SHELTER ISLAND II

Proceedings of the 1983 Shelter Island Conference on Quantum Field Theory and the Fundamental Problems of Physics

EDITED BY

Roman Jackiw, Nicola N. Khuri, Steven Weinberg and Edward Witten

INTRODUCTION TO THE DOVER EDITION BY

Silvan S. Schweber

Brandeis University
and Harvard University

Dover Publications, Inc.
Mineola, New York

Bibliographical Note

This Dover edition, first published in 2016, is an unabridged republication
of the work originally published in 1985 by The MIT Press, Cambridge,
Massachusetts. It is reprinted by special arrangement with The MIT Press. Silvan
S. Schweber has provided a new Introduction specially for this Dover edition.

Library of Congress Cataloging-in-Publication Data

Shelter Island Conference on Quantum Field Theory and the Fundamental
Problems of Physics (2nd : 1983 : Shelter Island, New York)
    Shelter Island II : proceedings of the 1983 Shelter Island Conference on
Quantum Field Theory and the Fundamental Problems of Physics / edited by
Roman Jackiw, Nicola N. Khuri, Steven Weinberg, and Edward Witten.
        pages cm
    This Dover edition, first published in 2015, is an unabridged republication
of the work originally published in 1985 by The MIT Press, Cambridge,
Massachusetts. It is reprinted by special arrangement with The Mit Press.
Silvan S. Schweber has provided a new Introduction specially for this Dover
edition.
    Includes bibliographical references.
    ISBN 978-0-486-79736-6
    ISBN 0-486-79736-8
        1. Quantum field theory—Congresses. 2. Nuclear physics—Congresses.
3. Cosmology—Congresses. I. Jackiw, Roman W., editor. II. Khuri, Nicola
N., 1933– editor. III. Weinberg, Steven, 1933– editor. IV. Witten, E., editor.
V. Title. VI. Title: Proceedings of the 1983 Shelter Island Conference on
Quantum Field Theory and the Fundamental Problems of Physics.
    QC174.45.A1S54 1983b
    530.14'3—dc23

                                                            2015006318

Manufactured in the United States by RR Donnelley
79736801   2016
www.doverpublications.com

Contents

II Historical Perspectives

Introduction to the Dover Edition

Silvan S. Schweber

Brandeis University and *Harvard University*

Together with Sheldon Glashow and Abdus Salam, Steven Weinberg was awarded the Nobel Prize in 1979 for his formulation of the electroweak theory, which unified the quantum field theoretic description of weak and electromagnetic interactions. He began his Nobel Prize lecture by stating

> Our job in physics is to see things simply, to understand a great many complicated phenomena in a unified way, in terms of a few simple principles. At times, our efforts are illuminated by a brilliant experiment, such as the 1973 discovery of neutral current neutrino reactions. But even in the dark times between experimental breakthroughs, there always continues a steady evolution of theoretical ideas, leading almost imperceptibly to changes in previous beliefs. (Weinberg 1980, 515).

Weinberg's remarks highlight often forgotten deep insights regarding the collective aspect of the physicists' enterprise, and, for that matter, of the scientific enterprise in general. Speaking as an outstanding physicist being recognized for his noteworthy contributions to physics, he indicated that he was part of a tightly knit community, that it is "*our* job in physics," that it is "*our* efforts," and that it is the *collective* set of beliefs that are imperceptibly changed by the steady evolution of theoretical ideas.

In this broad historical overview of "the steady evolution of theoretical ideas" from Shelter Island I in 1947 to Shelter Island II in 1983, I focus on certain aspects of the story of the establishment of the standard model by the high-energy physics community, and of some of the subsequent developments. It is my conviction that an adequate historical account must treat the collective as an essential component of the dynamics that tries to explain the growth of scientific knowledge; hence my emphasis on the community.

The standard model is the quantum field theoretical description of the interactions among quarks, gluons, leptons, and Higgs bosons. It combines the electroweak theory, the quantum field theoretic description of weak and electromagnetic interactions, with quantum chromodynamics (QCD), the field theoretic description of the strong interactions. QCD explains how quarks, though unobservable, can combine and give rise to observable mesons and nucleons (the hadrons), which in turn can combine to give rise to stable nuclei which, when binding with electrons, give rise to the observed atoms and molecules whose structure and properties are described by quantum electrodynamics (QED).

The basic working assumption of the standard model is that there are two "foundational" sets of building blocks of matter:

1. A fermion component consisting of
   a.) Six quarks (up, down; charm, strange; top, bottom), expressing the "flavor" degrees of freedom of the quark. Each quark also carries an additional degree of freedom expressed by its three "colors"; and
   b.) Six leptons, $(V_e, e^-; V_\mu, \mu; V_\tau, \tau)$, which only carry a flavor degree of freedom, identical with that of the quarks.

And

2. A bosonic component consisting of
   a.) The photon, the $W^\pm$, and the Z, the carriers of the electroweak forces between quarks and leptons;
   b.) The Higgs boson, which is responsible for assigning masses to the $W^\pm$ and the Z, and to the quarks; and
   c.) Eight color gluons. The gluons carry color degrees of freedom and are the carrier of the strong forces between quarks. They also interact strongly with one another,

with only the photon, the $W^\pm$, and the Z having flavor attributes. The electroweak interactions are described by a gauge theory based on an $SU(2)$x$U(1)$ symmetry that involves only the flavor degrees of freedom; the strong interactions are described by a gauge theory based on $SU(3)$ color degrees of freedom. The quarks are involved in both sets of interactions; each factor, $SU(3)$, $SU(2)$, and $U(1)$, has a single gauge-coupling constant.

The standard model, as expressed in the language of quantum field theory (QFT), is one of the most remarkable theories yet devised and one of the greatest accomplishments of the human species. The range of its applicability, from the subnuclear to the cosmological—some 20 orders of magnitude—is staggering. But the standard model's 20 or so parameters, and its inability to incorporate general relativity, indicate its limitations[1]: new concepts of interactions, objects, space and time, and geometry will have to be considered and probably introduced. It is widely believed that relativistic quantum field theory, the *foundational* mathematical language used so far in the quantum description of nature, which attributes a point-like character to quarks, gluons, and leptons and to their interactions with one another and with the Higgs bosons, reached the limits of its validity in the domain in which the standard model applies.

Establishing and corroborating the validity of the standard model was a *collective enterprise* by the high-energy physics community. On the theoretical side one can point to certain individuals—Schwinger, Feynman, Tomonaga, Weisskopf, Dyson, Salam, Marshak, Landau, Goldberger, Treiman, Gell-Mann, Chew, Low, T. D. Lee, Yang, S. Weinberg, Källèn, Sudarshan, Drell, Symanzik, K. Wilson, Sakata, Fubini, Nambu, Goldstone, Gottfried, Glashow, Higgs, Englert, Brout, Kibble, Penrose, Hawking, Georgi, Callan, Bjorken, Zeldovich, Fadeev, Gribov, Popov, Kirzhnits, Veltman, 't Hooft, Jackiw, Bell, Adler, B. Lee, Coleman, E. Weinberg, Politzer, Gross, Polyakov, Preskill, Wilzcek, Scherk, Schwarz, Witten, among others—whose contributions stand out. But every one of them drew on the *cumulative*, collective knowledge of the community. The new knowledge they produced was critically analyzed by the community, and the acceptance of their findings by the community asserted their validity and reflected a collective consensus.

Many more names could easily be added to the preceding list of theorists. Some of them were rewarded with Nobel Prizes—but it should be remembered that there is always an element of contingency in the selection. The high-energy community was—and is—populated by remarkable individuals, many of them "off-scale." For a host of cultural reasons, our Western civilization tends to mythologize certain individuals. It is my conviction that some of the above mentioned theorists were—are—as off-scale as Einstein, Maxwell, or Newton. This is not to make Einstein, Maxwell, or Newton less off-scale, but to note that there were but a handful of natural philosophers in Newton's time; that physics became a separate discipline only in Maxwell's time, with as yet no distinction between experimental and theoretical physics; and that there were only a few dozen theoretical physicists during the first two decades of the twentieth century, when Einstein made his outstanding contributions. This is in contrast to the several hundred distinguished theorists addressing problems in high-energy physics in the period between Shelter Island I and Shelter Island II. And perhaps even more important is the fact that with the plethora of new accelerators of ever higher energies emerging after Shelter Island I, the ever growing, ever more accurate, and robust sets of data that become evidence for a model or a theory are much more extensive and wide-ranging than those Einstein dealt with to corroborate his general theory of relativity. In addition, the tempo of changes to the collective set of beliefs as new data was produced and theoretical ideas evolved increased dramatically with the size of the high-energy community.

To corroborate my conviction about the role of the collective, and to exhibit the dynamics involved in the creation and acceptance of new knowledge, would necessitate an extensive analysis of the restructuring of physics departments after World War II; an analysis of the role played by preprints; by the summer schools and the lectures given there by outstanding contributors to the advances who assessed the collective advance, lectures that introduced the most recent developments to large numbers of graduate students and postdocs; by the ever greater number of conferences; and by a number of other such institutional factors—not to mention the political and economic factors that made all this possible. It thus would require a much more extensive presentation than is possible in this brief introduction. I will here only note the importance of what Robert Merton has called the Matthew effect[2], namely, that the rich in fame and status get richer and the poor poorer. It is not merely that an eminent scientist will often get more credit than a colleague who has done similar work independently but is not as well recognized; or that credit will usually be given to researchers who are already famous. It is also that the eminent scientist will usually be at a "richer" and "higher status" institution, will hence have more outstanding postdoctoral students working with him or her, will be exposed to more seminars and colloquia devoted to recent advances, and therefore will have more interactions with active researchers working at the frontiers of the field. Thus he or she will have the opportunity (and the ability) to work out, quickly and successfully, latent generative ideas he or she may have heard in such meetings. The net effect of all this is not just that his or her status is maintained by being at such an institution, but that the potentiality of its being enhanced is always there.[3] The converse is also true: those not in such positions, or at such institutions, are at somewhat of a disadvantage. The introduction of ArXiving one's papers has greatly democratized the process, but often when a paper is made accessible on ArXiv, its senior author is already busy extending the work.

Within the compass of this introduction, it is only possible to give a broad overview of the developments between Shelter Island I and Shelter Island II. I will limit myself to contrasting the states of theoretical understanding at the two endpoints, and adumbrate the change in perspective of the community over the time interval, thereby attempting to make understandable the rationale for the presentations at Shelter Island II.

Several observations about the change in beliefs stand out. The mid 1970s witnessed several outstanding developments:

In high-energy physics, the establishment of the standard model, but also the recognition that all our foundational quantum field theories are but effective field theories of limited range, not fundamental theories[4].

In cosmology, the establishment of the Robertson-Walker model that describes the present-day—when averaged—homogeneous and isotropic distribution of matter and energy in the universe, as its "standard model," and after the discovery in 1965 of the 3^0K cosmic background radiation by A. A. Penzias and R. W. Wilson, the general acceptance of a big bang as the beginning of the universe.

In the mid-1970s these two fields—high-energy physics and cosmology—became amalgamated. The amalgamation and its elaboration transformed the natural sciences. In physics, as Murray Gell-Mann observed in his opening lecture at Shelter Island II, "particle physics and cosmology [have] essentially merge[d] into one field, the field of fundamental physics which underlies all of natural science."[5] In high-energy physics the fundamental question addressed by the community became: "How did the standard model come to be?" Data for the empirical corroboration or suggestion of a more "foundational" theory came not only from SLAC, Fermi Lab, and CERN and its accelerators, or Kamiokande and its neutrino detectors, but also, and ever more frequently, from astrophysical and cosmological data, such as the data obtained from the Hubble space telescope, the COBE satellite, and the LIGO gravitational wave detector. The acceptance of big bang cosmology transformed nuclear physics into nuclear science, a subdiscipline that encompasses all aspects of the genesis of matter since the time after the big bang, when the standard model became a fairly accurate and robust description of the foundational entities that make up matter.[6] Similar issues are addressed in the planetary sciences, where investigations try to give answers to questions such as "How did all the chemical compounds that arise in the formation of planets come to be?," "How did the moon come to be?," and "What makes a planet habitable to forms of life?" In that general sense, the natural sciences have become evolutionary sciences, animated by a Peircian *Weltanschauung* that stipulates "Don't tell me what the law is. That closes the door to understanding. Tell me how it came to be."[7]

In the 1970s a second major confluence of developments took place, and resulted in a new symbiotic relationship between physics and mathematics. It turns out that the nonlinear Yang-Mills equations that are central components of the standard model have *non-perturbative* solutions that have deep, unsuspected properties—e.g., solitons, instantons, and monopoles. Understanding these aspects of non-Abelian gauge theories required understanding the topological properties of the theories. Jackiw's lecture at Shelter Island II is an elegant and beautiful presentation of some of these developments. Indicative of this new symbiotic relation is that in 1982, making use of the nonlinear space of instanton parameters as a new geometrical tool, Simon Donaldson (as a second year graduate student!) proved a result which, according to the eminent mathematician Sir Michael Atyah, "stunned the mathematical world." Donaldson's findings implied that there exist "exotic" 4-dimensional manifolds that are topologically but not differentially equivalent to the standard Euclidean 4-space. Subsequently, in the hands of Witten, Polchinski, Seiberg, and others, supersymmetric field theories and string theories turned out to be amazingly generative in the development of new results in topology and other mathematical fields.[8] Witten was awarded the Fields medal for his contributions to mathematics.

The intent of my survey is to point to these deep transformations of the metaphysics of high-energy theoretical physics. The practice was also deeply changed by the remarkable advances in computing and by the accuracy of the calculations required for the interpretation of the data that the novel methods and instrumentation in atomic physics brought about[9] and that accelerators at the Tev scale generated.[10]

Shelter Island II thus highlights the closing of an era and the beginning of a new one.

In closing this Introduction, I wish to reiterate that at best I have "pointed to" a fine-grained, scholarly account of all these developments and an assessment of their impact remains to be produced by competent scholars.[11]

My article is structured as follows:

1. Shelter Island I Revisited
2. From Shelter Island I to Shelter Island II
   2.1 The Metaphysics of Shelter Island I and Thereafter

1. Shelter Island I Revisited

The Shelter Island I conference was held from June 2 to June 4, 1947, at the Ram's Head Inn on Shelter Island at the tip of Long Island. How Shelter Island I came to be is told in the historical section of the present volume.[12] Here I only want to make a few brief, retrospective remarks concerning the sociology of Shelter Island I, and also to report on the new light that has been shed on what happened at Shelter Island I by the recent discovery of the detailed notes that Hans Bethe took at the conference.

The dominant figures at Shelter Island I were the physicists responsible for the wartime development of radar and of the atomic bomb: Oppenheimer, Bethe, von Neumann, Rabi, Teller, Weisskopf, Uhlenbeck, Wheeler. Also invited to the conference were some of the younger physicists they considered outstanding—Schwinger, Feynman, Bohm, Marshak, Nordseick, Feshbach, Lamb, Pais—who all, except for Pais, had worked at Los Alamos or at the MIT or Berkeley Rad Labs during the war.

Only two of the physicists invited were not Americans. One of them was Kramers, who was in the United States as the chair of the technical committee helping the United Nations formulate a plan for the international control of atomic energy. The other was Pais, who was a member of the Institute for Advanced Study.

Otherwise, all of the physicists invited were American. In fact, Frank Jewett, the head of the National Academy of Science, the institution that sponsored the conference, Duncan McInness, who formulated the idea of the conference, and Karl Darrow, the secretary of the American Physical Society, who had helped organize the conference, all wanted the conference to marshal the power of *American* physics to address and solve the most pressing problems in fundamental physics, just as it had solved urgent and deeply consequential wartime problems. All three wanted the conference to demonstrate that theoretical physics in the United States had come into its own with the "younger men" who had been born and trained there. The conference was to prove the strength of American theoretical physics not only in wartime activities, but also in "pure" physics.

In the United States, physicists had become admired and lavishly rewarded for their wartime efforts. The United States did not suffer the physical destruction the war had wreaked on Europe, the Soviet Union, and Japan. So after the war it prospered economically from the industrial goods it could manufacture and from the access to cheap Middle Eastern oil it had obtained from Great Britain in return for its Lend Lease help during the war. The *Servicemen's Readjustment Act* of 1944, usually called the G. I. Bill, granted returning veterans tuition and living allowances to enable them to continue their education, thereby causing universities, and their physics and engineering departments in particular, to expand dramatically. In addition, research contracts by the armed forces and by the AEC made it possible for physics departments to support large numbers of research assistants and postdoctoral fellows. Several generations of outstanding physicists were trained as a result of the expansion of physics departments. The magnitude of the enterprise is reflected in the fact that from 1950 to 1960 an average of 500 PhDs in physics were awarded annually by American universities. The number of physics PhDs granted in 1970 climbed to 1500. It declined sharply thereafter with the sharp curtailment of government funds to universities, reflecting the upheaval brought about by the civil rights movement, the Vietnam War, the student protest movements, and the Republican Party gaining extended control of the presidency, Congress, and the state legislatures. More generally, the 1970s marked the beginning of a deep change in the political and economic structure of the United States. Neoliberalism began assuming a dominant role in shaping economic policy; globalization deeply changed the structure of commerce and industry; fundamentalism began playing a more vociferous and influential role in shaping the political agenda as reflected in the limitations on teaching evolution in public schools; and a distrust of scientific expertise and of science became generally apparent. The cancellation of the Superconducting Super Collider (SSC) in 1993 is perhaps the clearest indication of the public's disenchantment with high-energy physics and with "fundamental" science in general.

Shelter Island II was organized by members of the physics department of Rockefeller University in part to substantiate the value and importance of the physics department to the university. The Rockefeller Institute, founded in

1901, had been devoted entirely to medical research. In 1959 it started graduate programs leading to PhDs in various disciplines, and in 1965 it became Rockefeller University. The University established a physics department in the early 1960s. In the late 1970s, because of government cutbacks and the changing national political and socio-economic context, Rockefeller University was restructuring itself and considering closing departments.[13]

This difference in economic, social, political, and cultural contexts is one of the sharp contrasts between Shelter Island I and II. Another is the size and cost of the two conferences. Some twenty-five scientists attended Shelter Island I, and its total cost was less than $1500, all paid for by the National Academy of Sciences. Some hundred theoretical physicists were invited to Shelter Island II. Its cost, close to $100,000 though many of the attendees paid their own expenses, was covered privately, by the support of the Richard Lounsbery Foundation, the Andrew W. Mellon Foundation, the Texaco Philanthropic Foundation, and Susan B. Borden.[14]

Shelter Island I was a landmark whose impact was comparable to that of the Solvay Congress of 1911. That Congress set the stage for the subsequent developments in quantum theory by virtue of Planck's and Einstein's interpretation of the precise measurements of blackbody radiation as a function of temperature. Similarly, Shelter Island I provided the initial stimulus for the post-World War II developments in quantum field theory— renormalization theory, effective, relativistically invariant computational methods, Feynman diagrams—this, by virtue of Lamb and Retherford's experiment on the fine structure of hydrogen and Nafe, Nelson, and Rabi's experiment on the hyperfine structure of hydrogen. The conference was also responsible for the elucidation of the structure of the mesonic component of cosmic rays. Rossi reported the experiment carried out by Conversi, Pancini, and Piccioni in Rome that indicated that the meson detected in cosmic ray showers near the surface of the earth could not be the meson postulated by Yukawa to be responsible for the nuclear forces between nucleons. This led to the two-meson hypothesis advanced at the conference by Marshak. He suggested that there were two kinds of mesons. The heavier ones were produced at the top of the atmosphere by cosmic rays in nuclear reactions. He identified these with the mesons Yukawa had postulated. The π-mesons then decay into the lighter μ mesons, which do not interact strongly with nuclei.

Shelter Island I, and the two subsequent conferences it engendered, the Pocono and Oldstone conferences, were small, closed, and elitist in spirit. In a sense they mark the postponed end of an era, that of the 1930s, and its characteristic style of doing physics: small groups and small budgets, as in the case of the Washington Conferences of Theoretical Physics. None cost more than $1500. The ravages of inflation are illustrated by the fact that the average expenditure to accommodate *all* the **25** attendees for one night at the comfortable and elegant inns where the conferences took place was $200.

What the newly found notes that Bethe took at Shelter Island I record is that on the first day of the conference Lamb presented the most recent data of his experiment with Robert Retherford. It established that contrary to the solutions of the Dirac equation for the energy levels of an electron in a Coulomb field, the 2s and 2p levels of the hydrogen atom do not degenerate. In the discussion which followed Lamb's presentation, Oppenheimer pointed out that if one calculated the *difference* between the 2p and 2s energy levels by using hole theory, a finite answer might be obtained given that the divergences encountered were logarithmically divergent (see Figure 1).

Lamb's talk was followed by Rabi's, who presented the data that Nafe, Nelson, and he had obtained on the hyperfine structure of H and D. That

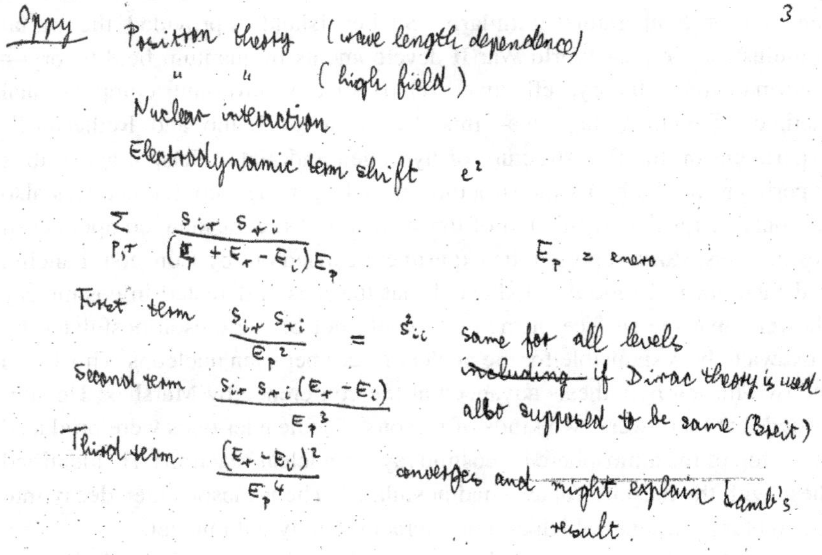
3

Oppy Positron theory (wave length dependence)
" " (high field)
Nuclear interaction
Electrodynamic term shift e^2

$$\sum_{p,r} \frac{S_{ir} S_{ri}}{(E_p + E_r - E_i) E_p} \qquad E_p = \text{energy}$$

First term $\frac{S_{ir} S_{ri}}{E_p^2} = S^2_{ii}$ same for all levels ~~including~~ if Dirac theory is used

Second term $\frac{S_{ir} S_{ri} (E_r - E_i)}{E_p^3}$ also supposed to be same (Breit)

Third term $\frac{(E_r - E_i)^2}{E_p^4}$ converges and might explain Lamb's result.

Figure 1 (From Bethe's notes of Shelter Island I)

afternoon Rossi reported on the experiment that had been carried out in Rome by Conversi, Pancini, and Piccioni on the absorption of mesons in the atmosphere.

The entire morning session of the second day of the conference was taken up by Kramers's extensive presentation of his version of the Lorentz theory of an extended charge, in which structural effects had been encapsulated in the experimental mass of the particle. He concentrated on the *classical* version of his nonrelativistic theory, although Bethe's notes make clear that in the last part of his talk he indicated what quantizing the theory would do (see Figure 2).

That his talk was influential is reflected by the fact that Kramers's use of the Hertz potential and his derivation of the potential that the dressed electron with experimental mass experiences became the point of departure for Schwinger in his quantum electrodynamic calculation of the magnetic moment of the electron and of the Lamb shift.

In the afternoon of the second day, Weisskopf reviewed the divergence difficulties in the hole-theory calculations of the self energy of an electron. In a paper that he had written in 1939 he had said that in hole theory the divergence would be logarithmic to all orders of perturbation theory. According to Breit's notes of the conference, Weisskopf agreed with Oppenheimer's hunch that a hole-theoretic calculation of the difference in the 2p and 2s energy levels would not diverge. During the ensuing discussion, Weisskopf and Schwinger indicated how a hole-theoretic calculation of the level shift might be attempted, and suggested further reasons why a finite result might result from the applications of Kramers's ideas.

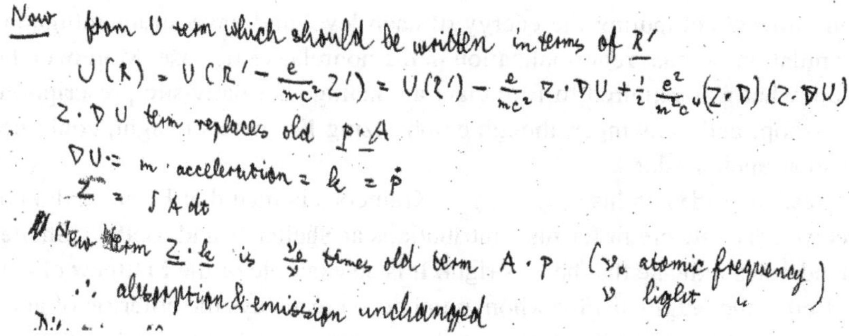
Now from U term which should be written in terms of R'

$U(R) = U(R' - \frac{e}{mc^2} Z') = U(R') - \frac{e}{mc^2} Z \cdot \nabla U + \frac{1}{2} \frac{e^2}{m^2c^4} (Z \cdot \nabla)(Z \cdot \nabla U)$

$Z \cdot \nabla U$ term replaces old $p \cdot A$

$\nabla U = m \cdot$ acceleration $= k = \dot{p}$

$Z = \int A \, dt$

New term $Z \cdot k$ is $\frac{\nu_0}{\nu}$ times old term $A \cdot p$ (ν_0 atomic frequency, ν light ")

∴ absorption & emission unchanged

Figure 2 (From Bethe's notes taken at Shelter Island I)

After the conference was over, Bethe performed his famous nonrelativistic calculation on the train ride from New York to Schenectady. The paper in which he proved that the level shift would be accounted for by quantum electrodynamics was completed by June 9 and thereafter circulated to the participants of the Shelter Island conference. In his paper (1947), Bethe acknowledged in a footnote the comments by Weisskopf and Schwinger that a hole-theoretic calculation of energy level differences would be finite. Their insight justified the high-energy cut-off he introduced in his calculation.

But in his paper Bethe *did not* acknowledge Kramers's talk, even though Kramers's presentation had been very important by stressing the importance of expressing observables in terms of the *experimental* mass of the electron. At Shelter Island II Bethe did interject during Lamb's talk that Kramers had "inspired him."

Why Bethe did not acknowledge Kramers's presentation in his paper was perhaps because Bethe thought that he had a much simpler way than Kramers to incorporate Kramers's insight. Bethe had noted that the quantum electrodynamically calculated self-energy of a free nonrelativistic electron could be ascribed to an electromagnetic mass of the electron, and. though divergent, had to be added to the mechanical mass of the electron. The only meaningful statements of the theory involve the sum of the electromagnetic and mechanical masses, which combination is the experimental mass of a free electron. In contrast to Kramers's approach, Bethe's was a *model-independent* formulation of mass renormalization that did not assume an extended charge distribution to the electron. And in contrast to Schwinger and Weisskopf's initial insight that a hole-theoretic calculation that computed the *difference* between the energies of two levels would be finite, Bethe's mass renormalization allowed computing the energy of each level and gave an unambiguous formulation of mass renormalization in the nonrelativistic case. Moreover, he knew what was required to formulate an analogous relativistic prescription. Weisskopf and Schwinger, though emphasizing Kramers's insight, could not do so at Shelter Island.

Dresden (1988), in his biography of Kramers, claimed that Kramers did not receive adequate credit for his contributions at Shelter Island. Bethe's Shelter Island notes indicate that he was right. It is an example of the Matthew effect! To Bethe, the pragmatist for whom numbers were always the criterion of good physics, and who had just been deeply and successfully involved in the war effort, calculating numbers that translated into physical effects and measurable empirical data, the challenge was to get the numbers out and account for the magnitude of the 2p-s shift in hydrogen and for the new values of the hyperfine splitting in H and D that Nafe, Nelson, and Rabi had measured. Accounting for the empirical data would become evidence for the correctness of the theory, or at least for the model that approximated it. Kramers's approach was too model-dependent, too theoretical, and too far removed from calculating numbers. For Bethe, the value of a novel idea was gauged by whether it could help you calculate numbers that could be compared with empirical data.

Bethe's calculation was a crucial one. By introducing in a simple manner the concept of mass renormalization and its associated meaning, and indicating how it is to be used in the case of a bound electron, Bethe's calculation gave a new perspective on how to address quantum electrodynamical calculations and obtain numbers that could be compared with experimental data. Furthermore, Bethe could convincingly justify the cut-offs he had to introduce in his nonrelativistic calculation and obtain a logarithmic expression for the Lamb shift that agreed with the observed level shift. It should be added that only Bethe could have evaluated the logarithmic contribution as quickly as he did. He had encountered similar logarithmic expressions when calculating quantum mechanically the energy loss of a fast charged particle traversing matter in his *Habilitationschrift* in 1929!

Thereafter Schwinger made another crucial calculation, perhaps a more influential one than Bethe's, by calculating the quantum electrodynamic contribution to the magnetic moment of the Dirac electron. His calculation was crucial because it was the first *field theoretical* perturbative calculation of quantum electrodynamic radiative corrections that was Lorentz and gauge-invariant. Up to this point all calculations had been based on hole theory.

Stated field theoretically, in hole theory one subtracted from the Hamiltonian for the electron field the expectation value of the Hamiltonian operator for the state representing the vacuum *in the absence of the electromagnetic field.* Thus

$$H_0^{matter} = \int \left[\psi*(r)(\boldsymbol{\alpha}\cdot\boldsymbol{p} + \beta m)\,\psi(r) - \left\langle \psi*(\boldsymbol{\alpha}\cdot\boldsymbol{p} + \beta m)\,\psi \right\rangle_{vac} \right]$$

Similarly, the charge current four vector of the quantized electron field was defined as

$$\{j(r), \rho(r)\} = e\left[\psi*(r)\{\boldsymbol{\alpha}, 1\}\psi(r) - \left\langle \psi*\{\boldsymbol{\alpha}, 1\}\psi \right\rangle_{vac} \right]$$

For the free electron field

$$\psi(r) = \sum_m a_m e^{ip \cdot r} u_m \qquad \psi^*(r) = \sum_m a_m^* e^{-ip \cdot r} \overline{u_m}$$

where u_m are the Dirac spinor functions, a_m^*, a_m are the creation and annihilation operators, and E_m is the absolute value of the energy for an electron in the state m. The symbol m denotes a definite value of the linear momentum p and a specification of the four possible spin-states corresponding to this momentum p.

Hole theory was not gauge-invariant, because the vacuum subtraction prescription labeled states according to p and only p-eA, and not p is gauge-invariant. The lack of gauge invariance led to arbitrariness and difficulties in all the prewar calculations. It had already arisen in the calculation of the electron self-energy in the lowest order of perturbation theory. The self-energy was divergent, but introducing a cut-off for the possible momenta p to obtain finite results destroyed the Lorentz invariance of the theory.

Schwinger's calculation indicated that a gauge- and Lorentz-invariant quantum field theoretic description of both electrons and radiation could be formulated, and that with it one could calculate observable consequences.

2. From Shelter Island I to Shelter Island II

The Shelter Island, Pocono, and Oldstone conferences[15] were the precursors of the Rochester Conferences on High-energy Physics. But they differed from these later conferences in important ways. The Rochester conferences were more professional and democratic in outlook, and had the imprint of the new era: the large group efforts and the large budgets involved in machine physics.[16] Also, whereas Shelter Island and Pocono looked upon quantum electrodynamics (QED) as a self-contained discipline, the Rochester conferences saw "particle physics" come into its own, with QED as one of its subfields— albeit one with a privileged, paradigmatic position.

Shelter Island II, which was held in June 1983, 36 years after Shelter Island I, in many ways marks the end of a period of remarkable advances in what was then considered "fundamental" physics, a period that culminated with the experimental verification, to a remarkable degree of precision, of the standard model, and of its electroweak component in particular. It can be said that Shelter Island II marks the end because

1. In the early 1970s it was shown that Yang-Mills gauge theories with broken symmetry were renormalizable and made acceptable the Glashow-Salam-Weinberg electroweak theory. By 1973 the existence of neutral currents was confirmed experimentally, thus corroborating a key prediction of Weinberg's electroweak theory.
2. In 1973 it was shown that non-Abelian gauge theories had the property of being asymptotically free and thus could explain Bjorken scaling. This opened the way for the acceptance of quantum chromodynamics (QCD), a Yang-Mills gauge theory based on an $SU(3)$ color symmetry, for the description of the strong forces.
3. In 1974 the J/ψ meson was identified as a bound state of a Glashow-Iliopoulos-Maiani charmed quark and anti-quark, cementing the acceptance of both electroweak theory and QCD.
4. In 1975 Kenneth Wilson's lattice gauge formulation of QCD gave strong support to the notion of quark confinement.
5. In 1976 the discovery of the τ meson completed the present-day version of the standard model with its three generations of quarks and leptons.

The subsequent detection of the Ws, the Z, the top and bottom quarks, and in 2012 of the Higgs boson, anchored the standard model as the effective field theory describing subnuclear phenomena up to the 1 Tev scale.

The lectures delivered at Shelter Island II reveal the concerns of the theoretical physicists about the shortcomings of the standard model, confronting "fundamental" problems *beyond* the standard model, and exploring and extending a new symmetry formulated in the mid-1970s: supersymmetry[17]. Thus Murray Gell-Mann, in his magisterial opening lecture entitled "From Renormalizability to Calculability?," outlined the theoretical advances that might make it possible to calculate fundamental constants with no need for infinite renormalizations. The same is true for the other major introductory lecture, Steven Weinberg's paper entitled "Calculations of the Fine Structure Constants." Shelter Island II thus indicates the new paths being charted to overcome the limitations of renormalizable relativistic quantum field theories and of the standard model.

My aim in the present section is to adumbrate the genesis of the standard model, and indicate some of the problems that were not resolved by the standard model. Addressing some of these problems required the use and development of fairly complex and difficult new mathematical structures, in fact mathematics being concurrently constructed by mathematicians. Shelter Island II gives proof of a new symbiotic relationship between physics and mathematics.[18]

Needless to say, the strong coupling between theory and experiment, and the cooperative, mutually beneficial relationship between theory and experiment were crucial elements making possible the genesis of the standard model. Bjorken has stated the dynamics succinctly:

> It is my credo that technological advances drive the progress in experimental physics and that experimental physics in turn drive the theory. Without these ingredients, the most brilliant theoretical constructs languish worthlessly. There is in my opinion no greater calling for a theorist than to help advance the experiments. It is not an easy thing to do (Bjorken 1997, 596).

Nonetheless, I will concentrate on the theoretical aspects because that is what Shelter Island II was concerned with. The title of its proceedings, *Shelter Island II*, edited by Jackiw, Khuri, Weinberg, and Witten, characterized Shelter Island II as a conference on quantum field theory and *the Fundamental Problems of Physics*. I will neither give a continuous account of the theoretical advances nor attempt to convey the entanglement of the various problems being addressed at any given time. I refer the reader to Hoddeson, Brown, Riordian, and Dresden's *The Rise of the Standard Model* and to Tian Cao's *Conceptual foundations of quantum field theory* for detailed accounts and interpretations of the history given by some of the physicists responsible for the advances and by historians, philosophers, and sociologists of science.[19] And since Hoddeson, Brown, Riordian, and Dresden in their introduction to *The Rise of the Standard Model* have given a masterful overview of the developments, I focus on what I consider certain key elements which highlight the "perceptible" changes in perspective by the high-energy community in the late 1970s. The reader should refer to Volume I of Gottfied and Weisskopf's *Concepts of Particle Physics*, published in 1983, for a detailed, wide-ranging, insightful, and readily accessible exposition of the standard model that was generally accepted when Shelter Island II convened.

I have made extensive use of their Volume II, published in 1986, for the more technical aspects of my presentation.

2.1 The Metaphysics of Shelter Island I and Thereafter

It was John Archibald Wheeler who, at a joint meeting of the National Academy of Sciences and the American Philosophical Society in the fall of 1945, noted that the experimental data in "elementary particle physics" that had been obtained during the 1930s made it possible to identify four "fundamental" interactions: the "strong" forces responsible for nuclear interactions, the electromagnetic forces between electrically charged particles, the "weak" forces responsible for beta decay, and gravitational forces. It was evident that the gravitational forces could be neglected when considering elementary particle processes, given their weakness and the distances being probed in experiments. At the first Rochester Conference in 1950, most physicists held the view that electrons, photons, protons, neutrons, π-mesons, μ-mesons, and neutrinos were "elementary particles," that the four interactions were independent realms, and that quantum electrodynamics, pseudoscalar meson theory, and Fermi's beta decay theory were "fundamental" theories of the electromagnetic, nuclear, and weak interactions. But what became clear very quickly was that the elucidation of the dynamics of the subnuclear world would require their simultaneous consideration.

Elucidating the properties of "mesons" was a case in point. The two meson hypothesis was confirmed by the discovery of the pion by Lattes, Muirhead, Ochialini, and Powell through its decay sequence π→μ→e. In 1949 Lederman found that the μ meson disintegrates into an electron and two neutrinos; this corresponds to an ordinary β decay. Immediately thereafter, Lee, Rosenbluth, and Yang established that μ decay and μ capture could be modeled by a four Fermi interaction similar to Fermi's original β decay. Fermi's theory of the neutron's and proton's weak interaction

$$n \rightarrow p + e^- + \bar{v}$$
$$p \rightarrow n + e^+ + v$$

had described the weak forces by an interaction Hamiltonian $G_{np}\left(\psi_n^*\gamma_0\gamma_\mu\psi_p\right)\left(\psi_e^*\gamma_0\gamma^\mu\psi_\nu\right)$ and was immediately generalized to the form $G_{npi}\left(\psi_n^*\gamma_0 O_i\psi_p\right)\left(\psi_e^*\gamma_0 O_i\psi_\nu\right)$, with O_i being one of the following: $1, \gamma_5, \gamma_\mu, \gamma_5\gamma_\mu, \sigma_{\mu\nu}$. Furthermore, Lee, Rosenbluth, and Yang noted that the coupling constants G_{np} and $G_{\pi\mu}$ in the four Fermi interaction for nuclear β-decay and that for μ decay and μ absorption were of the same order of magnitude. Klein, Puppi, and Tiomo and Wheeler arrived independently at the same conclusion. One thereafter spoke of a "universal" Fermi interaction. Given that the weak interactions, all seemed to have the same value of the coupling constant, Lee, Rosenbluth, and Yang suggested that in analogy with electromagnetism, where all charges, irrespective of their masses, interact with the same coupling constant, the weak interactions have the same property. And in analogy with the electromagnetic forces that are quantum field theoretically interpreted as mediated by the exchange of spin one photons (which are massless because electromagnetic forces have infinite range), Lee, Rosenbluth, and Yang suggested that the weak forces should be considered mediated by an intermediate massive spin one boson (since the weak forces are short-ranged). They had hoped that this might solve the problem of the non-renormalizabilty of the Fermi interaction. But spin one theories with massive quanta were likewise not renormalizable.

By the late 1950s, with the discovery of the plethora of new, "strange," "elementary particles," and the failure of meson theories to account for the nuclear forces, it was realized that the traditional view, which correlated a quantum field with each "elementary particle," was not tenable when trying to understand the hadronic world. In fact, Lev Landau and Geoffrey Chew went further and called for the abandonment of field-theoretic descriptions.[20] During the 1960s Geoffrey Chew and his students formulated a program for calculating S-matrix elements for strong interaction processes that had no reference to quantum field theory by stipulating certain analytic properties they should satisfy, and requiring that Lorentz invariance and unitarity be satisfied. However, Chew was not able to develop a consistent, reliable method for calculating the S-matrix for an arbitrary process.

Similarly, in the weak interactions the τ-θ puzzle challenged another widely accepted view. The puzzle consisted in the fact that two particles, the θ and the τ meson, which—to the extant experimental accuracy—had the same mass and lifetime and therefore seemingly were the same particle, decayed in two different modes the θ into two pions and the τ into three; therefore, and because the pion was known to have negative parity, it seemed possible that parity was not conserved. As a result of questions that Martin Block and Feynman posed at the 1956 Rochester Conference to Lee and Yang, who had constructed a model to explain the seeming violation of parity conservation, they made a careful study of all known experiments involving weak interactions. After reviewing them, Lee and Yang concluded that there was no evidence of parity conservation in any weak interaction (Lee and Yang 1956). In his Nobel lecture, Yang commented that although past experiments on the weak interactions gave no proof of parity conservation, "In strong interactions, . . . there were indeed many experiments that established parity conservation to a high degree of accuracy. . . ." Furthermore, Yang noted that "The fact that parity conservation in the weak interactions was believed for so long without experimental support was very startling. But what was more startling was the prospect that a space-time symmetry law which the physicists have learned so well may be violated." Wu, Ambler, Hayward, and Hobson with an experiment on the β-decay of polarized Co^{60} nuclei, and Garwin, Lederman and Weinreich, and Friedman and Telegdi with their experiments on the π-μ decay, proved that parity was violated in the weak interactions.[21]

Very shortly thereafter, Marshak and Sudarshan, and Feynman and Gell-Mann proposed that the operator O in the Fermi interaction $G_F\left(\psi_1^* \gamma_0 O \ \psi_2\right)\left(\psi_3^* \gamma_0 O_i \psi_4\right)$ is $O = \gamma_\mu(1-\gamma_5)$, which would account for the parity violation and most of the available experimental β-decay data. The interaction was immediately interpreted as implying that the neutrino is "left-handed," since only the operator $(1-\gamma_5)\psi$ v entered into the description of the neutrino field. Furthermore, the Fermi interaction was interpreted as a current-current interaction of the form $G_F J_\mu J_\mu^*$, where J_μ has both a vector and an axial vector component. The vector piece of that current was conserved, giving rise to a charge that guaranteed universality, while the axial vector current component was only partially conserved (PCAC) in the case that massive particles were part of the current-current description. The presence of masses broke the axial symmetry.

The role of an assumed underlying symmetry in the classification of the hadrons, and the efforts in trying to understand the manifestations of the breaking of that symmetry, also put symmetry considerations center stage in the analysis of strong interaction processes. It became clear that understanding the mechanisms of braking symmetries was necessary. This is the story of the various manifestations of an $SU(3)$ symmetry in the classification of the hadrons, of how quarks grew out of a simpler conceptualization of the "eightfold way" "three" more elementary entities, first as mathematical entities, and later reconceived as physical ones (see Hoddeson et al. 1997, Cao 2010). I will not retell that story here, but instead focus on developments in quantum field theory.

After the success of renormalized quantum electrodynamics, theorists turned their attention to quantum-field theoretical descriptions of the meson-nucleon interaction. It was believed that such interactions would be descriptions with a fixed coupling constant just as in the case of QED. In fact, to "derive" the value of the (dimensionless) fine structure constant, $e^2 / 4\pi\hbar c$, was considered a "fundamental" problem, considered by some, such as Pauli, to be the most challenging one. Its echo still resonated in the lectures of Gell-Mann and Weinberg at Shelter Island II.

It was recognized that the coupling constants of the nuclear and electromagnetic forces differed greatly in magnitude, with $g^2_{NN\pi} / \hbar c \approx 15$, whereas $e^2 / 4\pi\hbar c \cong 1/137$, and that this called into question the use of perturbation theory in the meson-nucleon case. General properties of quantum field theories, such as causality and relativistic invariance and their consequences, such as dispersion relations, were therefore investigated. Two features stand out in the researches of the field theorists: the role of symmetries and that of renormalization. A basic assumption was that the physical systems under consideration could be described by a Lagrangian, L, which is a local function of the *fields* $\phi_i(x)$ and of their first derivatives, $\partial_\mu \phi_i$. The fields $\phi_i(x)$ were associated with the "elementary particles" they were to describe, which were characterized by their spin, and the (bare) mass and the (bare) charge parameters that appeared in the Lagrangian. The action, $\int \mathcal{L}(\phi_i(x), \partial_\mu \phi_i(x)) d^4x$, was stipulated to be a scalar, invariant under Lorentz transformations.

As had been the case with Heisenberg coming to matrix mechanics by focusing on observables, so it was after the insights of mass and charge renormalization had been absorbed. The focus came to be placed on the scattering amplitude, i.e., the S-matrix, and its *perturbative* expansion and associated Feynman diagrams. However, if a Feynman diagram contains loops, the corresponding contribution to the S-matrix scattering amplitude contains integrals over the loop momenta, and these integrals (usually) are divergent. One must therefore introduce an algorithm—called a regularization—to modify the theory and render loop integrations finite in order to be able to operate with them, and possibly move them about. An example of such an algorithm is the introduction of a cut-off such that all momenta beyond some fixed value Λ are not integrated over.

When the scattering amplitudes are calculated in perturbation theory with the original Lagrangian, they are expressed in terms of the (bare) masses and (bare) charges parameters that appeared in the Lagrangian. But once radiative corrections are taken into account, these are not the observed physical masses and charges of the entities the theory is intended to describe. Some physical processes must be chosen in order to define the empirically determined masses and charges, and the S-matrix scattering amplitudes must then be re-expressed in terms of them. This process is called renormalization. Note that renormalization would be necessary even if no divergences were encountered. A renormalizable theory is one in which the scattering amplitudes remain well defined and consistent when the regularization is removed (as in the above example, where $\Lambda \to \infty$).

Dyson, at the 1948 Oldstone conference, had introduced the precise concept of renormalizabity, i.e., the possibility that a quantum field theory having all the infinities encountered in perturbation theory could be absorbable in a *finite* number of parameters that enter into the description of the theory. He pointed out that the requirement of renormalizabity required that Lagrangians, in turn, contain terms of mass dimension 4 or less (with $\hbar$ and $c = 1$), where bosonic fields have dimension 1 and fermionic fields have dimension 3/2.

Lagrangians encapsulate the equations of motion that a field system obeys. They have the further property of being able to readily incorporate in their formal structure the symmetries under which the field system is invariant. The consequences of the symmetry then become expressed by conserved (Noether) currents, $\partial_\mu J_i^\mu = 0$, where $J_i^\mu(x)$ is derived from the Lagrangian density by virtue of the invariance of action under the symmetry. The corresponding charges $Q_i = \int d^3x\, J_i^0(x)$ are conserved $dQ_i/dt = 0$. A given symmetry is always associated with an either discrete or continuous group of transformations. The invariance of the system under the symmetry operations translates in quantum theory to the statement that the Hamiltonian is invariant under such transformations, or equivalently that the Hamiltonian commutes with the generators of the group $[H, Q_i] = 0$, and hence that $dQ_i/dt = 0$. The most familiar example is the conservation of electric charge. If the Lagrangian is invariant under the symmetry operation $\phi_j \to e^{iq_j\varphi}\phi_j$, $\phi_j^* \to e^{-iq_j\varphi}\phi_j^*$ (with φ constant), there is a corresponding Noether current which is conserved. The Dirac Lagrangian

$$\mathcal{L} = \bar{\psi}\gamma_\mu \partial_\mu \psi + m\bar{\psi}\psi$$

is invariant under this transformation, and the associated charge operator is

$$Q = \int d^3x \, \bar{\psi}\gamma_0 \psi \, .$$

Using the equal time anticommutation relations between ψ and $\bar{\psi}$, one verifies that $e^{iQ\varphi}$ generates the phase transformation on the Dirac field.

Very soon after the discovery of the neutron, Heisenberg noted that the quantum mechanical models of nuclear structure for deuteron, helium, and other light nuclei indicated that the nuclear forces between protons, between neutrons, and between neutrons and protons were nearly equal. He therefore hypothesized that there exists a further symmetry between a neutron and a proton which have the same spin angular momentum: they could be considered two states of the same entity, the nucleon, where the proton has isospin +1/2 and the neutron isotopic spin −1/2, each carrying the same charge as far as the nuclear forces are concerned. Translated into field-theoretic language a Lagrangian describing a system of neutrons and protons interacting through nuclear forces was to be invariant under any rotation in isotopic spin space. This invariance would result in conserved Noether currents and in corresponding laws charges and conservation. If the nucleon field is considered a spinor in isotopic spin space,

$$\psi(x) = \begin{pmatrix} p(x) \\ n(x) \end{pmatrix}$$

then under a rotation therein ψ is transformed

$$\psi \rightarrow e^{i\varepsilon \cdot \tau}\psi$$

where the τs are the Pauli matrices. The Noether conserved currents are $J_\mu^i = \frac{1}{2}\bar{\psi}\gamma_\mu \tau^i \psi$. By using the equal time anticommutation rules, one verifies that the corresponding charges satisfy the $SU(2)$ algebra

$$[Q_i, Q_j] = i\varepsilon_{ijk} Q_k \, .$$

The $SU(2)$ isotopic spin symmetry is approximate, because the proton carries an electric charge whereas the neutron does not. The interaction Lagrangian that describes the electromagnetic interactions of nucleons

$$\mathcal{L}_{em} = \frac{e}{2}\bar{\psi}(x)\gamma_{\mu}(1+\tau_3)\psi(x)A^{\mu}(x)$$

is not invariant under isotopic spin rotation. Nonetheless, even though $\mathcal{L}_o + \mathcal{L}_{em}$ is not invariant under isotopic spin rotations—that is, the $SU(2)$ symmetry is broken—the now time-dependent charges $Q_i(t)$ still satisfy the $SU(2)$ algebra

$$[Q_i(t),Q_j(t)]=i\varepsilon_{ijk}Q_k(t).$$

This observation was the point of departure for Gell-Mann's important work on current algebras in the case of broken symmetries. Treiman, Jackiw, and Gross have given a full exposition of this approach.[22]

Another approach to the failure of the meson theory was Pauli's formulation of a non-Abelian gauge theory as a possible framework for dealing with the strong interactions. Yang and Mills in 1954, and independently Shaw, a student of Salam, in his Ph.D. dissertation submitted to Cambridge University in 1955, reached a similar end point by considering the consequences of demanding that like gauge transformations in QED, theories must be invariant under "local symmetries"—that is, the requirement that the symmetry transformation can be arbitrarily chosen at different space-time points. Local symmetries are introduced by making the group parameters functions of space and time. If G is a symmetry group whose elements are expressed in terms of its generators θ^a

$$G = e^{i\varepsilon^a\theta^a}$$

where the θ^as satisfy the Lie algebra of the group

$$[\theta^a,\theta^b]=if^{abc}\theta^c$$

with f^{abc} the structure constants of the Lie algebra, local symmetry means that the ε^a become functions of x, $\varepsilon^a(x)$.

Consider the set of spin 1/2 matter field described by $\psi(x)$ with $\psi(x)$ transforming as an irreducible representation of G, as follows:

$$\psi(x) \rightarrow e^{i\varepsilon^a(x)\theta^a}\,\psi(x) = S(x)\psi(x)$$

Under this transformation the kinetic energy term in the Lagrangian transforms as follows:

$$\overline{\psi}\,\gamma^\mu\,\partial_\mu\psi \rightarrow \overline{\psi}\,\gamma^\mu\,\partial_\mu\psi + \overline{\psi}\,\gamma^\mu e^{-i\varepsilon^a(x)\theta^a}(\partial_\mu e^{i\varepsilon^a(x)\theta^a})\psi$$

To make the Lagrangian invariant, it is necessary to introduce a spin one field $B_\mu(x)$ that transforms under G as follows:

$$B_\mu(x) \rightarrow S B_\mu S^{-1} + S(\partial_\mu S^{-1})$$

If the kinetic part of the Lagrangian is modified so that it becomes $\mathcal{L} = \overline{\psi}\,\gamma^\mu(\partial_\mu + B_\mu)\psi$, then the matter Lagrangian will be invariant. For $B_\mu(x)$ to be a full-fledged dynamical field, a G invariant kinetic energy term must appear for it in the Lagrangian. Since the function

$$F_{\mu\nu} = \partial_\mu B_\nu - \partial_\nu B_\mu + [B_\mu, B_\nu]$$

transforms covariantly under $G(x)$, $\mathrm{Tr}\,F^{\mu\nu}F_{\mu\nu}$ is G invariant. The full gauge-invariant Lagrangian is then

$$\mathcal{L} = \frac{1}{g^2}\mathrm{Tr}\,F^{\mu\nu}F_{\mu\nu} - \overline{\psi}\,\gamma^\mu(\partial_\mu + B_\mu)\psi.$$

(See Mohapatra 1986, Gottfried and Weisskopf 1984, 1986 for details.) The B field thus introduced, however, is massless, and thus gives rise to a long-range force, contrary to what is necessary for the description of the strong nuclear forces.

How to break the symmetry and give the B field quanta a mass was the problem that had to be solved.

The point of departure for the solution of the problem was the 1956 Bardeen, Cooper, Schrieffer (BCS) theory of superconductivity.[23] When charged particles are present, quantum electrodynamics can exhibit a superconducting phase, characterized by the condensation of the charged field

$$\langle 0|\psi(x)\psi(x)|0\rangle \neq 0$$

where the field $\phi(x) = \psi(x)\psi(x)$ describes the Cooper pair and $|0\rangle$ is the field-theoretical, degenerate ground state (which is not invariant under gauge transformations). The condensation of Cooper pairs creates a gap in the spectrum by making the photon massive, and makes electrical currents superconducting. Its effect on magnetic fields is known as the Meissner effect. Magnetic fields cannot penetrate the superconductor except in thin flux tubes. Thus, when two magnetic monopoles are inserted in a superconductor, a thin flux tube gets created between them. The energy stored in the flux tube depends linearly on the length of the flux tube, and therefore the potential between two magnetic monopoles is linear. Such a linear potential is known as a confining potential (Gottfried and Weisskopf 1986).

In 1959, in an influential paper, Nambu suggested that the masses of elementary particles might arise in a similar way as in BCS: the vacuum state might not respect the symmetries of the theory—the phenomenon known as a spontaneously broken symmetry—and like the quasiparticles of the BCS theory, elementary particles might acquire mass. In 1961 Nambu and Jona-Lasinio went on to construct a model of a massless fermion field $\psi(x)$ that exhibited these characteristics. It was based on the Lagrangian density

$$L = i\overline{\psi}(x)\gamma^\mu \partial_\mu \psi(x) + g\left[\left(\overline{\psi}(x)\psi(x)\right)^2 - \left(\overline{\psi}(x)\gamma_5\psi(x)\right)^2\right]$$

which is invariant under the phase changes

$$\psi(x) \to e^{i\alpha}\,\psi(x) \qquad \text{and} \qquad \psi(x) \to e^{i\beta\gamma_5}\,\psi(x).$$

By Noether's theorem, there would be a vector current, $j^\mu = \overline{\psi}(x)\gamma^\mu\,\psi(x)$, that is conserved and an axial vector current, $j_5^\mu = \overline{\psi}(x)\gamma^\mu\gamma_5\,\psi(x)$, that is conserved. Nambu and Joan-Lasinio then stipulated that the ground state, i.e., the vacuum state, does not respect the chiral symmetry and that the symmetry is broken spontaneously by a non-zero expectation value $\langle 0|\overline{\psi}(x)\psi(x)|0\rangle \neq 0$.

They then showed that this would imply a non-zero mass for the "quasiparticle" of that theory, which they identified with the nucleon.

However, their model also predicted the existence of a pseudoscalar, mass zero, spin zero particle—the "Goldstone" (1961) boson of that model.[24]

2.2 From Goldstone Bosons to the Standard Model

In 1961 Goldstone, in an important paper, described a simple *relativistic* field theory that manifested spontaneous symmetry breaking, i.e., its vacuum, the ground state of the theory, is not invariant under the symmetry that leaves the Lagrangian invariant. It consisted of a complex scalar field ϕ with the Lagrangian density

$$\mathcal{L} = \partial_\mu \phi^* \, \partial^\mu \phi - V(\phi)$$

$$V(\phi) = m^2 \phi^* \phi + \frac{1}{2} \lambda (\phi^* \phi)^2$$

where m and λ are the mass and the self-interaction coupling constant of the scalar field. The model is invariant under a global change of phase

$$\phi(x) \to e^{i\alpha} \, \phi(x)$$

which transformations define the Abelian symmetry group $U(1)$. When $m^2 > 0$ the model is that of a self-interacting scalar field, whose quanta are particles and antiparticles of mass m. However, when $m^2 < 0$ the potential V

$$V(\phi) = -\left| m^2 \right| \phi^* \phi + \frac{1}{2} \lambda (\phi^* \phi)^2$$

has minima on the circle $\left| \phi \right|^2 = \left| m^2 \right| / \lambda$. Therefore in the ground state, the vacuum state, the value of ϕ will be non-zero, with a magnitude close to $\left| \phi \right| = \sqrt{\left| m^2 \right| / \lambda}$ but with arbitrary phase α. There will be a *degenerate* family of vacuum states $\left| 0_\alpha \right\rangle$ labeled by the phase angle α, each $\left| 0_\alpha \right\rangle$ being the vacuum state of a distinct Hilbert space constructed by applying the field operators to it.[25]

Massless Goldstone bosons also appear in this model. This can be seen by choosing a particular minimum, such as, for example, the one where ϕ is real and positive, and expanding about that point. Upon defining the shifted real fields φ_1, φ_2 by

$$\phi = (v+\varphi_1+i\varphi_2)$$

and inserting them in the Lagrangian density, one finds that a term in φ_2^2 does not appear. The model thus describes two kinds of particles: massive φ_1 quanta and massless φ_2 quanta. The massless φ_2 quanta are **Goldstone** bosons. That their presence is required is a consequence of a theorem that Goldstone, Salam, and Weinberg (1962) proved: massless Goldstone bosons are present in any **manifestly Lorentz covariant** field theory in which a **continuous** symmetry is spontaneously broken. The requirement that it be a quantum system with an infinite number of degrees of freedom, such as a quantum field theory, is essential. The following is a heuristic proof of their theorem.

Let $U(\varepsilon)$ denote the unitary representation of the continuous symmetry group which leaves the Lagrangian invariant

$$\mathcal{L}(\phi_a',\partial_\mu\phi_a')=\mathcal{L}(\phi_a,\partial_\mu\phi_a)$$

and under which a field transforms as:

$$\phi_a \rightarrow \phi_a'=U(\varepsilon)\phi_a U^{-1}(\varepsilon)$$

Consider an infinitesimal transformation

$$U(\varepsilon)=1+i\varepsilon_j Q_j$$

where the Q_j are the generators of the transformation. If the vacuum state is degenerate and is not invariant under $U(\varepsilon)$, the non-invariance of the vacuum state implies that $Q_j|0\rangle \neq 0$.[26] Hence for some j, $Q_j|0\rangle$ is some state $|m\rangle$. Since $dQ_j / dt = 0$

$$\langle m|[H,Q_j]|0\rangle = E_m \langle m|Q_j|0\rangle = E_m \langle m|\int d^3x\rho(x)|0\rangle$$
$$= E_m \delta^{(3)}(p_m)\langle m|\rho(0)|0\rangle$$
$$=0$$

and therefore in the limit $p_m \rightarrow 0$, $E_m \rightarrow 0$, $|m\rangle$ is the state of a zero mass particle.

Although Nambu and Jona-Lasinio (1961) and Anderson (1963) had indicated that in the *nonrelativistic* context the Goldstone boson could have a mass, most field theorists believed that in a relativistic theory the Goldstone-Salam-Weinberg theorem required that the Goldstone bosons be massless.

The story of how that limitation was overcome has been told repeatedly (see Bernstein 1974, and in particular Brout 1997, Kibble 2009a, b, and Guralnik 2009). Higgs (1964), Englert and Brout (1964), and Guralnik, Hagen, and Kibble (1964) solved the puzzle of how to do so in relativistically invariant *gauge* theories.[27] They showed that when a gauge theory is combined with a scalar field that spontaneously breaks the symmetry group, the gauge bosons indeed consistently acquire a non-zero mass. Higgs pointed out how in a gauge theory the fixing of the gauge can break the relativistic invariance, rendering the Goldstone-Salam-Weinberg theorem inapplicable.

I have given this somewhat lengthy exposition of some of the elements of the Higgs mechanism because for all its singular features—for example, the use of scalar fields—it became the mechanism most often invoked for "breaking" symmetries, as we shall see when reporting on the exposition of inflation at Shelter Island II.

Also, in addition to providing masses for the gauge vector bosons, the Englert-Brout-Higgs-Guralnik-Hagen-Kibble mechanism generates the masses of the other particles that enter in the description of the theory. Any fermion that interacts with the scalar Higgs field ϕ via an interaction of the form $g\bar{\psi}\Gamma\psi\phi$ may acquire a mass of order $g\langle 0|\phi|0\rangle$. This is the mechanism that gives masses to the leptons and the quarks in the standard model. The masses remain arbitrary parameters, because they are determined by the arbitrary coupling constants g. Thus, the heaviness of the top quark is ascribed to its strong coupling, g, to the Higgs field.

These developments coincided with the developments of methods to quantize Yang-Mills theories—i.e., non-Abelian gauge theories (Fadeev and Popov 1967, DeWitt 1967)—and the proof of their renormalizability even in the broken symmetry phase (Hooft 1971, Lee and Zinn-Justin 1972a, b). This led to the acceptance of the Weinberg-Salam quantum field theoretical model of electroweak interactions (Weinberg 1967, Salam 1968).

The formulation of the theory of strong interactions proved to be more involved, and required the elucidation of how the cross section of deep inelastic electron-proton scattering depends on energy. The 1968 experiments at SLAC introduced the notion of scaling by Bjorken, and of essentially freely moving subnuclear constituents of hadrons by both Bjorken and Feynman, who called them partons. Their connection with the quarks that had been introduced by Gell-Mann and Zweig as fictitious entities to elucidate the $SU(3)$ symmetries of the hadron spectrum, and the incorporation of the color properties attributed to them by Greenberg, eventually led to the current theory of the strong interactions, the non-Abelian gauge theory known as quantum chromodynamic. Its crucial property of being asymptotically free, as proved by Politzer (1973) and by Gross and Wilczek (1973a, b 1974), explained the weakness of the interactions between quarks at short distances, and thus the success of the parton model in explaining scaling. Furthermore, the analysis of renormalizable relativistic quantum field theories by the renormalization group approach that Curtis Callan and David Gross (Callan 1968, 1970, 1972, 1973) and Kenneth Wilson (1973, 1984) had developed from 1968 to 1973 led to the deeply significant conclusion that only gauge theories of the Yang-Mills type could have the property of being asymptotically free.[28]

When Shelter Island II took place, key components of the standard model had been corroborated by experiments accurate to 5%. As then formulated, the model amalgamated the Glashow-Weinberg-Salam $SU(2)\times U(1)$ electroweak theory with quantum chromodynamics (QCD) based on the $SU(3)$ color group as the gauge group. A particularly attractive feature of the standard model is that all the fields are gauge theories. Once the gauge group and its representations are specified, the gauge interactions are uniquely determined; one coupling constant determines the strength of the interactions for a given group factor. Even though gauge invariance seemingly requires gauge bosons to be massless, which would result in long-range forces, the non-Abelian character of the $SU(3)$ color group, which manifests itself in interactions between the color charge carrying gluons, the QCD interactions are in fact short-ranged. Strong coupling effects at large distance scales—reflecting the duality of the color electric and magnetic forces—prevent the colored quarks and gluons from propagating as free particles. The weak interactions too are short-ranged, because the associated gauge bosons acquire masses by the introduction of the Englert-Brout-Higgs-Guralnik-Hagen-Kibble mechanism: that is, by the introduction of a Higgs scalar field which spontaneously breaks the gauge symmetry of the vacuum of the theory. The existence of the scalar Higgs boson was established in 2012.

The success of electroweak theory raised the question whether it is possible to "grand unify" the standard model—that is, whether it is possible to formulate a gauge field theory in which the strong, weak, and electromagnetic interactions would be described as a single simple gauge group with a single coupling constant, with the quarks and leptons and their antiparticles combined in the same multiplet. In 1974 Howard Georgi and Sheldon Glashow formulated an $SU(5)$ model that was the first full unification of $SU(3) \times SU(2) \times U(1)$ into a simple group with a single coupling constant. It yielded electric charge quantification, and thus explained why the electron and the proton have an electric charge of equal magnitude. It also predicted that the proton would decay, but would possess an extremely long lifetime, of the order of 10^{30} years. Though looked for, proton decay has not yet been observed, indicating that it has a lifetime greater than 10^{32} years. Note that proton decay implies that baryon number is not conserved in this grand unified theory (GUT).

In the standard model the gauge coupling constants of the $SU(3)$, $SU(2)$, and $U(1)$ groups are independent quantities. Because of the effects of quantum fluctuations and their renormalization, these coupling constants must be understood as running parameters whose values change with the distance scale being explored, or equivalently with the energy scale being explored. The running of the gauge couplings is described by renormalization group equations. In 1974 Georgi, Quinn, and Weinberg showed that the three gauge couplings would meet approximately at some high-energy unification scale $M_X \sim 3 \times 10^{16}$ GeV. This supports the notion of the existence of a GUT, in which the three interactions are unified and the GUT symmetry is broken at some scale of the order $M_X \sim 3 \times 10^{16}$ GeV by a Higgs mechanism, but above which breaking would not operate.

2.3 GUTs and Cosmology

Grand unified theories (GUTs) offered insights into what might lie beyond the standard model. However, the most dramatic predictions of GUTs occur at the energy scale of 10^{14} Gev. But the only situation in which such high energies existed was in the very early universe, in the immediate aftermath of the big bang.

The big bang cosmological hypothesis was corroborated in 1965.[29]

At the end of the 1920s Edwin Hubble had established that the universe was expanding, with each galaxy receding from every other galaxy with a velocity proportional to the distance between them. Hubble's law is usually written with our galaxy as singled out. It then states that

$v = Hl$

where v is the recession velocity of the observed galaxy, l is its distance from us, and H is Hubble's constant, a "constant" that varies as the universe evolves. This picture of a homogeneous expansion of the universe, when extrapolated backward in time, implies that at some instant in the past all the galaxies coalesced and the density of the universe and its temperature were infinite. Although there had been earlier speculations, by Friedmann and Lemaître, about the evolution of the universe from a big bang, it was George Gamov and his students, Ralph Alpher and Robert Herman, who, in the late 1940s, made use of the extant knowledge of nuclear physics to make the study of the early universe an integral part of cosmology. Gamov pointed out that the very early, very hot universe would have been permeated by blackbody radiation and that as the universe evolved from its hot initial state and expanded, this blackbody radiation would have cooled, and the universe at present would be bathed by the cooled blackbody radiation whose temperature he estimated to be around 7^0K. Gamov, Alpher, and Herman's prediction was confirmed in 1964 when Arno A. Penzias and Robert W. Wilson discovered an isotropic uniform background of microwave radiation with an effective temperature of about $3°K$.[30] It quickly became accepted that the uniform microwave background radiation (MBR) that Penzias and Wilson had discovered was the primordial fireball.[31]

The second set of phenomena that Gamov investigated to support the Big Bang theory was concerned with nucleosynthesis in the very early universe before galaxies and stars had formed. On the assumption that neutrons were the primordial elements,[32] such calculations were first undertaken by Alpher, Gamov, and Herman in the late 1940s. The modern theory of cosmological nucleosynthesis was formulated in 1966 by P. J. E. Peebles, who at the time was a student in Dicke's group studying cosmology. Dicke, in the early 1960s, had started a program of experimental gravitational research at Princeton, and in 1966 had set up an experiment to detect the cosmic background radiation that he believed would be the remnant of conditions of the early universe in the oscillating model of cosmology that he believed in, in which the universe expanded and contracted.

Peebles assumed the validity of the big bang theory. Given the ambient temperature during the first few minutes after the big bang, there were virtually no nuclei present, since nuclei would not be able to remain bound in that environment. The universe then consisted of a hot gas of photons and neutrinos, with a very much smaller density of protons, neutrons, and electrons. As the universe cooled, the protons and neutrons combined to form nuclei. Extending earlier calculations by Alpher, Follin, and Herman, Peebles calculated the expected abundances of the light chemical elements—deuterons, He^3, He^4, and Li^7. He found that most of the matter in the universe would remain in the form of hydrogen, with about 25% by mass being converted into He^4, and He^4 being 10^8 times more abundant than lithium 7. These numbers depended on the present-day abundance of protons and neutrons in the universe, whose value was not known with great accuracy, but they were in general agreement with the observed cosmic abundance of these elements.

Going back in time in the history of the universe correlates temporal changes with temperature changes. One uses the temperature changes to conjecture a historical process for the evolution of the universe in a big bang scenario. Thus, if one goes back sufficiently in time one arrives at an era when the temperature and density were so high that radiation and matter were in thermal equilibrium. But as the universe expanded and the temperature dropped, the radiation went out of equilibrium with matter. However, a detailed analysis indicates that the blackbody spectrum that existed when matter and radiation were in equilibrium was preserved as the photons went out of equilibrium with matter. The story of the evolution of the universe from the time is that it was a very hot plasma of electrons, positrons, and neutrinos, and a very small density of neutrons, protons, deuterons, and helium atoms, can be told with some confidence. Steven Weinberg did so in his popular and very influential exposition *The First Three Minutes: A Modern View of the Origin of the Universe,* which he published in 1977 (see also Weinberg 2007).

After QCD was formulated, the question arose whether the standard model could be incorporated in that framework, with quarks, gluons, and leptons being the "elementary" constituents of the primordial soup. Also of great importance was to understand how the details of the symmetry-breaking mechanism would manifest itself. Kirzhnits and Linde in 1972 had qualitatively investigated whether a *global* broken symmetry would be restored if the temperature were sufficiently high. Weinberg (1974) took up the problem and investigated the case of *local* broken symmetries in relativistic quantum field theories—and, in particular, in gauge theories—and showed how to calculate the critical temperature at which a phase transition occurs for general renormalizable fields by using a Feynman diagrammatic approach. At Weinberg's suggestion, Jackiw also took up the problem of spontaneous symmetry breaking at finite temperature and showed how to calculate the critical temperature more generally by functional methods (Dolan and Jackiw 1974; see also

Kirzhnits and Linde 1976). In Weinberg and Jackiw the spontaneous breaking of symmetry is accomplished by a Higgs-like scalar field that acquires a non-zero vacuum expectation value. The phase transition occurs because thermal fluctuations at high temperatures destroy the non-zero expectation value of the Higgs field.

To better understand the details of GUTS symmetry breaking by the Higgs mechanism, Coleman in 1977 investigated the quantum field theory of a scalar field whose Lagrangian density is of the form

$$\mathcal{L} = \partial_\mu \phi^* \, \partial^\mu \phi - U(\phi)$$

where U has two relative minima, $\phi_\pm$, but only one of them, ϕ_-, is an absolute minimum. The state ϕ_- corresponds to the vacuum state of the quantum theory. The ϕ_+ state, which is a stable equilibrium state in the classical theory, is rendered unstable by barrier penetration at the quantum level. Coleman called it a "false vacuum." The qualitative features of the decay process closely parallel the nucleation processes of statistical mechanics, the crystallization of supersaturated solution, or the boiling of a superheated fluid, which had previously been investigated by Langer (1967). Thus one can think of the false vacuum as corresponding to the superheated fluid phase, and the true vacuum to the vapor phase. In that situation, thermodynamic fluctuations continuously cause bubbles of the vapor phase to materialize in the fluid phase. What happens to a bubble depends on its size and on the gain (or loss) in volume energy as compared to the gain (or loss) in surface energy. Coleman invoked an identical picture to describe the decay of the false vacuum. "Once in a while, a bubble of false vacuum will form so that it is classically energetically favorable for the bubble to grow. Once this happens, the bubble spreads throughout the universe converting false vacuum to true." The issue becomes computing the probability of decay of the false vacuum per unit time per unit volume, with Coleman stressing that "such a computation would be bootless [that is, pointless] were it not for cosmology."

2.4 Inflation

The history of the universe before the period of radiation domination is much more conjectural than the history of the period after. It is generally assumed that the universe immediately after the big bang was homogeneous and

isotropic, had negligible curvature, and can be described generally relativistically by a Robertson-Walker metric

$$ds^2 = -dt^2 + R^2(t)(dx^2 + dy^2 + dz^2).$$

The expansion of the universe is described by the scale factor $R(t)$. The
Einstein field equation

$$\left(\frac{\dot{R}}{R}\right)^2 = \frac{8\pi}{3}G\rho$$

where G is Newton's gravitational constant and ρ is the energy density, determines the behavior of $R(t)$, which in this regime grows as $\sqrt{t}$. The quantity
$H = \dot{R}/R$ is the Hubble constant.

Several authors had speculated that before this Robertson-Walker era there
was an earlier period of inflation, during which $R(t)$ grew exponentially (See
Smeenk 2005). However, the idea attracted little attention until Alan Guth
"incited interest in the possibility of inflation by noting what it was good for"
(Weinberg 2007, 201).

In the late 1970s Guth was working with Henry Tye on seeing what happens
to $SU(5)$ magnetic monopoles in the early universe. They made use of Coleman's
model for the potential of the Higgs-like scalar fields that spontaneously broke
the $SU(5)$ symmetry to $SU(3)$x$SU(2)$x $U(1)$. They also made use of Coleman's
calculations for the decay rate of the false vacuum; that is, when Higgs-like scalar fields get caught in a local minimum of the potential, which is a state with
unbroken $SU(5)$ symmetry. This false vacuum has a huge negative pressure. It is
this peculiar feature of the false vacuum that is responsible for the dramatic effect
that it can have on the evolution of the universe. It produces a constant rate of
expansion—that is, an $R(t)$ that grows exponentially. Barrier penetration eventually stops the inflation, with the scalar fields rolling down the potential toward
the global minimum, which corresponds to the present universe. Guth then noted
that inflation could solve not only the monopole problem (see Smeenk 2005) that
depended on the $SU(5)$ GUT, but also three problems in cosmology, and could
give answers that would depend only weakly on any GUT, answers that non-
inflationary models could not provide. These three problems were

1. How did the universe become so homogeneous on large scales?
2. Why is the mass density of the early universe so extraordinarily close to the critical density?
3. Can one find a physical origin for the primordial density perturbations which led to the evolution of galaxies and clusters of galaxies?

Guth's original formulation of inflation had a major flaw. It was soon replaced by a new inflation model advanced by Andrei Linde and by Andreas Albrecht and Paul Steinhard (Linde 1990, Guth 1997). Both Guth and Linde made extensive presentations at Shelter Island II.

It is surely one of the startling facts about the conference that inflation—a theory that makes conjectures about what happened from 10^{-36} to 10^{-33} seconds after the big bang—should receive that much attention. The fact that inflation is a falsifiable theory—it made a prediction of some of the properties of the fluctuations in cosmic microwave background that can be experimentally tested—is surely a factor for its respectful reception. In 2007, Weinberg would write: "So far the details of inflation are unknown, and the whole idea of inflation remains a speculation, though one that is increasingly plausible" (Weinberg 2007, 202).

2.5 Supersymmetry and Strings

If one construes that an "ultimate" GUT would unify all particles and all interactions, then gauge theories that incorporate the symmetries exhibited by QCD and the electroweak theory in a larger group are examples of how this could be accomplished for the *interactions*. However, the "foundational" entities present in QCD and the electroweak theory are of *two* kinds: fermions (quarks, leptons), and bosons (photons, $W^{\pm}$s, Zs, gluons). One way a more extensive "unification" could be achieved is if all the elementary entities were composites of a set of "fundamental" *fermions*, which would exhibit the symmetry of some "fundamental" group.

Another way would be to look for a new symmetry that transforms bosons into fermions, and thus puts them on equal footing. This latter kind of symmetry is dubbed "supersymmetry" (SUSY). A supersymmetry transformation turns a bosonic state into a fermionic state, and conversely. The operator Q that generates such transformations must be an anticommuting spinor, with

$$Q|Boson\rangle \rightarrow |Fermion\rangle$$
$$Q^{\dagger}|Fermion\rangle \rightarrow |Boson\rangle$$

Because Q and $Q^{\dagger}$ are fermionic operators, they carry spin angular momentum 1/2. Supersymmetry must therefore be a space-time symmetry, and the generators Q and $Q^{\dagger}$ satisfy an algebra of anticommutation and commutation relations with the generators of the Lorentz group (see Mohapatra 1986, Martin 2011). The single-particle states of such a supersymmetric theory fall into irreducible representations of the supersymmetry algebra, called supermultiplets, the members of which all have the same mass. Each supermultiplet contains both fermion and boson states, which are commonly known as superpartners of each other, with each supermultiplet containing an equal number of fermion and boson degrees of freedom. Because of the mass constraint, supersymmetry must be a broken symmetry.

Another formulation of supersymmetry is based on the notion of superspace. (See Wess and Zumino 1974, Gates *et al.* 1983, Seiberg and Wilczek 1999.) Ordinary space-time is given four additional dimensions θ_{α} ($\alpha = 1, \cdots, 4$) that anticommute with one another: $\theta_{\alpha}\theta_{\beta} = -\theta_{\beta}\theta_{\alpha}$. They can be considered quantum dimensions that have no classical analog.

Soon after the formulation of supersymmetric theories, it was recognized that they had remarkable renormalization properties, rendering finite some of the Feynman loop diagrams. How this happens, in which supersymmetrized theories, and why this happens was reviewed by West in his presentation at Shelter Island II (see also Freedman and van Nieuwenhuizen 1978, West 1986).

The fact that in the supersymmetric version of $SU(5)$ the three couplings meet at a ***point*** gave support not only to supersymmetry, but to the notion that a more unified foundational theory might exist. The problem of how to unify the electroweak and the strong forces, and how to incorporate gravity into such a unified theory, was considered one of the most fundamental unsolved problems. However, it was recognized that the usual approach to "quantizing" field systems, whether by Feynman's path integral or by more conventional approaches, could not overcome the difficulties encountered in attempts to formulate a quantum theory of gravity. By the early 1980s, it had become apparent through the work of John Schwarz and others that string theory offered a framework for overcoming the difficulties encountered in quantizing gravity. The theory arose in the late 1960s as an alternative to quantum field theory in which the fundamental entities are considered oscillations of one-dimensional strings rather than zero-dimensional points; a particle is just a string in a particular state of vibration. In string theory, interactions are spread out over the strings rather than concentrated at a single point, thereby preventing infinities. The existence of a spin 2 graviton in the theory came to be seen as a necessary feature of string theory, and string theorists came to characterize gravity as an essential component of string theory and a "prediction" of it .

Supersymmetry likewise came to be considered another general prediction of string theory. It was at the center of many of the presentations at Shelter Island II. In addition to P. C. West's lecture on "Supersymmetry and Finiteness" of quantum field theories, Bruno Zumino lectured on the application of supersymmetric models to the proof of the Atyah-Singer index theorem. In the attempts to unify gravitation and gauge theories, Kaluza-Klein theories and their supersymmetric extensions attracted a great deal of attention in the late 1970s. At Shelter Island II, M. J. Duff lectured on the consequence of making Kaluza-Klein versions of gravitational theory supersymmetric; as did Ed Witten in his lecture on "Fermion Quantum Numbers in Kaluza-Klein Theory."

String theory was also very much a center of interest of many of the participants of Shelter Island II. String theory's point of departure had been in S-matrix theory, but it has become clear that quantum field theories and string theories are deeply intertwined and related by dualities. John Schwarz, one of the developers of the field and an important contributor to it, has come to see string theory "as the logical completion of quantum field theory" (Schwarz 2012).

I refer the reader to Schwarz's beautiful presentation of the subject in Henley 2012. The history of the birth of string theory has recently been narrated and analyzed by Cappelli et al. (2012) and by Rickles (2014).

2.6 Computing

Computers and computing made possible remarkable advances in the calculations of higher-order Feynman diagrams in QED. Kinoshita, Veltman, Strubbe, Hearn, Lautrup, and others wrote extensive programs to do the algebraic operations in reducing the Feynmam diagrams and to efficiently carry out multidimensional integrations by Monte Carlo methods. The size of this subdiscipline can be inferred from the presentations they gave at the Second Colloquium on Advanced Computer Methods in Theoretical Physics, held in Marseilles in June 1971.

At Shelter Island II, Toichiro Kinoshita reviewed what had been accomplished in the calculations of the anomalous magnetic moments of the electron and of the muon to achieve the impressive precision and accuracy experimentalists had obtained since Shelter Island I.

In 1981 Dehmelt et al. made measurements of the anomalous magnetic moment of an electron and of a positron in a Penning trap. They obtained the following values:

$$a_{e^-} = 1,159,652,200(40) \times 10^{-12}$$

and
$$a_{e^+} = 1,159,652,222(50) \times 10^{-12}$$

One usually writes the QED calculation of the anomalous magnetic moment of the electron as a power series in α/π, thus

$$a_e = C_1(\alpha/\pi) + C_2(\alpha/\pi)^2 + C_3(\alpha/\pi)^3 + C_4(\alpha/\pi)^4 + \ldots$$

Kinoshita reported that reliable calculations had been carried out to the sixth order. That is, the values of C_1, C_2, C_3 had been calculated, and after four years of hard work he and Linquist had obtained a preliminary value for C_4 after evaluating some 891 Feynman diagrams! The value of ae calculated from QED is

$$a_e = 1/2\ (\alpha/\pi) - 0.328478966(\alpha/\pi)^2 + 1.1765(13)(\alpha/\pi)^3$$
$$- 0.8(1.4)(\alpha/\pi)^4.$$

To this must be added contributions stemming from the vacuum polarization effects due to muons, tauons, and hadrons, as well as certain weak interaction effects. Kinoshita indicated that there were discrepancies between the calculated and empirical values of a_e stemming from the lack of knowledge of α to the same precision.

Kinoshita's lecture on the calculations and measurements of a_e and a_μ gave proof that they had become tests of electroweak theory and its radiative corrections.

In the Introduction, I suggested that Shelter Island II should be seen as indicative of both the closure of an era and the manifestation of a new era in physics. A dramatic illustration of this is that Kinoshita and Lundquist in 1981 could come up with a reliable *estimate* of the contributions of some 891 Feynman diagrams in a three-loop calculation of QED after renormalization. In 2012 Zvi Bern and his colleagues could undertake a calculation in a supersymmetric gravitational theory that included seven loops contributions that generated tens of thousands of Feynman diagrams—and this to prove that the theory is finite! (See Bern, Dixon, and Kosover 2012.)

2.7 Physics and Mathematics

That mathematicians at times construct mathematical entities and structures, seemingly to satisfy strictly internal needs of the discipline, yet later find useful and important applications in physics is well known. Riemannian geometry and its Levi-Civita tensor calculus formulation, and their appropriation by Einstein as the framework and the mathematical language of general relativity, is an oft cited example of the phenomenon.

In the 1960s and 1970s mathematicians and physicists witnessed a similar confluence of concepts and ideas. On one side were the Abelian and non-Abelian gauge theories describing the electroweak and strong interactions; and on the other was an internally motivated extension of Riemannian geometry involving mathematical structures known as fiber bundles. When studying the topology and differential geometry of surfaces in an arbitrary number of dimensions, mathematicians found it useful to introduce auxiliary spaces, such as the space consisting of the tangent planes to the surface being studied, or the space of lines normal to that surface. Such spaces are called fiber bundles, the fibers being the auxiliary spaces and the bundle the totality of fibers that are fitted together. The classification of bundles and their invariants was studied in the 1930s and 1940s by Whitney, Pontrjagin, Chern, and others. They obtained integral formulas for the invariants of bundles that generalize the Gauss-Bonnet formula. The latter states that the integral of the Gaussian curvature over an entire surface is a topological invariant—that is, the value of the integral is not changed by local deformation of the surface—and is a multiple of 2π.

The mathematician Isidore Singer, who with Michael Atyah had formulated the famous Atyah-Singer index theorem, made a presentation at Shelter Island II on the content of that theorem. The proof of the theorem makes use of elliptic differential operators on a manifold to study its geometry and topology.

Unfortunately, the content of that lecture was not included in the proceedings. However, a year earlier he had written an article in *Physics Today* which gives a very informative and accessible overview of the connections among gauge theories, differential geometry, and topology. Roman Jackiw, in his lecture at Shelter Island II, gave a more intuitive presentation as to why non-Abelian gauge fields divide themselves into topologically distinct, disconnected classes determined by an integer k, called the topological charge, given by

$$k = -(\frac{1}{8\pi^2})\int d^4x \, \mathrm{Tr}\,(\varepsilon^{\mu\nu\alpha\beta} \, F_{\mu\nu} \, F_{\alpha\beta})\,.$$

He pointed out that some quantum numbers and conservation laws may be topological in nature, and stressed the important role topology played in nonperturbative solutions of gauge theories. He illustrated this with the charge fractionalization of polyacetaline.

Zumino in his presentation made clear how Clifford algebras, and hence Dirac operators, make the proof of the Atyah-Singer theorem more transparent to physicists.

To understand why the turn towards geometry and topology after the discovery of string theory, think of the motion of a string through space as sweeping out a two-dimensional surface. One can characterize string theory as the theory of these two-dimensional surfaces. Solving the dynamical problems of strings by using the Feynman integral-over-paths requires summing over all possible two-dimensional surfaces. Mathematicians had classified all possible two-dimensional surfaces according to their topology. Physicists made use of these results.

Edward Witten has been at the center of contributions to geometry stemming from supersymmetric quantum field theories and string theory that have elucidated many aspects of algebraic topology and of topology in general. In 1994 Atyah could state:

> [T]he whole area between Quantum Field Theory and Geometry . . . has now produced a wealth of new results. . . . Witten's work . . . greatly extended the scope of the Jones knot invariants. Other results of this type

. . . include formulae for the volume of moduli spaces, computations of their cohomology, and information about rational curves on three-dimensional Calabi-Yau manifolds. A new and much simpler proof of the positive energy theorem of Schoen and Yau emerged from ideas of Witten, based on the Dirac operator and a super-symmetric formalism. . . ."[33]

I am not competent to analyze these developments. What I want to emphasize is the symbiotic nature of the connection that has evolved between physics and mathematics, a bond that was already manifest at Shelter Island II. What is novel about this interaction, and why I have labeled it a *symbiosis,* is that it involves ideas at the forefront of theoretical physics and of mathematics, that it stems from seeing mathematics as being more unified than previously thought and seeking further unification, and similarly in physics that it stems from attempt at unifying the description of the known forces in nature, and that it involves the leading geometers and the leading field theorists and string theorists.

Epilogue

Shelter Island I assembled the group of leading young American physicists who had been endowed with self-confidence by the war-time work they had done, work that had produced the means to give the Allies victory in World War II. These young physicists went on to produce a usable quantum electrodynamics and laid the foundations for making possible the formulation of the standard model later on.

Shelter Island II brought together leading young physicists who had been endowed with self-confidence by the work they, and other members of their generation, had done to create the standard model. That is why inflation could be taken so seriously at the conference and why, despite the fact that no detailed experimental predictions could be made from string theory, they nonetheless would invest huge efforts into advancing the theory during the next two decades until the LHC went into operation.

Bibliography

Atiyah *et al.* (1994). Responses to "Theoretical Mathematics: Toward a cultural synthesis of mathematics and theoretical physics" by A. Jaffe and F. Quinn. *Bulletin of the American Mathematical Society* 30/2: 178-207.

Atiyah, M. (2002). The unreasonable effectiveness of physics in mathematics. In Fokas *et al.* (2002), pp. 25-38.

Bardeen, J., Cooper, L.N. and J.R. Schrieffer. (1957). Theory of Superconductivity. *Physical Review* **108**: 175-1204.

Bernstein, J. (1974). Spontaneous symmetry breaking, gauge theories, the Higgs mechanism and all that. *Reviews of Modern Physics* **46**/1: 7-48.

Bernstein, J. and G. Feinberg. (1986). *Cosmological Constants. Papers in Modern Cosmology.* (New York: Columbia University Press).

Bern, Z., Dixon, L. J., and Kosower, D. A. (2012). Loops, trees and the search for new physics. *Scientific American, 306*(5), 34-41.

Bjorken, J. 1997. Deep-Inelastic Scattering: From Current Algebra to Partons. In Hoddeson *et al.* (1997), pp. 589-599.

Brout, R. (1997). Notes on spontaneously broken symmetry. In Hoddeson *et al.* (1997), 485-500.

Callan, C.G. 1970. Bjorken scale invariance in scalar field theory. *Physical Review* **D2**: 1541-1547.

Callan, C.G. 1972. Bjorken scale invariance and asymptotic behavior. *Physical Review* **D5**: 3202-3210.

Callan, C.G. and D. Gross. 1973. Bjorken scaling in quantum field theory. *Physical Review* **D8**: 4383-4394.

Callan, C. G., and Coleman, S. (1977). Fate of the false vacuum. II. First quantum corrections. *Physical Review D, 16*(6), 1762.

Cao, T.Y. ed. (1999). *Conceptual Foundations of Quantum Field Theory.* New York: Cambridge University Press.

Cao, T.Y. (2010). *From Current Algebra to Quantum Chromodynamics: a case for structural realism.* Cambridge; New York: Cambridge University Press.

Cappelli, A., Castellani, E., Colomo, F. and P. Di Vecchia, eds. (2012). *The Birth of String Theory.* (Cambridge: Cambridge University Press).

Coleman, S. (1977). Fate of the false vacuum: Semiclassical theory. *Physical Review D, 15*(10), 2929.

DeWitt, B.S. 1967. Quantum theory of gravity. I. The canonical theory. *Physical Review* **160**: 1113.

Dolan, L. and R. Jackiw. 1974. Symmetry behavior at finite temperature. *Physical Review D* **9**: 3320-3341.

Dresden, M. (1987). *H.A. Kramers*. (Berlin: Springer Verlag).

Fokas, A. *et al.* eds. 2002. *Highlights of Mathematical Physics*. (Providence R.I.: American Mathematical Society).

Englert, F. and R. Brout. 1964. Broken Symmetry and the Mass of Gauge Vector Mesons. *Physical Review Letters* **13**: 321-323.

Fadeev, L.D. and V.N. Popov. 1967. Feynman diagrams for the Yang-Mills field. *Physics Letters B* **25**: 29-30.

Forman, P. (1982). The fall of parity. *Physics Teacher* **20/5**: 281-88.

Freedman, D. Z., and Van Nieuwenhuizen, P. (1978). Supergravity and the unification of the laws of physics. *Scientific American*, *238*(2), 126-43.

Gates, S.J. Jr., M.T. Grisaru, M.T., Rocek, M. and W. Siegel. (1983). *"Superspace: or one thousand and one lessons in supersymmetry,"* (Reading, Mass.: Benjamin/Cummings).

Glashow, S.L. (1961). Partial-symmetries of weak interactions. *Nuclear Physics* **22**: 579-588.

Glashow, S.L. (1998). *Interactions: A Journey Through the Mind of a Particle Physicist and the Matter of this World*. With Ben Bova. (New York, NY: Warner Books).

Goldstone, J. (1961). Field Theories with Superconductor Solutions. *Nuovo Cimento* **19**: 154-164.

Goldstone, J., Salam, A. and S. Weinberg. (1962). Broken Symmetries. *Physical Review* **127**: 965-970

Gottfried, K. and V.F. Weisskopf (1984-1986). *Concepts of Particle Physics*. (Oxford: The Clarendon Press; New York: Oxford University Press). 2 volumes.

Gross, D.J. and F. Wilczek. (1973a). Ultraviolet behavior of non-Abelian gauge theories. *Physical Review Letters* **30**: 1343.

Gross, D.J. and F. Wilczek. (1973b). Asymptotically free gauge theories. I. *Physical Review D* **8**: 3633.

Guralnik, G.S., C.R. Hagen and T.W.B. Kibble. (1964). Global Conservation Laws and Massless Particles. *Physical Review Letters* **13**: 585-587.

Guralnik, G.S. (2009). The history of the Guralnik, Hagen and Kibble development of the theory of spontaneous symmetry breaking and gauge particles. *International Journal of Modern Physics A* **24**: 2601-2627.

Guth, A.H., Huang, K. and R.L. Jaffe. eds. 1983. *Asymptotic Realms of Physics: Essays in Honor of Francis E. Low.* (Cambridge, Mass.: MIT Press).

Guth, A. H. (1997). *The Inflationary Universe: The Quest for a New Theory of Cosmic Origins.* With a foreword by Alan Lightman. (Reading, Mass.: Addison-Wesley Publishing, 1997).

Henley, E.M. and S.D. Ellis, eds. (2013). *100 Years of Subatomic Physics.* (Hackensack, NJ: World Scientific).

Higgs, P.W. (1964). Broken Symmetries and the Masses of Gauge Bosons. *Physical Review Letters* **13**: 508-509.

Higgs, P.W. 1997. Spontaneous breaking of symmetry and gauge theories. In Hoddeson et al. (1997), pp. 506-510.

Höche, S. *et al* (2013). Computing for pertubative QCD. A Snowmass White Paper. ArXiv 1309.3598v1.

Hoddeson, L., Brown, L., Riordan, M. and M. Dresden, eds. 1997. *The Rise of the Standard Model. Particle Physics in the 1960s and 1970s.* (Cambridge University Press, Cambridge).

Hooft, G.'t. 1971. Renormalizable lagrangians for massive Yang-Mills fields. *Nuclear Physics B* **35**: 167-188.

Hooft, G.'t and Veltman, M. (1972). Regularization and renormalization of gauge fields. *Nuclear Physics* B 44/1: 189–213.

Hooft, G.'t. 1997a. Renormalization of Gauge Theories. In Hoddeson et al. (1997), pp. 179-198.

Hooft, G.'t. 1997b. *In Search of the Ultimate Building Blocks.* (Cambridge, England; New York: Cambridge University Press).

Jackiw, R., Khuri, N. N., Weinberg, S., and Witten, E. eds. (1985). *Shelter Island II. Proceedings of the 1983 Shelter Island Conference on Quantum Field Theory and the Fundamental Problems of Physics.* (Cambridge, Mass.: MIT Press).

Jaffe, A. and F. Quinn. (1993). Theoretical mathematics: toward a cultural synthesis of mathematics and theoretical physics. *Bulletin of the American Mathematical Society* **21**: 1-13.

Kibble, T.W.B. (2009a). Englert-Brout-Higgs-Guralnik-Hagen-Kibble mechanism. *Scholarpedia* **4**: 6441.

Kibble, T.W. (2009b). Englert-Brout-Higgs-Guralnik-Hagen-Kibble mechanism (history). *International Journal of Modern Physics A* **24**: 6001-6009.

Kirzhnits, D.A. and A.D. Linde. (1972). Macroscopic consequences of the Weinberg model. *Physics Letters B* **42**: 471-474.

Kirzhnits, D.A. and A.D. Linde. (1976). Symmetry behavior in gauge theories. *Annals of Physics* **101**: 195-238.

Kragh, H. (1999). Quantum generations: a history of physics in the twentieth century. (Princeton, N.J.: Princeton University Press).

Kragh, H. (2007). *Conceptions of Cosmos: From Myths to the Accelerating Universe: A History of Cosmology.* (Oxford; New York: Oxford University Press).

Krieger, M. H. (2016). *Doing Mathematics: Convention, Subject, Calculation, Analogy.* 2nd edition. (River Edge, N.J.: World Scientific).

Langacker, P. (2010). *The Standard Model and Beyond.* (Boca Raton, FL: Taylor and Francis).

Langacker, P. (2012). Grand Unification. *Scholarpedia* 7(10): 11419.

Langer, J.S. (1967). Theory of the condensation point. *Annals of Physics*, *41*(1): 108-157.

Lee, T. D., and C. N. Yang. (1956). Question of Parity Conservation in Weak Interactions. *Physical Review* **104:** 1671-77.

Linde, A.D. (1990). *Inflation and Quantum Cosmology.* (New York: Academic Press).

Martin, S. P. (2011). A Supersymmetry Primer. hep-ph/9709356 version 6.

Merton, R. (1968). The Matthew effect in science. *Science* 159, No. 3810: 56-63.

Mohapatra, R.N. (1986). *Unification and Supersymmetry: The Frontiers of Quark-Lepton Physics.* (New York: Springer-Verlag).

Nambu, Y. (1960). Quasi-particles and gauge invariance in the theory of superconductivity. *Physical Review* 117: 648.

Nambu, Y. and G. Jona-Lasinio. (1961). Dynamical model of elementary particles based on an analogy with superconductivity. I, II. *Physical Review*, 122: 345 and 124: 246.

Nambu, Y. (2008). A Superconductor Model of Elementary Particles and its Consequences. *International Journal of Modern Physics A* 23: 4063-4079.

Peirce, C.S. (1881). The architecture of theories. *The Monist* 1/2:161-176. Reprinted in Peirce, C. S. (1963). *Selected Writings: (Values in a Universe of Chance).* Edited, with an introduction and notes, by Philip P. Wiener. (New York, Dover Publications).

Politzer, H.D. (1973). Reliable perturbative results for strong interactions. *Physical Review Letters* 30: 1346-1349.

Politzer, H.D. (1974). Asymptotic freedom: An approach to strong interactions. *Physics Reports* 14: 129-180.

Polkinghorne, J. (1989). *Rochester Roundabout: the Story of High-Energy Physics.* (New York: W.H. Freeman).

Rickles, D. (2014). *A Brief History of String Theory: From Dual Models to M-Theory.* (Berlin, Heidelberg: Springer Verlag).

Salam, A. (1968). In Svartholm, N. ed. *Elementary Particle Physics: Relativistic Groups and Analyticity.*" Eighth Nobel Symposium. (Stockholm: Almquvist and Wiksell), p. 367.

Schwarz, J.H. and N. Seiberg. (1999). String Theory, Supersymmetry, Unification, and All That. The American Physical Society Centenary issue of *Reviews of Modern Physics* 71 (1999) S112. hep-th/9803179.

Schwarz, J.H. (2013). "String theory and M-theory." In Henley and Ellis (2013), pp. 519-550.

Schweber, S.S. (1994). *QED and the Men who Made It*. (Princeton: Princeton University Press).

Schweber, S.S. (2015). Hacking the Quantum Revolution. *European Journal of Physics-H.*

Singer, I. (1982). Differential geometry, fiber bundles and physical theories. *Physics Today* 3: 41-44.

Smeenk, C. (2005). False Vacuum: Early Universe Cosmology and the Development of Inflation. (Boston: Birkhäuser); 223-257.

Treiman, S.B., Jackiw, R. and D.J. Gross (1972). *Lectures on current algebra and its applications*. (Princeton, N.J.: Princeton University Press).

Weinberg, S. (1972). *Gravitation and Cosmology: Principles and Applications of the General Theory of Relativity*. (New York: Wiley).

Weinberg, S. (1967). A Model of Leptons. *Physical Review Letters* **19**: 1264-1266.

Weinberg, S. (1974). Gauge and global symmetries at high temperature. *Physical Review D* **9**: 3357-3378.

Weinberg, S. (1977). *The First Three Minutes: a Modern View of the Origin of the Universe*. (New York : Basic Books). Updated ed. (New York: Basic Books 1993).

Weinberg, S. (1980). Conceptual foundations of the unified theory of weak and electromagnetic interactions. *Reviews of Modern Physics* 52(3), 515-523.

Weinberg, S. (2008). *Cosmology*. (Oxford; New York: Oxford University Press).

Weinberg, S. (2009). Effective Field Theory, Past and Future. arXiv: 1964v3 [hep-th] 26 Sept 2009.

Wess, J., and Zumino, B. (1974). Supergauge transformations in four dimensions. *Nuclear Physics B*, *70*(1), 39-50.

West, P. C. (1986a). *Introduction to Supersymmetry and Supergravity.* (Singapore: World Scientific).

West P.C. ed. (1986b). Supersymmetry: a decade of development. (Bristol; Boston: Adam Hilger).

Wilson, K. (1974). Confinement of quarks. *Physical Review D* **10**: 2445.

Wilson, K.G. (1977). Quantum chromodynamics on a lattice. In *New Developments in Quantum Field Theory and Statistical Mechanics Cargèse 1976* (pp. 143-172). (New York: Springer US).

Yang, C. N. and Mills, R. (1954). Conservation of Isotopic Spin and Isotopic Gauge Invariance." *Physical Review* **96**(1): 191–195.

Yang, C. N. (1959). The law of parity conservation and other symmetry laws of physics. Nobel Lectures Physics: 1942–1962, 1964. Nobel Lecture.

Yang, C.N. (1991). Symmetry and Physics. In G. Ekspong ed. *The Oskar Klein memorial lectures. Vol. 1: Lectures by C.N. Yang and S. Weinberg.* (Stockholm: Stockholm U).

Endnotes

[1]   See Langacker 2010.

[2]   It takes its name from verse 25:29 in the Gospel of Matthew: "For unto every one that hath shall be given, and he shall have abundance: but from him that hath not shall be taken even that which he hath." See Morton 1968.

[3]   See in this connection the recollections of Glashow (1998) and that of Guth (1997).

[4]   I want to emphasize the difference between "foundational" and "fundamental." The quantum theory as formulated in terms of quantum field theories has hierarchesized the physical world. The atomic, nuclear, and subnuclear domains each have a set of "elementary" entities that populate that domain and each has a "foundational" theory, an effective field theory, which governs the dynamics of these entities. See Schweber 2015. See Stephen Adler's lecture at Shelter Island II, which makes the case for considering "Einstein's theory of gravitation as a long-wavelength Effective Field Theory." See also Weinberg's Nobel Prize lecture (Weinberg 1980) and his talk at the 1996 conference on quantum field theory that Cao had organized at Boston University. (Weinberg 1999.)

[5]   Gell-Mann, "From Renormalizability to Calculability," in *Shelter Island* II, p. 33.

[6]   See Schweber 2015.

[7]   Peirce (1881), The architecture of theories.

[8]   See Atyah 2002.

[9]   See Kinoshita's article in Shelter Island II.

[10] See Höche *et al.* 2013.

[11] For the mathematical component see Kreiger (2016).

[12] Schweber, A short history of Shelter Island I. See also Schweber 1994.

[13] I thank Roman Jackiw for sharing some of his recollections of Shelter Island II.

[14] A detailed narrative of how Shelter Island II came to be, its cost, and sociology is in preparation.

[15] All three were sponsored by the National Academy of Sciences (NAS). The second, the Pocono conference was held from March 30 to April 2, 1948 and the third, Oldstone, was held for four days from April 11 to April 14, 1949.

[16] See Polkinghorne 1989.

[17] See Freedman and Van Nieuwenhuizen 1978.

[18] The lecture that Isadore Singer delivered at Shelter Island II on the Atyah-Singer index theorem was unfortunately not included in the book.

[19] See also Cao 1999, 2012.

[20] See Cao 1999 and Kaiser 1999 for a review of these developments.

[21] See Forman 1982.

[22] See also Cao 2010 and also Weinberg 2009 for his involvement with current algebra.

[23] For a historical account of these developments see Schweber 2015.

[24] Interestingly, Nambu and Jona-Lasinio noted that in the BCS theory the particle that plays the role of the Goldstone boson is not massless because of the existence of the long range Coulomb interaction.

[25] Furthermore, the representation of the canonical commutation relations on this Hilbert space is inequivalent to the usual Fock-space representation.

[26] If the vacuum is invariant under $U(\varepsilon)$ the particles in a given supermultiplet will all have the same mass. Similarly, there will be relations among scattering amplitudes.

[27] See in particular Bernstein 1974 for a contemporaneous, historically sensitive account of these developments.

[28] See the lectures by David Gross (1976) on the "Applications of the renomalization group to high-energy physics" that he gave at the 1975 Les Houches Summer School in Theoretical Physics.

[29] For the history of the Big Bang theory and of its rival, the steady state theory, see Kragh 2007.

[30] A detailed history of these developments is given in Weinberg's *The First Three Minutes* that was published in 1979. See also his *Cosmology* published in 2007 for the detailed calculations. Bernstein and Feinberg (1986) also narrate that history, and in addition present the original papers. See also Guth (1999).

[31] Penzias and Wilson had measured the cosmic background radiation at only one wavelength. By the early 1980s further measurements indicated that at least for long wavelengths the distribution of wavelengths agreed with what was expected from blackbody radiation at a temperature of 2.9^0K.

[32] Neutrons would decay into protons, neutrons and protons would combine to form deuterons, deuterons would combine with neutrons to produce He^3, and the latter combining with neutrons would produce He^4. Gamov had hoped to produce all the elements this way, but the

non-existence of stable nuclei with mass 5 and mass 8 implied that Gamov's process of nucleo-synthesis could not account for the formation of the heavier elements.

[33] Witten's work gave rise to a controversy within the mathematics community. Jaffe and Quinn (1993) were concerned about the rigor displayed in Witten's "conjectures" and suggested that these results be confined to a new subdiscipline branded as "theoretical mathematics" to safeguard the standards and purity of the traditional mathematical practice. See Atyah *et al* (1994) for the response of leading mathematicians and mathematical physicists to Jaffe and Quinn's proposal.

Introduction

The Conference on the Foundations of Quantum Mechanics held at Shelter Island, June 1–3, 1947, was destined to have a significant position in the history of modern physics. It was the first major postwar conference on theoretical physics in the U.S. The list of participants reads today like an honor role of modern physics. These two facts would have been enough to single it out. But the Lamb shift was first reported at Shelter Island, and the resolution of the two-meson dilemma began there. These two unplanned events have marked Shelter Island I as the starting point of a series of developments that have radically changed our ideas on the foundations of physics and given us a new cosmology. Many young physicists do not know where Shelter Island is, but all know what happened there in 1947.

The initiative for holding the first Shelter Island Conference should be credited to Duncan MacInnes, a distinguished physical chemist and then a member of the Rockefeller Institute for Medical Research, the predecessor of what is today Rockefeller University; the invitations that the 25 participants received were typed on stationery headed "The Rockefeller Institute for Medical Research." S. Schweber's article (see pp. 301–343) gives a history of Shelter Island I and its origins.

Rockefeller University took the lead in organizing Shelter Island II, and it was fortunate that many of the participants in the first conference could attend the second. The second Shelter Island Conference had three objectives: first, to honor the living members of the class of 1947 and listen to their comments on that event and what preceded it; second, to have a critical review of the developments of the past 36 years; finally, to assess future directions. Shelter Island II consisted of five scientific sessions and one historical session. It was held at The Ram's Head Inn in exactly the same room as the first conference, 36 years later (June 1–3, 1983). This book includes almost all the papers delivered at the conference as well as a summary of the panel discussion on the 1947 conference.

The organizing committe wishes to thank Rockefeller University and its administration for sponsoring the conference and supporting its realization. We are also grateful to Brookhaven National Laboratory for providing audiovisual equipment. Finally, we especially recognize the diligent and excellent work of Judy E. Cuozzo and Gladys Roberts in preparing and running the conference.

N. N. K.

W. Lamb, Jr., and V. Weisskopf

I. Singer and B. Allen

L. Pauling

R. Feynman

T. D. Lee

S. Drell, D. Gross, and H. Bethe

R. Jackiw, F. Seitz, and I. Rabi

S. Hawking and N. Khuri

I SCIENTIFIC PROCEEDINGS

From Renormalizability to Calculability?

Murray Gell-Mann

It is a great honor to be called on to speak at the beginning of this meeting of very distinguished colleagues, celebrating the thirty-sixth anniversary of that remarkable gathering here at Shelter Island. I hope you'll forgive my poorly organized remarks and especially the randomness of my references: I shall attribute research to the authors only occasionally (and no doubt wrongly), but the serious technical lectures that follow will take care of the errors and omissions, in both the physics and the references.

Thirty-six years ago I was just starting to learn particle physics, and it struck me very forcibly that theoreticians were in disgrace. They were in trouble on two grounds: the experimental muon didn't agree with the theoretical meson in its properties; and all radiative corrections in the one theory available—quantum electrodynamics—gave infinity. Just at that time, in the year of Shelter Island, things righted themselves. Marshak's two-meson hypothesis, followed by the experimental discovery of the pion, fixed one difficulty. And the measurement and calculation of the Lamb shift fixed the other one.

Since then, we theoreticians have been riding high. Every once in a while, of course, experiments give us a jolt, as in the case of CP violation or the existence of the third family of spin $\frac{1}{2}$ fermions. (And of course we never knew where the second one came from, or the first one either, for that matter!)

In the meantime—in fact, a decade ago—we theoreticians in partnership with our experimental colleagues have largely solved the problems that confronted us thirty-six years ago, while replacing them with others. We seem to have a renormalizable theory of strong, electromagnetic, and weak interactions that works. In this talk I shall assume that it is correct, with perhaps minor modifications, at least for energies below 1 TeV. It is of course a Yang-Mills theory, based on color $SU(3)$ and electroweak $SU(2) \times U(1)$, with 3 families of spin $\frac{1}{2}$ leptons and quarks, their antiparticles, and some spinless Higgs bosons in doublets and antidoublets of the weak isotopic spin to break the electroweak group down to U_1 of electromagnetism. I shall assume also that the t quark will be discovered by our experimental friends to complete the third family. The Higgs dynamics is, of all this, the part least tested by experiment and the most artificial looking. Otherwise the theory has considerable confirmation, especially now that the intermediate boson, $X^{\pm}$ (which some people still call $W^{\pm}$, although I'm sure that will stop after a while) has been found, twenty-five years after we predicted it.

Essentially, QED has been generalized. QED was invented around 1929 and has never changed; but in the years after the Shelter Island meeting it was understood to be renormalizable, to give definite predictions to every order in perturbation theory. Now, QED has merely been generalized to include the strong and weak interactions along with electromagnetism, the quarks and neutrinos along with the electrons, and the mysterious 3 families. The new theory has similar characteristics, but it has many dimensionless parameters instead of one—very many, especially in connection with the Higgs bosons, which couple to themselves, and then couple separately to all the fermions, with coupling constants adjusted to give all the fermion masses.

The renormalizability means that we can absorb all the infinities into these parameters and into the mass scale, but it doesn't help us to calculate. This is a generalization of the uncalculability of 1/137 in QED. Now, that was always an issue, but people hesitated to take it up in the old days. (I think that was largely because they didn't want to get mixed up with Eddington and his "intricule" and "extricule," and thus be subjected to ridicule.)

When I say "renormalizability" I mean perturbative renormalizability. After thirty-five years of trying, theorists have been unable to invent any scheme for rendering finite or renormalizable a theory that isn't so in perturbation theory. Now, of course some day someone may solve that problem, but I shall assume today that finiteness and renormalizability apply to perturbation theory. I remember many occasions when it had been announced that somebody had solved the problem. The day I arrived at the Institute for Advanced Study as a very young man, I was told that Ning Hu had shown the pseudovector meson theory to be nonperturbatively renormalizable, and I've been told the same kind of thing many, many times since, just the way mathematicians are told all the time that somebody has proved Riemann's conjecture.

(I heard last night from André Martin that Hilbert sent a telegram before he gave a speech in Berlin, saying that he had proved Riemann's conjecture. Then he arrived and the audience was very disappointed in the speech, because he said nothing about such a proof. It turned out he had sent the telegram in case he was killed on the way, on his first airplane trip, of which he was terrified. Disappointed as you may be in my remarks today, it probably won't be as bad as that.)

As usual, solving the problems of one era has shown up the critical questions of the next era. The very first ones that come to mind, looking at the standard theory of today, are

• Why this particular structure for the families? In particular, why flavor-chiral, with the left- and right-handed particles being treated differently, rather than, say, vectorlike, in which left and right are transformable into being treated the same?

• Why 3 families? That's a generalization of Rabi's famous question about the muon, which I'll never forget: "Who ordered that?") The astrophysicists don't want us to have more than 3 families. Maybe they would tolerate a 4th, but no more, with massless or nearly massless neutrinos; it would upset them in their calculations of the hydrogen and helium isotope abundances. Of course, if the neutrinos suddenly jumped to some huge mass in going from known families to a new one, then they would be less upset.

• How many sets of Higgs bosons are there in the standard theory? Well, the Peccei-Quinn symmetry, which I'll mention later, requires at least 2, if you believe in that approach. If there's a family symmetry group, there may be more, because we may want a representation of the family symmetry group: maybe there are 6 sets of 4 Higgs bosons; nobody knows.

• Why $SU(3) \times SU(2) \times U(1)$ in the first place? Here, of course, there have been suggestions. We note that the trace of the charge is zero in each family, and that suggests unification with a simple Yang-Mills group at some high energy, or at least a product of simple groups with no arbitrary $U(1)$ factors. If the group is simple or a product of identical simple factors, then we can have a single Yang-Mills coupling constant.

I should like to emphasize in this talk the thread of renormalizability, finiteness, and calculability. It runs all through our work from Shelter Island I to Shelter Island II—through QED, $SU(3) \times SU(2) \times U(1)$, unified Yang-Mills theories, attempts to unify with gravity, and super-strings. Hence the title: "From Renormalizability to Calculability?"

After the triumphs of thirty-five years ago, many theoreticians—although not my colleague Richard Feynman—changed their minds about the significance of renormalizability. In QED, after all, e and m are arbitrary. Only one of them is dimensionless (and as I said before, nobody wanted to get mixed up with Eddington and his calculation of alpha). So it seemed perfectly all right just to say that the renormalized theory was *the* theory and that we must put in the renormalized e and m. Feynman persisted in regarding renormalization in a different way, as it's treated in the cartoon, courtesy of Pierre Ramond and the artist Joan Cartier.*

Now—is it really like that? Well, in the 1970s I would say that Richard Feynman was led into error by this idea in the sense that he disbelieved charm and neutral weak currents on the grounds that they were needed for a renormalizable theory, and that he thought the requirement of a renormalizable theory was not a good criterion. However, at a deeper level he may still be right. Many theorists are now raising the question of whether our renormalizable broken Yang-Mills theories [the $SU(3) \times SU(2) \times U(1)$ and some higher unified one] are not, after all, just renormalizable phenomenological theories in which the cutoff dependence is replaced by the existence of a huge number of lumped parameters—renormalized dimensionless parameters calculable only in a more fundamental theory of a different kind that would really be finite.

Right in the $SU(3) \times SU(2) \times U(1)$ theory we're faced with the situation that quantum chromodynamics has a hidden uncontrollable parameter, the mixing angle between chromoelectric and chromomagnetic fields that violates CP unless it can be rotated away by a γ_5 transformation on some flavor of quark field. Well, if the u quark actually had 0 ultraviolet mass, that would be OK—but it seems to have roughly half the ultraviolet mass of the d quark rather than 0 times it. We seem to have three more choices: we dial the angle to 0 and look for a theory where the radiative corrections are somehow finite and small enough to be compatible with the experimental levels of CP violation; or we put in a cutoff and say that with the cutoff the corrections are small; or else we find a symmetry that would rotate the angle away. This is the Peccei-Quinn symmetry, which involves at least doubling the higgsons; the associated $U(1)$ symmetry gives us a light modified Goldstone Higgs boson called a higglet, with a very small mass. In connection with the higglet, what happens is that the product of the coupling constant of the higglet to each fermion times the vacuum expected value of some Higgs field is fixed, and if you make one very large, the other one becomes small. It has been suggested in the last few years that maybe the expected value is very large and each coupling constant very tiny, and that

---

* Editors' note: The cartoon referred to shows a cleaning woman tidying up the office of the Field Theory Group by sweeping infinities under the rug; the caption has the cleaning woman saying that she's "got this one renormalized."

then you would get an invisible higglet that no experimental or astrophysical test yet devised could find. Also it has been claimed that if you make the expected value too big, and the coupling constants too small, then the universe is somehow unhappy, so that one must have an expected value somewhere around 10^{12} GeV. You may ask whether that is an important energy in some other connection, and actually it has turned up in some recent other research as a possibly fundamental energy scale. By the way, some people have called the higglet by another name, in which case it's extremely easy to discover in any supermarket.*

Here, as elsewhere, we seem to have to dial various renormalized quantities to small values. The situation of having numerous arbitrary dimensionless parameters is even more humiliating when some of them are very small. Now there's a hierarchy of remedies for this. (If you want to use that word—it means "sacred rule" and has to do with priestly government. I don't know what it's doing in our subject!) First of all, we'd like to interpret small or nearly symmetric quantities as coming from a slightly broken symmetry; otherwise they don't make any sense to us. The small quantities would approach zero in the limit of perfect symmetry. Second, becoming more ambitious, we would like them if dialed to small values to stay there and not acquire uncontrollable radiative corrections resulting in unknown values, which could again be large. Third, and even better, we would like to be able to calculate them and know why they are small. We should distinguish among these three objectives. When we can't avoid dialing some renormalized quantity to a small value (knowing that if the unrenormalized quantity were small there would be infinite and therefore uncontrollable corrections reparable only by renormalization), that situation has recently been described as a problem of "naturalness." In practice each theoretician seems to dial one or more quantities to a small value in his own work and then attacks other theorists for performing unnatural acts when they do the same thing. In fact, the idea of avoiding dialing renormalized quantities to symmetric or nearly symmetric values is really a very old one. Many of us have thought for thirty years (and said) that simplicity and symmetry lay typically in the direction of higher energies, and that effective coupling constants and effective masses as followed upward in energy via the renormalization group would show their true underlying symmetries in the limit of high energy. We showed thirty years ago that symmetry in the limit

* Editors' note: At this point a box of Axion laundry presoak is held aloft by Gell–Mann.

of high energy was essentially equivalent to symmetry applying to unrenormalized quantities. We indicated then that we should not dial renormalized quantities to symmetric or nearly symmetric values, or to zero.

Now, in today's standard Yang-Mills theory, or its unified generalization, because the broken symmetries are violated by the expected values of Higgs fields, which give ultraviolet fermion masses and vector boson masses, the symmetries do show up as you go to energies large compared with those masses.

It has been suggested in the last few years (and not followed up very much) that there may be other kinds of symmetry. For example, authors including Iliopoulos, Tomaras, and Maiani have shown that coupling constants obeying the renormalization group equations, as in other problems in nonlinear systems dynamics, can tend toward symmetry merely by virtue of the properties of the equations. That can happen if the energy is reduced, and such symmetries would never be fully realized. The equations would try to approach a symmetric situation as energy approaches zero, but they would encounter the symmetry-violating masses and the symmetries would never achieve full expression. However, I shall not pursue this fascinating heresy. If we use the usual description of broken symmetry, then what we would like best when we see a small or nearly symmetric renormalized quantity is to have it reflect a broken symmetry, with a finite, calculable correction that will explain the small number. This is actually not easy to do in our broken Yang-Mills theories.

Now all of you know that it's tempting to try to include the $SU(2) \times U(1) \times SU(3)$ theory in a unified Yang-Mills theory, and that what happens if you follow the three coupling constants of the three groups up in energy by the renormalization group equations, assuming fermion families like the ones we see today, is that the coupling constants approach one another at around 10^{15} GeV. The fact that all three come together is nontrivial and suggests that maybe there really is some kind of unification with, say, a simple Yang-Mills group that is badly broken. If you have just 3 fermion families, then the value of the coupling constant near the unification energy is something like 1/40, and that would replace 1/137 as the dimensionless parameter that requires explanation.

The simplest version of this kind of theory uses $SU(5)$, as you know, and employs a large breaking, which smashes the group down to $SU(3) \times U(1) \times SU(2)$ at around 10^{15} GeV by means of one or more adjoint representations of higgsons, and a weak breaking based initially on 5's and $\bar{5}$'s. Unfortunately such a weak breaking presents some difficulties; while the masses of the tau lepton and the b quark have the right ratio when the finite renormalizations are taken into account, the ratio of the muon and s quark masses (if we understand the ultraviolet mass of the s quark, which I happen to think we do) doesn't come out so good. You don't even try to get agreement in the case of the electron and the d quark; you assume that their masses are so small that they must arise from some other mechanism. The discrepancy can be corrected if we complicate the theory by introducing μ-s additional higgsons belonging to other representations of $SU(5)$.

Another possible unified theory makes use of $SO(10)$ and puts each family of fermions into a 16, which on reduction to $SU(5)$ gives a 5 and a $\overline{10}$ and a singlet that is an inactive left-handed antineutrino. A further generalization uses the graph E_6, with each fermion family in a 27.

So ... what I'll call "quantum unified dynamics" (some people call it "tripe" or "guts" or something of that kind), while it has very much to recommend it, is still complicated; and it still doesn't explain the fermion representations and multiplicities or the very large number of coupling constants; it has some of the same difficulties as the $SU(3) \times SU(2) \times U(1)$ theory. The word "guts" is appropriate in this discussion in that it takes enormous guts to suppose that we can extrapolate our ideas from 100 GeV, where we are experimentally, to 10^{15} GeV. Fortunately, there are consequences of these theories that don't involve experiments at 10^{15} GeV.

It should be mentioned that 10^{15} GeV is so close to the Planck mass (2×10^{19} GeV, where the strength of quantum gravity is around unity) that we may consider two possibilities:

One is to regard that ratio as being very important, so that there are two stages—a stage of unification with gravity around the Planck mass, and a stage where gravity can be ignored and Yang-Mills theories unified.

The other is to suppose that there is only one stage and that 10^{15} is just another name for 10^{19}, so to speak.

I would say we don't really know which of those to believe, although the former certainly looks tempting.

One famous important feature of quantum unified dynamics is that without special measures one tends to get proton decay at a rate comparable to the experimental limits. That is one probe of very high energy phenomena at low energies. Another possible probe is connected with neutrino masses, as we shall discuss shortly. It may be that in the future we shall find some other such tests or probes at moderate energies attainable by experiment, say 1 TeV or so. That would be very useful, and also provide additional justification for constructing our next machines. Astrophysical and cosmological tests are also of some value. Basically, what we're doing is looking for rules that obtain in $SU(3) \times SU(2) \times U(1)$, but are violated in a unified theory in a detectable manner.

Anyway, Sakharov's ingenious suggestion many years ago that proton decay, a suitable CP violation, and nonequilibrium conditions in the very early universe could explain the predominance of matter over antimatter was sharpened by many clever theorists after quantum unified dynamics suggested that the proton decay be taken seriously. And it looks now like an idea that we would hate to lose, especially since a rough prediction of the number of baryons over the number of photons can be given, and is not that far off. So quantum unified dynamics that has proton decay is desirable; and although scenarios are available for complicating the theory to avoid proton decay, we should probably not invoke them. An analogous scenario is found naturally in the simplest version of $SU(5)$ and gives rise to conservation of baryon number minus lepton number. As you know, that restricts proton decay so that it can yield a positron plus mesons but not an electron. It also has another effect; neutrino masses are made impossible. That result depends, though, on the group and on the choice of transformation properties of the particular Higgs field components that are nonzero in the vacuum. If you go to $SO(10)$, which is a rather nice generalization of $SU(5)$, not only do you add to each fermion family a right-handed neutrino and a left-handed antineutrino, but also you can easily arrange the representations so that you violate the conservation of the number of baryons minus the number of leptons.

Neutrino masses then arise in the following ways: you can get a huge Majorana mass m_M for the right-handed neutrino; you can have a normal-size Dirac mass m_D connecting the left- and right-handed neutrinos; and to order m_D^2/m_M you can get a small left-handed neutrino mass as well. Now, these left-handed neutrino masses are quite interesting astrophysically and cosmologically, because we need dark matter to bind the galaxies and clusters gravitationally. Dark matter is important to the universe as well; here not enough is known about galactic evolution for astronomers to be sure, but it looks as if enough dark matter is needed to make the universe at least approximately flat asymptotically. One possibility for that dark matter is for it to consist of elementary particles, and one possibility for the elementary particles is to have neutrinos with tiny masses. The sum of those left-handed neutrino masses would have to be of the order of tens of eV if all the dark matter consists of neutrinos. Larger masses for the neutrinos would give trouble: the universe would be overclosed (except in the case of enormous masses, which would lead to a totally different regime). If there are masses for these left-handed neutrinos, then there will also be comparable or somewhat smaller transition masses among them, and those are being actively sought in the laboratory. Other candidate particles for dark matter are now envisaged by theoreticians, including higglets.

Let us return now to the question of mysterious small numbers. In unified Yang-Mills theory, we have a number of them and they are very small. The renormalization-group-invariant mass, which is sometimes called Λ_{QCD}, in units of the unification mass, gives something like 10^{-16}. The mass of the weak bosons, which comes from e times a Higgs expected value, in units of the unification mass, comes out around 10^{-13}. The masses of the various fermions, which come from g coupling constants multiplied by expected values of the various Higgs φ's, where those g's are Higgs couplings to the fermions — range from 10^{-13} for the t quark (if they find it around present energies), down to 10^{-18} for the electron. We want to know why these are so small, and also why they are of the same general order of smallness: after all, one of them could be 10^{-100}! In the case of Λ_{QCD}/m_{UNIF}, we know that it comes from the logarithmic change of coupling constants as we move up from Λ_{QCD} to the unification mass, and so it can be represented as $\exp(-C/\alpha_{UNIF})$, where C is a known number and α_{UNIF} is the unified fine structure constant. We are then led to the question of why the fine structure constant is small compared with 1—but at least we don't have to explain an exponential smallness any more! What about the other small constants? Again, we can be slightly ambitious, moderately ambitious, or very ambitious. We can try to fix them so that if we dial them to small values, they stay there; or we can try to explain why they're nearly zero—for example, by showing that they also correspond to such exponentials (in certain models that's true); or we can actually calculate all of the small numbers, and that has so far eluded everybody. But the last is what we really want: to be able to calculate the constants. To express some of them as $\exp(-C/\alpha)$ is OK, but then we ought to be able to calculate the coupling constant involved.

Now, in $SU(3) \times SU(2) \times U(1)$ Yang-Mills theory, and especially in unified Yang-Mills theory, a great deal of effort has gone into trying to make modest improvements in this "naturalness' situation. In quantum unified dynamics, as far as I know, no one has ever given a thorough explanation, free of difficulties, of the small ratios I've just listed—why they're all small and why they're all of the same general order of smallness. Certainly no one has calculated all of them!

However, in some schemes there has been partial success, especially in showing that if the constants are dialed to small values they stay small. The greatest difficulties of "naturalness' occur in connection with keeping the masses associated with the Higgs bosons low, as required, for example, to keep the weak breaking tiny in quantum unified dynamics (the ratio of 10^{-13} or so). We can regard the standard $SU(3) \times SU(2) \times U(1)$ theory as a phenomenological renormalizable theory within quantum unified dynamics—if there is a quantum unified dynamics. The well-known tendency of spinless bosons in field theory to get quadratic self-masses emerges here as a difficulty for the weak breaking higgsons to avoid acquiring large masses from the cutoff, so to speak, when we consider the small theory as being a renormalizable effective field theory within the big one. Most of the nostrums of the last decade—fashions that last a year or two, typically—have been connected with fixing certain Higgs boson masses to be low, first tying them down to zero, and then, with corrections, to small values more or less calculable depending on the nostrum. One type of prescription had to do with discrete symmetries. More recent attempts can be classified in two ways: according to the degree of success in explaining the small parameters (success that's never been complete, of course) and according to which "elementary" particles are treated as composite, if any.

The proper name, I would suggest, for an elementary field in quantum field theory is "haplon." I encourage every one to use that name. It comes from the Greek for "single" or "simple." In fact it's cognate in Indo-European to "simple" because the Indo-European "s" becomes an "h" in Greek or Welsh. Sun, for example, is "helios" in Greek and "haul" in Welsh. The "haploid" generation is very familiar in biology, implying a single set of chromosomes instead of two, and so forth. This choice of name has already received the approval of many physicists—Dimopoulos, Nanopoulos, and Iliopoulos, and for the benefit of my French friends I add Rastopopoulos.

The question is, which of our familiar "elementary" particles are kept on as haplons? And which ones are sacrificed to complexity? Everyone's favorite candidate for being thrown to the wolves is the higgson. One idea was to fabricate them out of new fermions and antifermions possessing a new, exactly conserved color called "supercolor," "primed color," "hypercolor," "technicolor," or some such, with a Λ' (renormalization-group-invariant mass) much bigger than Λ_{QCD}. Quarks and leptons and the new fermions would all lack ultraviolet masses. They would have only infrared masses, which would look ultraviolet to us at energies much smaller than Λ'. This scheme hasn't been too popular lately, perhaps because a generalization of the primed color group to a larger group, approximately conserved, that would explain the concentration of mass in the highest family of quarks and leptons, gave rise to excessively large neutral current interactions connecting one family with another.

Some schemes have made the quarks and leptons composite. In other schemes, the $X^{\pm}$ bosons, which some people call $W^{\pm}$, and the Z^0 are easily sacrificed, since they are not massless gauge bosons for exact symmetries. The final step would be to make the photon and gluons composite, so that the whole $SU(3) \times SU(2) \times U(1)$ theory would contain only phenomenological fields, all composite objects. In the case of photon and gluons, one should check in any particular theory whether their masslessness and the exact conservation of charge and color are preserved when they are composite.

The favorite method nowadays for gaining control of corrections to small quantities is to apply $N = 1$ supersymmetry either to the standard theory or to quantum unified dynamics; as you know, $N = 1$ supersymmetry connects each state of a given J_z to one with a value of J_z differing by a half-unit. The spin 0 higgsons are thus connected to otherwise unwanted spin $\frac{1}{2}$ superpartners, which one can tie down near zero mass by using approximate chiral invariance. In this way (or in other related ways) one can arrange for the light higgsons to stay light. Their spin $\frac{1}{2}$ superpartners, like other unwanted superpartners, can then be pushed up to higher masses where they would not have been detected.

There is a whole "slanguage" that has evolved recently to describe new particles required by supersymmetry. I must admit to having been present when Glennys Farrar and Pierre Fayet coined names like "photino" and "gluino," for the spin $\frac{1}{2}$ partners of the spin 1 bosons, and "goldstino." The "goldstino" is the spin $\frac{1}{2}$ Goldstone fermion with zero mass that we need in order to allow spontaneous breaking of supersymmetry; and supersymmetry, if it is an exact symmetry, has to be broken spontaneously because we do not see a universal degeneracy of fermions and bosons. In supergravity, which we will discuss soon, there is a gravitino (spin $\frac{3}{2}$) that accompanies the graviton (spin 2). The gravitino is ideally poised for eating the goldstino and thereby acquiring mass, and undoubtedly that's what happens if supersymmetry is right. The goldstino can be either elementary or composite: it can occur in the list of haplons or it can arise dynamically; either way it is there for the hungry graviton to eat. The rest of the slanguage is needed because the spin $\frac{1}{2}$ quarks and leptons are accompanied by unwanted spin 0 partners called squarks and sleptons, and the higgsons, as we indicated before, have unwanted spin $\frac{1}{2}$ partners called, I suppose, shiggsons. (It sounds awful, but some people can say all this with a straight face!) All of these are accompanied by new spin $\frac{1}{2}$ and spin 0 particles, superpartners of each other, which nobody wants otherwise, but which are there in order to break the supersymmetry in the manner prescribed by O'Raifertaigh. (His name, by the way, is written in a simplified manner; the "f" should really be "thbh.")

There has been some success along these lines, but more in the way of quantities staying small when dialed to be small than in the way of small quantities genuinely explained. Meanwhile, experimentalists are delighted that at least some theoreticians have made the desert bloom, and in fact turned it into a jungle.

Supertheories tend to have reduced divergences, and in some cases are completely finite in perturbation theory. Most applications exploit those properties, but one recent area of research ignores them, in what turns out to be a very interesting way. That is to say, some theorists take $N = 1$ supergravity, with one graviton and one gravitino, and couple it to the $N = 1$ super-Yang-Mills super-matter that we've just been describing. Now this is a crime, from the point of view we are adopting, because when gravity is accompanied by external matter, or when supergravity is accompanied by external supermatter, the divergences in perturbation theory are severe. However, one introduces a cutoff around the Planck mass, and then one calculates. New quantities of order unity occur, but one can only guess their values, so there is a good deal of freedom. One notices that a lot of familiar, important calculations are considerably altered if these numbers are correctly estimated. For example, the rate of proton decay and the preferred decay modes are changed (avoiding trouble with present experiments, by the way). Small neutrino masses, if they exist, can be altered. Quark-to-lepton mass ratios, which give some difficulties, as we indicated, can be improved if the unknown quantities come out right, and so on. Here supergravity is essentially a wild card that enables us to play the game in a much more exciting manner. However, since the quantities can only be estimated rather than calculated, it is a bit dangerous.

To return to our principal theme, what we want is, of course, a fundamental underlying theory. Whether the haplons include all, some, or none of our familiar elementary particles, except perhaps the graviton, we want everything to come out finite with no dimensionless parameters to start with, or at most one dimensionless parameter that gets renormalized, and a rest from the adjustability of numerous dimensionless quantities, of which, I think most of us are tired. Whether our impatience is justified, we don't know; it may be hundreds of years before we can get such a theory, if ever, but it's certainly what we would like.

West will, I think, tell us about $N = 4$ super-Yang-Mills and related theories that exhibit finiteness (no infinite charge renormalization anyway),* but they have some other problems, such as that all particles have to be put in the the adjoint representation of the gauge group, which is conceivable but not terribly nice unless the graph is exceptional. Getting scale-invariance violation off the ground is also rather difficult in those theories, but still they are interesting.

It is also possible that there are ordinary $N = 1$ super-Yang-Mills theories that are finite; if there are any that agree with phenomenology, that would be exciting.

But the best candidates we have for fundamental theories in which there is unification of all interactions and all particles, at most one dimensionless parameter, and possible renormalizability or finiteness are theories related to $N = 8$ supergravity, where we have 8 supersymmetries, 8 gravitinos, and so forth. In fact, we have a list of haplons with 1 graviton, 8 gravitinos, 28 spin 1 bosons, 56 spin $\frac{1}{2}$ fermions, 70 spin 0 bosons, and a partridge in a pear tree. In this theory, the number of dimensionless parameters is zero or one. There is κ, the square root of Newton's constant, which is dimensional; if that's all there is, then the theory is characterized by a chiral $SU(8)$ symmetry. However, there is also the option of introducing a dimensionless self-coupling parameter e, which gives a Yang-Mills character to those 28 spin 1 bosons; they gauge $SO(8)$, and the symmetry is thereby reduced from $SU(8)$ to $SO(8)$. This theory is only mildly divergent, if at all. In fact, ordinary gravity without external matter is only mildly divergent if at all! Gravity is suspected (but has not been found guilty) of a logarithmic divergence in two loops. The various supergravities are suspected (but have not been found guilty) of logarithmic divergences at higher numbers of loops. For $N = 8$ supergravity, there might be a divergence at three loops; some investigators think there will be no trouble before seven loops; and perhaps there is no trouble at all. There is no room for external supermatter

---

* Editors' note: The paper read to the conference by West appears in this volume.

in connection with $N = 8$ supergravity. All haplons are in the same supermultiplet.

There are variations of $N = 8$ supergravity. One is $N = 7$ supergravity, which has the same list of particles and a slightly reduced symmetry. Then there are generalizations to higher numbers of spatial dimensions, which are extraordinarily interesting, and, on restriction to four dimensions, permit reductions of symmetry and the generation of coupling constants. And finally there are superstring theories, which are the most elegant candidates for a unified description of Nature.

First, I would like to describe the most serious problem with this whole set of theories, $N = 8$ supergravity and all the beautiful modifications thereof; namely, how to relate it to what we know. The work on Yang-Mills and super-Yang-Mills theories that I described was unsatisfactory in some respects, but at least it dealt in part with familiar objects, and we were able to relate all of it in some way or other to experiment. Here we go into a different realm where everything is very hopeful, very beautiful, exceedingly promising, probably renormalizable or maybe even completely finite, free of parameters—but with no obvious connection with experiment! We have somehow to make a bridge between these two domains.

Yuval Ne'eman and I looked at this question a long time ago, before supergravity was written down, but after $N = 8$ supergravity could be envisaged. We studied the quantum numbers involved; in particular, we assumed that the $SO(8)$ would be gauged, and we noticed, obviously, that $SO(8)$ is too small to include $SU(3) \times SU(2) \times U(1)$. Well, we tossed out the $SU(2)$ bosons—they're expendable, as we said, because they are not massless and their symmetries are violated in some mysterious manner by higgsons. That leaves $SU(3) \times U(1)$, which we tried to identify with color and electric charge. We found that the spin $\frac{1}{2}$ fermions, while they look almost right, don't come out quite the way we want. The fundamental **8**, in which the gravitino appears, would be a triplet and an antitriplet of color and two singlets, with charges q, $-q$, q', and $-q'$, respectively. The easiest choice to take is $q' = 0$ and $q = -\frac{1}{3}$; then the fermions have some integral and some fractional charges, so there are things that look like leptons and things that look like quarks. That is very attractive, but one also gets a **6**, a $\overline{\textbf{6}}$, and an **8** of color, which we don't see, and too few quarks and leptons. The total number of spin $\frac{1}{2}$ fermions is all right, but some are wasted on sextets and octets.

Now let me show you a curiosity. I apologize for wasting your time on this scheme, which has probably nothing to do with physics; I offer it in the hope that somebody here can figure out something to do with it, because I can't. It is a last-ditch effort to salvage the identification of the spin $\frac{1}{2}$ haplons of $N = 8$ supergravity with the fermions that we know. There are the right number, in a certain sense. If we assume that the goldstinos are haplons and take 8 of them out of the 56, then we are left with 48, which is 3 times 16, and, as we mentioned in connection with $SO(10)$, there may be 16 left-handed fermions in each family if an inactive left-handed antineutrino is included. The symmetry is reduced from $SO(8)$, to $SO(7)$, breaking **56** into **48** + **8**. Of course $SO(7)$ contains $SU(3) \times U(1)$. We introduce values of the $U(1)$ generators that permit the same assignments for the 8 goldstinos and the 8 gravitinos, and then we examine the $SU(3) \times U(1)$ assignments of the 48 spin $\frac{1}{2}$ fermions. We find $(\mathbf{8 + 1}, \frac{1}{2})$; $(\mathbf{6 + \overline{3}}, -\frac{1}{6})$; $(\overline{\mathbf{3}}, -\frac{1}{6})$; and $(\mathbf{3}, -\frac{5}{6})$. These correspond to the observed quarks and leptons if we assign the charge $\frac{2}{3}$ quarks and the neutrinos to a $\overline{\mathbf{3}}$ of family $SU(3)$; we assign the charge $-\frac{1}{3}$ quarks and the negatively charged leptons to a **3** of family $SU(3)$; we identify the $U(1)$ generator with electric charge minus $\frac{1}{6}$ for the $\overline{\mathbf{3}}$ of family $SU(3)$ and with the electric charge plus $\frac{1}{6}$ for the **3** of family $SU(3)$; and we identify the $SU(3)$ of the theory with the diagonal $SU(3)$ of color $SU(3)$ times family $SU(3)$. These are a great many ifs and I have no idea how to justify them. Under family $SU(3)$, the charged weak interaction would be part of a **6**, the part that becomes a singlet when $SU(3)$ is reduced to $SO(3)$. I mention this scheme mainly to show to what lengths one has to go to associate the spin $\frac{1}{2}$ haplons of $N = 8$ supergravity with quarks and leptons. In this theory, we probably have to admit that quarks and leptons are composite.

Such a proposal was made by Ellis, Gaillard, and Zumino in 1980. I was working on similar ideas in the next office, but I didn't believe all their conclusions. What they did was very ingenious. They put $e = 0$, so that $SO(8)$ is not gauged, thus allowing the full global symmetry of chiral $SU(8)$. Then they supposed, as Cremmer and Julia had conjectured, that the $SU(8)$ somehow gauges itself dynamically. Next they assumed that the charge and the $SU(3)$ of color are not inside the $SO(8)$, but half in and half out of it. That way they didn't have to make the assignments symmetrical. For example, for the eight $J_z = \frac{3}{2}$ gravitinos, they were able to take the charges to be $-\frac{1}{3}$, $-\frac{1}{3}$, $-\frac{1}{3}$, $+1$, and four 0's. Then the first five constitute the **5** of $SU(5)$, and the last three constitute the **3** of a sort of highly broken family $SU(3)$, which is supposed to account (because 3×3 antisymmetric is $\overline{\mathbf{3}}$) for the three families of fermions.

Then they went through some less convincing arguments in order actually to drag out $SU(5)$ with three families; I pass over those in silence. A very important idea, however, is the suggestion that all the spin 1, spin $\frac{1}{2}$ and spin 0 particles in quantum unified dynamics are fake. None of them is a haplon. Of the familiar particles, only the theoretical graviton is a haplon. Quantum unified dynamics (including, of course, the standard Yang-Mills dynamics) is a phenomenological renormalizable field theory, and all the parameters in it are lumped parameters that express the dependence on the cutoff somewhere up near the Planck mass. Thus an explanation is offered of the complexity of unified dynamics, with its numerous parameters and apparently arbitrary representations.

This last idea they attributed to Veltman; I don't know whether Veltman admits having invented it. There is also 't Hooft who has worked on it, I am told, and Norton and Cornwall at UCLA, and perhaps others. I think it is an intriguing notion that none of today's elementary particles is elementary except the graviton. We will hear, by the way, I think from Steve Adler,* about the opposite notion that the graviton is composite and Yang-Mills fields are elementary. Maybe they are made up of each other and we're back to the bootstrap! That is not really a joke, as we shall see.

To return to $N = 8$ supergravity, I've mentioned the various attempts to connect it to experiment. Now let us describe some of its generalizations. $N = 8$ supergravity can be derived by dimensional reduction from two distinct $N = 2$ supergravity theories in 10 dimensions called A and B. We note also that A, but not B, can be derived from $N = 1$ supergravity in 11 dimensions. It is very tempting, therefore, as suggested by Scherk and Schwarz about eight years ago, to take 6 or 7 extra dimensions seriously and give them nonzero size. These dimensions probably have to be small, although it would be very interesting to know what the limits are on such dimensions from experimental physics and cosmology. I would prefer not to be constrained by the argument that a size R around the Planck length is required. That argument stems from the fact that the reciprocals of effective coupling constants arise from κ (the square root of Newton's constant) times these Rs, and that is a possible way, of course, of ultimately calculating coupling constants. If we try to identify these coupling constants with familiar ones that are no smaller than 10^{-2}, we must have R no larger than 10^2 times the Planck length. But perhaps those coupling constants are not the familiar ones and are very small.

* Editors' note: The paper read to the conference by Adler appears in this volume.

Many theorists have followed up the idea of extra dimensions within the framework of "Kaluza-Klein theory." The extra dimensions, taken seriously, tend to give internal symmetries, even broken internal symmetries—and they are really internal! The extra dimensions are wrapped up in a tiny structure, and for the first time the name "internal symmetry," which I always thought was silly, is justified. The isometries of the structure form a gauge group. One can use not only a linear representation of some symmetry group, like $SO(7)$ in 7 dimensions, or $SO(6)$ in 6 dimensions, but also (as I'm sure we'll hear at great length at this conference) a nonlinear quotient space representation—for example, $SU(3) \times SU(2) \times U(1)$ divided by $SU(2) \times U(1) \times U(1)$, which gives 7 dimensions; or $SU(3) \times SU(2)$ divided by $SU(2) \times U(1) \times U(1)$, which gives 6 dimensions. It may seem that $SU(3) \times SU(2)$ is too small, but remember, in 10 dimensions we have $N = 2$ supergravity, and so we have an additional $U(1)$ left over, and we can get $U(1) \times SU(3) \times SU(2)$ after all. (This remark is due to John Schwarz.)

Ideally the "compactification" of the extra dimensions will occur spontaneously, as a result of the equations of motion in the large space. Happily, mechanisms of spontaneous compactification are being found and will surely be discussed at this meeting.

Finally we go to string theories, which are particularly beautiful. You remember what a string theory is: it has an infinite number of states with increasing angular momenta lying on an infinite number of initially linear Regge trajectories. Early string theories, starting with the Veneziano model, were invented to describe hadrons according to the bootstrap idea. But then we developed QCD, and as far as hadrons were concerned, strings were seen as providing only a crude approximation to QCD. For hadronic purposes, of course, the slope of each Regge trajectory was around $1 \; (\text{GeV})^{-2}$.

The Veneziano model contains only bosons, and it also predicts a state of negative mass squared, a difficulty that has never been convincingly overcome by any explanation in terms of an unstable vacuum. But around 1970, Neveu and Schwarz found a string theory with bosons and fermions; important work on its properties was done at that time also by Ramond. The "critical dimension" for that string theory is 10, which means it is known to work in 10 dimensions, and might possibly exist in fewer dimensions if some conjectures of Polyakov and others turn out to be right; I shall assume here that it requires 10 dimensions.

A few years later, Scherk and Schwarz did further work on the Neveu-Schwarz string theory and adapted it to an entirely different task. They made a slight modification in the slope of the Regge trajectory, by a factor of 10^{38}, and adapted it for use in gravitation. It was later shown that the theory didn't have any problems: that there was no negative mass squared and no negative probability, assuming a modification suggested by Gliozzi, Scherk, and Olive.

The Neveu-Schwarz theory is supersymmetric, and indeed their work and that of Ramond had anticipated in a sense the invention of supersymmetry. Their old superstring, when restricted to massless states in 4 dimensions, was now shown to give $N = 4$ supergravity with $N = 4$ super-Yang-Mills matter. But, treated in 10 dimensions, it was renormalizable—string renormalizable to at least one loop. I say string renormalizable because the string theory is so far treated only on the so-called mass shell. It hasn't yet got the regular Lagrangian formalism, although there's no reason to believe that there isn't one, and Green and Schwarz are working on it. The full Lagrangian formalism may be available soon. Anyway, although string renormalizable in 10 dimensions, the theory is hideously divergent, of course, when we restrict to 4 dimensions, because we have supergravity with supermatter. But this $N = 4$ super-Yang-Mills theory by itself is finite, as you will hear from West.

More recently, Brink, Green, and Schwarz have brought us "Superstrings II." This theory was not considered years ago because it doesn't have any open strings, and at that time people wanted open strings and closed strings for hadron purposes. Because Superstrings II has no open strings, it has no external supermatter when trivially reduced to initially massless states in 4 dimensions, and lo and behold, it becomes $N = 8$ supergravity! So this theory, which started from nothing but the idea of a string with bosons and fermions, fixes 10 dimensions, and then when you restrict it to zero mass and to 4 dimensions, fixes $N = 8$ supergravity as the limit. If you stop on the way, with the massless states in 10 dimensions, you get $N = 2$ supergravity. And just as there are two forms, A and B, of $N = 2$ supergravity in 10 dimensions, one of which can be obtained by reduction from 11 dimensions and one not, so there are two forms of Superstrings II—IIA and IIB. These theories are miraculously finite to one loop and may well be finite to every order, and that's rather exciting. Thus we have as advantages that the number of dimensions is fixed, that supergravity comes out and doesn't have to be put in, and that the theory may be finite to all orders. But we're left with the fundamental question of how superstrings are related to the real world.

Let me add two remarks, one on the cosmological constant and one on the inflationary cosmology. You'll hear about both of these from lots of speakers, so I won't have to say much. I think the cosmological constant problem is one of the key issues of fundamental physics, and it's another question of renormalization. Renormalization, finiteness, and calculability are the whole story at our meeting here, as far as I can see, and that's entirely appropriate for a sequel to the 1947 Shelter Island Conference. Here we deal with renormalization of zero-point energy; that's something we always used to pass over very quickly in our books and classes: we put in some funny little points, and said it was gone. But GRAVITY IS WATCHING! It notices when we take away the zero-point energy, including the higher order corrections to that energy, and it says, "They've taken away a term in $\delta_{\mu\nu}$ from the $\Theta_{\mu\nu}$ tensor, and therefore they've added a term in $\delta_{\mu\nu}$ to Einstein's equation for me"—and that's a cosmological constant. The astronomical cosmological constant, which may or may not be identically the same quantity as the one in the equation (that's a problem of renormalization also, in a certain sense) is known to be zero or very close to zero. In natural units, it's 10^{-118} or smaller, natural units being those involving the Planck scale. That's the largest discrepancy in physics; it's large even for astrophysics.

(Gamow used to make fun of the big mistakes that astrophysicists tolerate. He wrote a letter about something or other in astrophysics and cosmology and deliberately made an error, preparing an erratum in advance. He sent the letter to the *Physical Review*, and then after it was printed he sent in the erratum, which said roughly, "In Equation so-and-so, there is a mistake by a factor of 10^{24} on the right-hand side. This does not affect the result.")

Anyway, 10^{118} bothers even astrophysicists. In supersymmetry theory (and this was originally one of the great things about supersymmetry) zero-point energy can be made to vanish to all orders if you include a constraint called R symmetry. However, when you put in spontaneous violation of supersymmetry, you get back a cosmological constant. Then, in supergravity theory, if you put in a self-coupling e, so that you have gauging of $SO(N)$, you get another contribution with the opposite sign. Maybe they cancel—but nobody has yet found a way to cancel these terms naturally. If one simply dials the algebraic sum to zero, one is introducing the largest fudge factor ever.

Of course, if supersymmetry is explicitly violated, the algebra is altered. That's another possible source of the cosmological constant.

Then Hawking has developed a whole new way of looking at these matters, which I'm sure he will discuss,* in which the quantum fluctuations of gravity or supergravity, including the topology of the space, result in bubbles in space-time of Planck length size, forming a foam, which, he says, disguises a huge fundamental cosmological constant as a zero effective cosmological constant for the universe as a whole, thus divorcing the astronomical from the fundamental. Well, we'll see how well that works; it's an extremely ingenious idea.

Then we shall hear from Linde and from Guth about the remarkable new explosive or inflationary cosmology that accomplishes a gigantic increase in entropy of the universe during its early moments and dilutes out all sorts of unwanted things: monopoles, asymptotic curvature, inhomogeneity, and anisotropy, while apparently explaining the horizon paradox as well. All this by a mere addition of 10^{87} to the entropy. A very beautiful idea, in my opinion. The transition is accomplished specifically by a phase change in the vacuum, which is usually attributed to the quantum unified dynamics around 10^{15} GeV, although there might be other possibilities. Perhaps people will want to explore the idea that it occurs at an earlier era, or at least at a higher energy, and involves gravity more intimately, or supergravity. Possibly there are even extra dimensions—sizable extra dimensions comparable with the others—at the moment when this happens. Or superstrings: since these have huge entropies, they might furnish an alternative way to generate the big jump in entropy.

I understand that if the phase transition uses quantum unified dynamics, there is a slight discrepancy today with the usual $SU(5)$ theory, involving a difficulty with the size of the fluctuations producing galaxies. But that's highly technical and might be reparable by some variation in the theory.

Let me raise one question that may be answered at this meeting, and that is whether there is some inconsistency between Hawking's idea of divorcing the cosmological constant in the equation from the one in astronomy and the idea in the exploding universe theory that the cosmological constant jumps from a finite value to zero. (One doesn't explain the zero, only the jump.) For the exploding universe it's both the fundamental constant and the astronomical one that jump, and perhaps we can understand that better

---

* Editors' note: The paper read to the conference by Hawkings appears in this volume, as do the papers by Linde and Guth mentioned in the following paragraph.

after a few days. Most likely there is actually no problem, but I don't understand it very well.*

Let me say in conclusion that I'm delighted that, during the last decade or more, an old prediction of mine has been confirmed; that particle physics and cosmology would essentially merge into one field, the field of fundamental physics, which underlies all of natural science. In our attempts to understand the basic structure of the universe, we theorists of fundamental physics, even though our day-to-day labors are often frustrating and petty like anyone else's, are engaged in a magnificent quest, along with our experimental friends, for a kind of Holy Grail; a universal theory. Will it prove as elusive as the Holy Grail? Will there be a Lancelot who almost grasps it? A Galahad to whom it is fully revealed? Whether or not we achieve the quest, each time we slay a dragon or rescue a maiden along the way, each splendid adventure is an accomplishment in itself; like the writing of a poem or a symphony, part of the soaring of the human spirit.

---

* Editors' note: Hawkings makes the following interjection: "There is no problem!" Gell-Mann: "That's what you told me a few weeks ago; I'll be delighted to hear it explained again. Anyway, some day we may have a consistent picture of a very early universe, somehow trading its non-zero cosmological constant for zero to an accuracy of 10^{-120}, and producing a lot of matter, including us."

Calculation of Fine Structure Constants

Steven Weinberg

The title of my talk may seem a bit ambitious, but please note the plural "constants." To calculate *the* fine structure constant, 1/137, we would need a realistic model of just about everything, and this we do not have. In this talk I want to return to the old question of what it is that determines gauge couplings in general, and try to prepare the ground for a future realistic calculation.

As far as I know, the only theories that have a chance of predicting the gauge couplings are those that get Yang-Mills fields out of gravity in more than 4 dimensions. Everyone knows that in the original model of Kaluza and Klein, now over 60 years old, one begins with pure gravity in 5 space-time dimensions. One of the spatial dimensions is assumed to be compact, a circle of circumference $2\pi\rho$, while the other 4 remain flat. Correspondingly, the metric is

$$\bar{g}_{LM}(x, y) = \begin{bmatrix} \eta_{\mu\nu} & 0 \\ 0 & \rho^2 \end{bmatrix} + \sum_{l=-\infty}^{\infty} \exp(iyl) \begin{bmatrix} g^l_{\mu\nu}(x) & A^l_\mu(x) \\ A^l_\mu(x) & \phi^l(x) \end{bmatrix}.$$

In the notation we will be using here, a bar denotes $4 + N$-dimensional quantities; indices L, M, etc., run over all $4 + N$ coordinates; indices $\mu\nu$, etc., run over space-time coordinates; and $\eta_{\mu\nu}$ is the Minkowski metric (diagonal, with elements $+1, +1, +1, -1$). Also, in the Kaluza-Klein model y is an angular variable running from 0 to 2π, and l is an integer running over all positive and negative values. In 4 dimensions, the metric excitations described by the fields $g^l_{\mu\nu}(x)$, $A^l_\mu(x)$, and $\phi^l(x)$ appear as particles of mass $|l|/\rho$ and spin 2, 1, and 0, respectively. Since ρ turns out very small, the only particles we observe experimentally are those with $l = 0$: a graviton described by $g^0_{\mu\nu}(x)$, a photon described by $A^0_\mu(x)$; and a scalar $\phi^0(x)$. Also, the electric charge of these particles is given by

$$q_l = l\sqrt{16\pi G}/\rho.$$

The Kaluza-Klein model thus provides an alternative to grand unification as an explanation of the quantization of electric charge. The fact that charge is quantized here is one of the most attractive features of the Kaluza-Klein model, although even before it was pointed out by Klein in 1926, Einstein had helped Kaluza get a professorship after years as a lowly *Privatdozent*. (Incidentally, I am told that the *l* in *Kaluza* has the same Polish pronunciation as the l in *Walesa*, so whatever misprononciation one uses for Walesa is equally appropriate for Kaluza.)

Unfortunately, the radius ρ of the fifth dimension is not dynamically fixed in the original Kaluza-Klein model, so it is not possible in this model to calculate the unit of electric charge from first principles. Instead, the experimentally determined value of the electronic charge e was historically used to calculate the size of the fifth dimension:

$$\rho = \sqrt{16\pi G}/e = 3.8 \times 10^{-32} \text{ cm.}$$

This is so small that one could well understand why the fifth dimension is unobservable, but the value of e remained mysterious.

In the last 20 years these ideas have been embodied in models of dimensionality $4 + N$ higher than 5. I believe the first step was taken by Bryce de Witt in his lectures at the 1963 Les Houches Summer School (or perhaps I should say by de Witt and his students, since Bryce assigned the derivation as a take-home problem). In the early work the extra dimensions were taken to form the manifold of a compact Lie group, or (as in the review of Salam and Strathdee) an arbitrary compact homogenous manifold, but the key point does not even depend on homogeneity: The gauge group of the spin 1 massless fields that are observed in 4 dimensions at low energy is identical with the group of symmetries of the compact manifold of extra dimensions; for each Killing vector of the manifold (i.e., each infinitesimal isometry) there is one massless gauge field. Thus, for instance, to get a low energy $SU(3) \times SU(2) \times U(1)$ gauge group we need a compact manifold with an $SU(3) \times SU(2) \times U(1)$ isometry group, which, as shown by Witten, requires a space-time of at least $4 + 7 = 11$ dimensions.

Recently I gave a simple prescription for calculating the gauge coupling constants in terms of the geometry of the compact manifold. Out of the Lie algebra of symmetries of the manifold, one can always choose a complete set of symmetry generators, each of which has the special property that if one starts at an arbitrary point of the manifold and follows the direction dictated by the symmetry transformation (i.e., follows the Killing vector), one comes back to the same point. The gauge coupling constant for the vector field associated with such a "Magellan curve" is given by

$$g = 2\pi\sqrt{16\pi G_0}/s, \tag{1}$$

where s is the root-mean-square circumference of the manifold along these

Magellan curves, the average being taken over starting points on the manifold, and G_0 is Newton's constant, apart from "induced gravity" corrections to which I will come back later. For instance, for the N-dimensional spherical surface S^N the isometry group is the group $O(N + 1)$ of rotations in $N + 1$ dimensions. Each of these rotations generates a family of Magellan curves, the "small circles," whose rms circumference is $2\pi\rho\sqrt{2/(N + 1)}$, so the $O(N + 1)$ gauge coupling constant is (for S^N radius ρ)

$$g = \sqrt{\frac{N + 1}{2}} \frac{\sqrt{16\pi G_0}}{\rho}. \tag{1'}$$

(Of course there are radiative corrections, and as we shall see they can be very significant.) For the sphere all the gauge coupling constants are equal, as a result of the special symmetry of the sphere. In general, the gauge couplings within any one simple gauge group will automatically be predicted by equation (1) to have the ratios required by the group structure, while the ratios of couplings of different simple groups will be expressed in terms of dynamically determined circumferences. In any case, we cannot calculate the overall scale of the gauge couplings unless we know how to calculate the size of the compact manifold.

Starting with the work of Cremmer and Scherk, a number of authors have developed models in which one can calculate the metric of the compact manifold as a solution of Einstein's field equations in $4 + N$ dimensions, with the energy-momentum tensor supplied by topologically nontrivial field configurations. (This is the case in particular for much of the work on higher-dimensional supergravity, which is covered by Duff's talk at this conference.* I will not go into supergravity here.) These models lead in some cases to values for the ratios of various circumferences of the compact manifold, which can be used along with the general rules mentioned earlier to predict the ratios of various gauge coupling constants that are not related by any simple group-theoretic considerations. The ratios come out to be square roots of rational numbers, just as in grand unified theories. Unfortunately, the strengths of the topological singularities in these models are free parameters (at least at the classical level), so the overall scale of the compact manifold and of the gauge coupling constants cannot be predicted, even if we know all parameters in the Lagrangian.

This is a serious problem, even apart from our natural ambition to be

---

* Editors' note: The paper read to the conference by Duff appears in this volume.

able to calculate fine structure constants. If the size of the compact manifold is not fixed by the underlying field equations, then we may expect it to evolve along with the cosmological expansion of ordinary space-time. Then the ordinary fine structure constant α would have been rather different 10^{10} years ago from its present value. But we know experimentally that this is not the case: the spectrum of quasars shows that α was just about the same when the light was emitted as it is in our laboratories today. (I understand that this problem has been under consideration by Gross and Perry.) Hence we have some experimental evidence that the size of the compact manifold is actually locked by the field equations into a fixed value.

There is in fact a class of models in which the size ρ of the compact manifold is not only fixed but calculable. One may suppose that the energy-momentum tensor responsible for the compactification arises not from topologically nontrivial classical field configurations, but from the quantum fluctuations in a large number n of matter fields that in $4 + N$ dimensions are free and massless. Now, this may seem like an unpromising approach, because you would think that if we include quantum matter fluctuations, we also have to include quantum gravity fluctuations, and this gets us into all the unsolved difficulties of quantum gravitation. However, for sufficiently large numbers of matter fields the quantum fluctuations (really, the Casimir pressure) of the matter can balance the classical action of gravitation, and the quantum effects of gravitation can simply be ignored. The quantum matter fluctuations of n massless matter fields give an energy density in $4 + N$ dimensions of order $n\rho^{-4-N}$. The Einstein tensor $\bar{R}_{LM} - \frac{1}{2}\bar{g}_{LM}\bar{R}$ is of order ρ^{-2}, and the gravitational constant $\bar{G}$ in $4 + N$ dimensions is given (at least in order of magnitude) by Newton's constant G times the volume of the compact manifold, and hence is of order $G\rho^N$. Einstein's field equations then give, in order of magnitude,

$$\frac{1}{\rho^2} \approx 8\pi G\rho^N \times n\rho^{-4-N},$$

and therefore

$$\rho \approx \sqrt{8\pi Gn}. \tag{2}$$

The rms circumferences s are of order $2\pi\rho$, so that (1) and (2) give gauge coupling constants roughly of order

$$g \approx 1/\sqrt{n}. \tag{3}$$

This is highly satisfactory for a number of reasons. First, for a large number n of matter fields the gauge coupling constants turn out small, in agreement with observed values. Also, and not unrelatedly, the size of the compact manifold comes out larger by a factor $\sqrt{n}$ than the Planck length $\sqrt{8\pi G}$, and in consequence one can show that the energy-momentum tensor is dominated by the one-loop matter terms, all other terms being suppressed by powers of $1/n$. Finally, as shown by Duff and Toms, these one-loop contributions to the energy-momentum tensor are actually finite in $4 + N$ dimensions if N is odd, so at least for odd N we can calculate the ρs and gs without having to know the coupling constants of counterterms [e.g., $R^{(4+N)/2}$] that might be needed to cancel infinities.

The need to add many matter fields here reminds me of the old childrens' story about stone soup. You probably remember the story. A poor peasant family hears a knock at the door of their hut one evening, and when they open it they find a starved-looking traveler. The peasants explain that they have no food at all, but the traveler tells them not to worry, he has a magic stone, which only has to be put into boiling water to make the most wonderful soup. Sure enough, when he puts the stone into boiling water, the traveler sniffs the steam, and exclaims that the soup will be really delicious. However, he says, it would be even better if a few potatoes were added. You know the rest: the peasants supply potatoes, then meat, and so on, until the soup is ready, and everyone has some, and they all agree that it is amazing how one can make such good soup from just a stone. These theories are like stone soup: we can get everything from just higher-dimensional gravity, but it would be even better if there were also some matter fields. (The problem raised by Witten, of getting low-mass fermions in nonreal representations of the low-energy gauge group, may require that we add yet more ingredients to the soup.)

Philip Candelas and I have been working on the calculation of one-loop potentials and manifold sizes in this sort of model. The vacuum metric is taken to satisfy Poincaré invariance in a 4-dimensional subspace, which constrains its form to be

$$\bar{g}_{MN}^{\text{VAC}}(x, y) = \begin{bmatrix} \eta_{\mu\nu} & 0 \\ 0 & \tilde{g}_{mn}(y) \end{bmatrix}, \tag{4}$$

where $\tilde{g}_{mn}(y)$ is the metric of the compact N-dimensional manifold, with coordinates y^n. The total action then takes the form

$$I = -\int d^4x \, V_{\text{eff}}, \tag{5}$$

$$V_{\text{eff}} = \int d^N y \sqrt{\tilde{g}} [\tilde{R}(y) + \bar{\Lambda}]/16\pi\bar{G} + V[\tilde{g}]. \tag{6}$$

Here $V[\tilde{g}]$ is the "potential" (in the sense, e.g., of Coleman and E. Weinberg) of the quantized matter fields in a classical background metric $\bar{g}_{MN}^{\text{VAC}}$. Also, $\bar{\Lambda}$ is a cosmological constant in $4 + N$ dimensions, which we include here in order to be able to find a solution with a flat 4-dimensional space-time. Of the Einstein field equations in $4 + N$ dimensions, the nm components just say that V_{eff} is stationary with respect to $\tilde{g}_{nm}$; the $n\mu$ and μn components are automatically satisfied; and the $\mu\nu$ components require that $\bar{\Lambda}$ be fine tuned to make V_{eff} vanish at its stationary point. (Of course it is very unpleasant to have to adjust $\bar{\Lambda}$ in this way, but I don't know of any remotely realistic model that does not have this problem.)

It is especially easy to solve these field equations for a class of one-parameter homogeneous manifolds, for which the symmetries of the manifold dictate that the metric takes the form

$$\tilde{g}_{nm}(y) = \rho^2 \gamma_{nm}(y), \tag{7}$$

where ρ is a free radius parameter, and γ is fixed (up to a choice of coordinates) by the symmetries of the manifold. This is the case, for instance, for spheres S^N, complex projective spaces CP^N, and the manifolds of simple compact Lie groups (and more generally, for homogeneous spaces G/H for which the generators of G that are not in H form a representation of H that is not reducible into real representations). In general, the relative normalization of ρ and γ can be fixed by specifying that the curvature scalar $\tilde{R}$ of the compact manifold is

$$\tilde{R} = -N(N-1)\rho^{-2}. \tag{8}$$

[The factor $N(N-1)$ is inserted so that for spheres ρ will be the usual radius. Note that $\tilde{R}$ is constant because by assumption the manifold is homogeneous; that is, any point can be carried into some other point by a symmetry of the manifold.] The potential (6) then takes the form

$$V_{\text{eff}} = \frac{1}{16\pi G_0}\left(\frac{-N(N-1)}{\rho^2} + \bar{\Lambda}\right) + V(\rho). \tag{9}$$

We have here introduced a constant G_0:

$$G_0 \equiv \bar{G} \Big/ \int d^N y \sqrt{\tilde{g}(y)}. \tag{10}$$

This equals the observed Newton constant, apart from "induced gravity" corrections I'll come to later. Note, however, that it is $\bar{G}$ that must be regarded as a fixed constant, with G_0 varying like ρ^{-N}. Therefore the condition that V_{eff} be stationary in ρ yields

$$0 = \frac{1}{16\pi G_0}\left(\frac{-N(N-1)(N-2)}{\rho^2} + N\bar{\Lambda}\right) + \frac{\rho\, dV(\rho)}{d\rho}. \tag{11}$$

Eliminating $\bar{\Lambda}$ by requiring also that $V_{\text{eff}} = 0$, we find

$$NV(\rho) - \frac{\rho\, dV(\rho)}{d\rho} = \frac{N(N-1)}{8\pi G_0 \rho^2}. \tag{12}$$

We can calculate the size of the compact manifold for any set of matter fields by simply calculating the matter potential $V(\rho)$ in one-loop order and solving (12) for ρ.

This is especially simple if the matter fields are massless and N is odd. There is no one-loop conformal anomaly for odd N, so in this case we can use ordinary dimensional analysis to see that $V(\rho)$ is proportional to $1/\rho^4$:

$$V(\rho) = C_N/\rho^4. \tag{13}$$

The constant C_N depends on N, on the number of various types of massless matter fields, and on the symmetries of the manifold, but need not depend on any continuous parameter. Using (13) in (12) then yields our result for ρ:

$$\rho^2 = \frac{(N+4)8\pi G_0 C_N}{N(N-1)}. \tag{14}$$

Note that this solution breaks down for the case of the original Kaluza-Klein model, with $N = 1$. This is because in this case the curvature $\tilde{R}$ automatically vanishes, so in order to allow $V_{\text{eff}} = 0$, the sign of $\bar{\Lambda}$ must be opposite to that of C_N, but in this case V_{eff} would have no stationary point for any finite ρ. (This case was treated earlier by Appelquist and Chodos.)

The constant C_N is roughly (as it turns out, *very* roughly) of the order of the number n of matter fields, so for large n, the radius ρ is indeed larger than the Planck length by a factor $\sqrt{n}$, as anticipated earlier. However, we can have a valid solution only for C_N *positive*; otherwise $\bar{\Lambda}$ cannot be adjusted to make V_{eff} vanish at its stationary point.

Now, to the calculation of C_N and ρ. Each of the $4 + N$-dimensional matter fields is manifested in 4 dimensions as an infinte tower of particles with masses proportional to $1/\rho$. To calculate the contribution of each of these particles to the potential, we use dimensional regularization, replacing the dimensionality 4 of space-time by a complex number d. The one-loop potential of a particle of mass M in d dimensions is

$$\frac{-ih}{2(2\pi)^d} \int d^d k \ln(k^2 + M^2 - i\varepsilon) = \frac{-h}{2}(4\pi)^{-d/2} \Gamma\left(\frac{-d}{2}\right) M^d, \tag{15}$$

where h is the number of helicity states, with an extra minus sign for fermions. The total potential is thus proportional to a sum over particle states of M^d (weighted by their degeneracies), which diverges for all $\text{Re } d > -N$, times a factor $\Gamma(-d/2)$, which diverges for $d \to 4$. Fortunately the sum can be analytically continued from $\text{Re } d < -N - 1$, where it converges, to $d = 4$, where its analytic continuation for odd N has a simple zero that cancels the poles in $\Gamma(-d/2)$. With b massless minimally coupled scalar bosons and f massless Dirac fermions in $4 + N$ dimensions, the result is

$$C_N = bC_N^{(0)} + fC_N^{(1/2)}, \tag{16}$$

where $C_N^{(j)}$ are ρ-independent numerical constants. In the special case of a spherical compact manifold, we find

$$C_N^{(0)} = \frac{1}{4\pi v} \int_0^{\pi/2} \frac{d\theta}{[2\cosh(\frac{\pi}{2}\tan\theta)]^{2v}}$$

$$\times \left\{ \left(\frac{v^3}{\pi^3}\cos\theta\sin 3\theta - \frac{15v}{\pi^5}\cos^3\theta\sin 5\theta\right)\sinh(v\pi\tan\theta) \right. \tag{17}$$

$$\left. + \left(\frac{6v^2}{\pi^4}\cos^2\theta\cos 4\theta - \frac{15}{\pi^6}\cos^4\theta\cos 6\theta\right)\cosh(v\pi\tan\theta)\right\},$$

$$C_N^{(1/2)} = \frac{(-)^{v+1}3\cdot 2^{v+1}}{\pi^6} \int_0^{\pi/2} \frac{\cos^3\theta\cos 5\theta\, d\theta}{[2\cosh(\frac{\pi}{2}\tan\theta)]^{2v+1}}, \tag{18}$$

where $v \equiv (N - 1)/2$. Here are some numerical results:

N	$C_N^{(0)}$	$C_N^{(1/2)}$
1	-5.05576×10^{-5}	$+2.022304 \times 10^{-4}$
3	$+7.56870 \times 10^{-5}$	$+1.945058 \times 10^{-4}$
5	$+4.28304 \times 10^{-4}$	-1.140405×10^{-4}
7	$+8.15883 \times 10^{-4}$	$+5.958744 \times 10^{-4}$
9	$+1.13389 \times 10^{-3}$	-2.992172×10^{-5}
11	$+1.32932 \times 10^{-3}$	$+1.477709 \times 10^{-5}$
13	$+1.37403 \times 10^{-3}$	-7.242740×10^{-6}
15	$+1.25249 \times 10^{-3}$	$+3.537614 \times 10^{-6}$
17	$+9.55916 \times 10^{-4}$	-1.725405×10^{-6}
19	$+4.79352 \times 10^{-4}$	$+8.412070 \times 10^{-7}$
21	-1.79909×10^{-4}	-4.101970×10^{-7}
$\geqslant 23$	Negative	Alternates

The value of $C_N^{(0)}$ for $N = 1$ is the one found (for untwisted scalars) in the original Kaluza-Klein model by Appelquist and Chodos. By itself, it might lead to pessimism about our chances of satisfying the condition $C_N > 0$, necessary for a satisfactory solution of Einstein's equations. Fortunately, $C_N^{(0)}$ changes sign between $N = 1$ and $N = 3$, and remains positive for 9 cases, from $N = 3$ to $N = 19$, after which it becomes and remains negative for all higher N. Also, $C_N^{(1/2)}$ alternates sign between $N = 3$ (mod 4) and $N = 5$ (mod 4). We conclude that the necessary condition $C_N > 0$ is satisfied for any mix of bosons and fermions when the total dimensionality $N + 4$ is equal to 3, 7, 11, 15, or 19; for suitable boson-fermion mixes when $N + 4 = 5, 9, 13, 17, 23, 27, 31, \ldots$; and not at all when $N + 4 = 21, 25, 29, \ldots$.

For spheres S^N the gauge group is $O(N + 1)$. The formula I gave earlier for the $O(N + 1)$ coupling constant can be derived by noting that the Einstein-Hilbert action in $4 + N$ dimensions when expanded in eigenmodes of definite 4-dimensional mass contains the term

$$-\frac{1}{4} \frac{g^2}{16\pi G_0} \left(\frac{s}{2\pi}\right)^2 F_{\mu\nu} F^{\mu\nu},$$

so (1) is obtained as the requirement that the Yang-Mills curl $F_{\mu\nu}$ be canonically normalized. But one must not forget (though at first we did) that with a large number of matter fields there is also an "induced Maxwell" term in the effective 4-dimensional action of the form

$$-\frac{1}{4}g^2 D_N F_{\mu\nu} F^{\mu\nu},$$

where D_N is a numerical coefficient like C_N, proportional to the number of matter fields. Since s^2/G_0 is of the order of the number of matter fields, these two terms are comparable, and we must write the normalization condition for the gauge fields as

$$g^2 = \left[\frac{(s/2\pi)^2}{16\pi G_0} + D_N\right]^{-1}. \tag{19}$$

For spheres $s/2\pi = \rho\sqrt{2/(N+1)}$, and ρ is given by (14), so the $O(N+1)$ coupling is

$$g_N^2 = \left[\frac{(N+4)}{N(N^2-1)} C_N + D_N\right]^{-1}. \tag{20}$$

This is what we wanted—a formula for a gauge coupling with no unknowns, except for the dimensionality N and the numbers of various types of matter fields that enter in C_N and D_N. As anticipated, we can make g_N^2 as small as we like by having enough matter fields, but one now has a new consistency condition: that (20) should be positive. We have not had time yet to calculate D_N for $N > 1$. (D_N is calculated for $N = 1$ in a recent paper of Toms.) Unfortunately the coefficients $C_N^{(0)}$ and $C_N^{(1/2)}$ are very small, and if the same is true of D_N, then we shall need a very large number of matter fields to get a reasonable gauge coupling. For instance, in 11 total dimensions (and neglecting D_N), we need 10^4 massless scalars to get $g_N^2/8\pi$ down to 0.23. It may be that the extreme smallness of $C_N^{(j)}$ that produces this problem is due to the great degree of symmetry of a sphere, and will not persist when we consider other less symmetric compact manifolds. At any rate, it seems to me that the important thing here is not to get the right value of the gauge coupling—the models studied so far are far from realistic—but to be able to calculate gauge couplings at all.

As everyone today knows, in specifying the value of a gauge coupling constant it is necessary to say not only what its value is but *where*—that is at what renormalization scale—it has that value. For the models discussed here, the answer is easy. In using the one-loop approximation, we were really assuming not only that there are many matter fields, but also that there are no large logarithms that could compensate for the factors $1/n$. (n is b or f or some linear combination of them.) If we set out to calculate the gauge coupling constants at a renormalization scale of, say, 1 eV, then the higher loop contributions to the potential could be characterized by factors not only of $1/n$, but also of $\ln(\rho \times 1\,\text{eV})$, and our one-loop calculations would therefore be invalid. Our calculations are only valid if we interpret them as giving the gauge couplings at energies roughly of order $1/\rho$, about 10^{17}–10^{18} GeV. In order to calculate the gauge couplings observed at accessible energies, we would have to use the results of our one-loop calculation as the initial condition for a set of Gell-Mann–Low differential equations, and integrate down from 10^{17} GeV to oridinary energies. Even if all this were feasible, in order to calculate the fine structure constant 1/137 from first principles we would also know how to calculate the electron mass (in units of $1/\rho$), and for this as yet we have no theory at all.

Up to now I have emphasized the calculation of the manifold size ρ in units of $\sqrt{8\pi G_0}$, because this is what we need to know in calculating gauge couplings. However, it is also of some interest to know what ρ is in terms of the experimentally determined Newton constant G. Matter loops induce a term in the effective 4-dimensional action of the form

$$-\frac{E_N}{\rho^2}[\sqrt{g}\,R]_{4\,\text{dim}},$$

where E_N is another numerical constant, proportional, like C_N and D_N, to the number of species of matter fields. (This is the "induced gravity" term used by Sakharov, Adler, Zee, etc., in calculations of the Newton constant.) Putting this together with the familiar term $-\sqrt{g}\,R/16\pi G_0$ from the classical action, we see that the true Newton constant G is given by

$$\frac{1}{16\pi G} = \frac{1}{16\pi G_0} + \frac{E_N}{\rho^2}. \tag{21}$$

Using (14) to eliminate G_0, we find

$$\rho^2 = 8\pi G\left[\frac{(N+4)C_N}{2N(N-1)} + E_N\right]. \tag{22}$$

E_N has been calculated for $N = 1$ by Toms. We have not yet had time to calculate E_N for $N > 1$, but we expect it to be comparable in magnitude with C_N. We see that one more consistency condition that will need to be satisfied is that

$$\frac{(N + 4)C_N}{2N(N - 1)} + E_N > 0.$$

I should also say a word about the stability of the compact manifolds. The calculation of ρ that we we did earlier should make it clear that these manifolds for $C_N > 0$ are stable against a uniform dilation or compression. The potential (9) is the sum of three terms: the matter potential is positive and varies like ρ^{-4}, the cosmological constant term is positive and varies like ρ^{+N} (recall that $G \propto \bar{G}/\rho^N$), and the curvature term is negative and varies like ρ^{N-2}. Thus the manifold is stabilized against implosion by the Casimir pressure of the matter quantum fluctuations and stabilized against explosion by the cosmological constant. This does not say, however, that a particular manifold like the sphere is necessarily stable against deformation. Page has recently studied the stability of spheres (with Casimir pressure supplied by massless scalars) against deformation into a simple kind of homogeneous 2-parameter manifold. He finds stability for $N = 5, 7, 9, 11, 13,$ or 15, but instability for $N = 3, 17, 19$. Of course, instability here is not necessarily a bad thing—we do not know that the compact manifold of the physical vacuum is a sphere, or, in other words, that the gauge group at 10^{17} GeV is an $O(N + 1)$.

There is also an instability of a rather different sort. So far, everything here has been for the case of zero temperature. Now, at finite temperature the 4-dimensional space-time has a Robertson-Walker rather than a Minkowski metric, and the time-independent treatment given earlier is inappropriate. However, if we shut our eyes to this problem, we can try to judge the stability of the compact manifold at finite temperature by simple including temperature-dependent effects in the static matter potential. These add a negative term to V_{eff}, of order $-T^{4+N}\rho^N$. Hence for small T the cosmological term still dominates for large ρ, but for T above a certain critical value the thermal term wins and V_{eff} decreases without limit as $\rho \to \infty$. This suggests (though it does not prove) that there is a critical temperature at which the compact manifold explodes.

In closing I want to return to the problem of ultraviolet divergences. I don't think that there is any profound significance to the fact that there are no one-loop divergences in odd dimensions. This just provides us with a chance (if there are many matter fields and N is odd) to calculate the gauge couplings without worrying about the difficult problems of quantum gravity. In this respect, the situation here is like the use of general relativity in astronomy. Even when we study a compact object like a neutron star, where the gravitational field is so strong that we have to take the nonlinearities in Einstein's equations into account, we are comfortable in ignoring effects of quantum gravitation. Why is this? It is not that G is small; G is not dimensionless, and the dimensionless quantity GM/r for neutron stars is of order unity. Rather, it is that $\sqrt{G}M$ is large; that is, there are a large number (about 10^{38}) of Planck masses in the star. When we ignore quantum effects in our calculations of compact manifold sizes, we are neglecting the same sorts of $1/n$ corrections as when we ignore quantum gravity in astronomy— only for us n is presumably a few hundred, not 10^{38}. But the ultraviolet divergences are still there, and will eventually have to be dealt with.

My own guess is that there is no renormalizable theory of gravitation, and that the final theory (whether in 4 or $4 + N$ dimensions) will be infinitely complicated, with a Lagrangian (or something like it) containing all possible terms allowed by symmetry principles. There is no problem with infinities in such a theory; for every ultraviolet divergence there is a counter-term ready to absorb it. The real problem is rather to understand why the infinite number of coupling constants have any specific values.

There is a possible answer to this problem, one that I have been advocating at every opportunity for some years past. The renormalized coupling constants depend of course on renormalization scale μ, and trace out trajectories in coupling constant space as μ is varied. It seems to me very likely that the generic trajectory runs off to infinity as μ increases, and that in consequence the theory develops diseases of one sort or another: tachyons, Landau ghosts, etc. One way, and perhaps the only way, to save the consistency of relativity and quantum mechanics is for the coupling constants (scaled by powers of μ to make them dimensionless) to lie on a trajectory that hits a fixed point of the renormalization group equations for $\mu \to \infty$. The trajectories that hit a given fixed point will form an "ultraviolet critical surface" in coupling constant space, and the requirement that the couplings lie on this surface will leave us with a smaller number of free dimensionless parameters, equal to the dimensionality of the critical surface minus one, plus one scale parameter.

This is very speculative, but there are solid reasons to believe that complicated field theories have ultraviolet critical surfaces of finite (and in fact small) dimensionality. Consider, for instance, the effective field theory that is used to study critical phenomena in water. This theory has an infinite number of coupling parameters, which depend on the temperature and pressure and all the microscopic properties of water molecules. Nevertheless, in order to bring about a second-order phase transition it is only necessary to adjust 2 parameters—say, the temperature and pressure. If there were some sort of external field that would allow us to change a microscopic parameter like the mass of the water molecules, we could instead adjust that parameter and the temperature *or* pressure. The important thing is that just two quantities need to be adjusted. It is now understood that this means that there is a fixed point in coupling parameter space, and that the surface of trajectories attracted to this fixed point as $\mu \to 0$ has dimensionality ∞ minus 2; that is, just 2 parameters need to be adjusted to place us on the *infrared* critical surfaces. But a fixed point is a fixed point, and if its infrared critical surface has dimensionality (in the above sense) of ∞ minus 2, then its ultraviolet critical surface has dimensionality 2. In other cases the dimensionality is even smaller; for instance, for ferromagnets it is 1. (Even though 2 parameters, T and H, are at our disposal, only 1 needs to be adjusted to produce a second-order phase transition.) But in any case, the ultraviolet critical surface seems always to be finite dimensional. Thus there are grounds to hope that in this way we will wind up with only a small number of free dimensionless parameters—perhaps none at all.

Of course, if one studies short range or high energy phenomena in real water or real ferromagnets, the description of the system in terms of an effective field theory for pressure or magnetization fluctuations eventually breaks down as μ increases, and we must go over to a description in terms of water molecules or iron atoms, or even their constituents. The requirement that the true theory must lie on an ultraviolet critical surface makes sense only when we describe nature in terms of the final short range degrees of freedom, whatever they are. In the end I think that all physical constants will be determined in this way, but if we are lucky—say, if there are many matter fields in higher odd dimensions—we may be able to calculate some of them long before we get to the fixed point.[1]

Acknowledgment

This work was supported in part by the Robert A. Welch foundation and NSF contract PHY-82-15429.

1. Candelas and I have now calculated the coefficients D_N and E_N for spinors and minimally coupled scalars. Our results show that the necessary conditions for g^2 and G/ρ^2 as well as $V(\rho)$ to be positive are satisfied in $N = 5, 9, 13, 17,$ and 21 compact dimensions, for a variety of suitable mixes of fermions and scalars.

Time as a Dynamical Variable

T. D. Lee

I Introduction

In this lecture I would like to discuss several closely related topics:

1. time as a dynamical variable,
2. discrete mechanics, and
3. random lattice field theory.

The quantum version of discrete mechanics [1, 2] was developed in collaboration with Richard Friedberg, and the random lattice field theory [3–5] with Norman Christ, Richard Friedberg, and H. C. Ren.

Let me start with the first. As we all know, time has always been regarded as a continuous parameter in physics. Even in general relativity, although the metric is a dynamical variable, the continuous four-dimensional space that the metric is embedded in is not. This concept can be traced to Newtonian mechanics.

As illustrated in table 1, in the usual continuum theory the position $\mathbf{r}(t)$ of a particle is a dynamical variable in classical mechanics, but the time t is a parameter. When we go over to the nonrelativistic quantum mechanics, the observable $\mathbf{r}(t)$ becomes an operator, while t remains a parameter. In the relativistic theory, $\mathbf{r}$ and t have to be treated on an equal basis. Two choices are open. Either regard t as an operator or $\mathbf{r}$ as a parameter. Our traditional course is to opt for the latter: Only the fields are operators or observables. The space-time coordinates are merely parameters. An alternative route is to see whether we can regard t as an operator; this is then the essence of this new approach.

Table 1

	Continuum theory		Discrete theory	
Classical mechanics	$\mathbf{r}(t)$	Dynamical variable	$\mathbf{r}, t$	Both dynamical variables
	t	Parameter		
Nonrelativistic quantum mechanics	$\mathbf{r}(t)$	Operator (observable)	$\mathbf{r}, t$	Both operators (observables)
	t	Parameter		
Relativistic quantum theory	Field $\phi(\mathbf{r}, t)$	Operator (observable)	$\mathbf{r}, t, \phi$	All operators (observables)
	$\mathbf{r}, t$	Parameters		

Thus, in the discrete version of relativistic quantum theory, the space-time position, as well as the field, is considered a dynamical variable. For example, in a collision experiment of, say, $e^+ e^- \to \mu^+ \mu^-$, the place and the time of the collision are regarded as part of the measurement, on the same footing as the electric field, magnetic field, ... In order to incorporate such a view, let us start with the discrete theory in its classical form.

II Classical Mechanics

Consider the example of a nonrelativistic point particle of unit mass moving in a potential $V(\mathbf{r})$. In table 2 we give the familiar formulation of classical continuum mechanics in the left column with A_c = action. The corresponding discrete version is given in the right column.

A fundamental postulate of discrete mechanics is that within a time interval T a particle can only assume N space-time positions

$$(\mathbf{r}_n, t_n),$$

with $n = 1, 2, \ldots, N$. The ratio

$$N/T \equiv \rho \tag{1}$$

is a fundamental constant of the theory. For convenience and without loss of generality we have arranged $t_1, t_2, \ldots, t_N$ in ascending order. The discrete action A_d is then given in table 2. Unlike the continuum case [where only $\mathbf{r}(t)$ is the dynamical variable], we regard $\mathbf{r}_n$ and t_n both as dynamical variables. Consequently there are two sets of equations:

$$\frac{\partial A_d}{\partial \mathbf{r}_n} = 0, \tag{2}$$

which gives the discrete version of Newton's law, and in addition

$$\frac{\partial A_d}{\partial t_n} = 0, \tag{3}$$

which yields the energy conservation. There are altogether $4N$ unknowns:

$$\mathbf{r}_n = (x_n, y_n, z_n) \qquad \text{and} \qquad t_n.$$

Exactly the same number of equations is supplied by (2) and (3). In continuum mechanics conservation of energy is a consequence of Newton's equation. This is not so in discrete mechanics.

Table 2

Continuum mechanics	Discrete mechanics[a]
$$A_c = \int_0^T \left(\tfrac{1}{2}\dot{\mathbf{r}}^2 - V \right) dt$$	$$A_d = \sum_n \left\{ \tfrac{1}{2}\mathbf{v}_n^2 - \tfrac{1}{2}[V(\mathbf{r}_n) + V(\mathbf{r}_{n-1})] \right\} (t_n - t_{n-1})$$ $$\mathbf{v}_n = \frac{\mathbf{r}_n - \mathbf{r}_{n-1}}{t_n - t_{n-1}}$$
fix $\mathbf{r}(0) = \mathbf{r}_0$ $\mathbf{r}(T) = \mathbf{r}_f$	fix $(\mathbf{r}_n, t_n) = \begin{cases} (\mathbf{r}_0, 0) & \text{when } n = 0 \\ (\mathbf{r}_f, T) & \text{when } n = N + 1 \end{cases}$
$\dfrac{\delta A_c}{\delta \mathbf{r}(t)} = 0$ gives $\ddot{\mathbf{r}} = -\nabla V$	$\dfrac{\partial A_d}{\partial \mathbf{r}_n} = 0$ gives $$\frac{\mathbf{v}_{n+1} - \mathbf{v}_n}{\tfrac{1}{2}(t_{n+1} - t_{n-1})} = -\nabla V(\mathbf{r}_n)$$
$\mathbf{r}(t) =$ dynamical variable	$\dfrac{\partial A_d}{\partial t_n} = 0$ gives
$t =$ parameter	$E_n \equiv \tfrac{1}{2}v_n^2 + \tfrac{1}{2}[V(\mathbf{r}_n) + V(\mathbf{r}_{n-1})] = E_{n+1}$

a. $N/T \equiv \rho =$ fundamental constant.

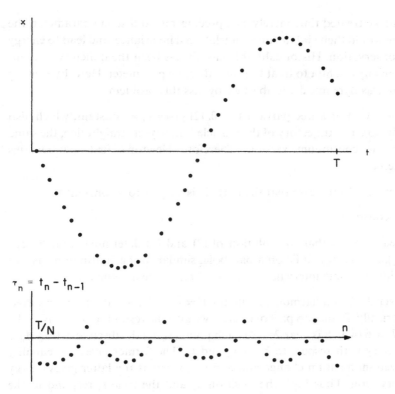

Figure 1
(Top) Numerical solution for a harmonic oscillator potential $V = \frac{1}{2}\omega^2 x^2$ in discrete mechanics, where $\omega = 1$, $N = 40$, and $T = 6$. (Bottom) Plot of the consecutive time spacings $\tau_n = t_n - t_{n-1}$ for n running from 1 to 40. The horizontal line indicates the average $\tau_n (= T/N = 6/40)$. The τ_n for $n = 8, 17, 28$, and 36 are missing in the plot because they are very large and therefore off-scale.

Had we treated time merely as a predetermined discrete parameter, the system would then violate time-translational invariance, and lead to energy nonconservation. Historically this has always been the difficulty encountered in any attempt to treat time as a discrete parameter. Here, by viewing $\mathbf{r}_n$ and t_n as dynamical variables we bypass this problem.

EXAMPLE 1   For a free particle $V = 0$, (2) gives $\mathbf{v}_n =$ constant, which also satisfies (3). The trajectory of the particle is always a straight line, the same as in the continuum version. Therefore Newton's first law remains unaltered.

EXAMPLE 2   In the case that the particle is subject to a constant force,

$\nabla V =$ constant,

it is easy to see that the solution of (2) and (3) determines that $(\mathbf{r}_1, t_1)$, $(\mathbf{r}_2, t_2), \ldots, (\mathbf{r}_N, t_N)$ all lie on a parabola, similar to the continuum case. In addition, the time intervals $t_2 - t_1, t_3 - t_2, \ldots$ are all equal.

EXAMPLE 3   For a harmonic oscillator $V = \frac{1}{2}\omega^2 x^2$, (2) and (3) can be solved numerically. In the top part of figure 1 we give the result for $\omega = 1, N = 40$, and $T = 6$ (which is near 2π). At each point (x_n, t_n) the discrete action A_d is stationary with respect to both x_n and t_n. The former gives the equality between momentum change and impulse, whereas the latter gives energy conservation. Thus both the position x_n and the time t_n respond to the potential. Both depend on the dynamics.

The time spacing $\tau_n \equiv t_n - t_{n-1}$ is not uniform and is plotted in the bottom part of figure 1. As we can see, τ_n is almost a periodic function in n, but with a frequency $\cong 4\omega$.

III Nonrelativistic Quantum Mechanics

Extension of the above classical system of a point particle to quantum mechanics is straightforward, as shown in table 3. In the left column of table 3 we give the usual continuum Feynman path integration formulation. For the Green's function $e^{-iH_{op}T}$, each path carries an amplitude e^{iA_c}, where A_c is the continuum action. Because t is just a parameter, the integration is only over the continuous path $[d\mathbf{r}(t)]$, which is defined to be

$$\lim_{\substack{T = N\varepsilon \\ \varepsilon \to 0}} J \prod_n d^3 r_n,$$

Table 3

Nonrelativistic continuum quantum mechanics	Nonrelativistic discrete quantum mechanics
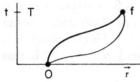	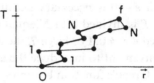

$H_{op} = -\frac{1}{2}\nabla^2 + V$

$(\hbar = m = 1)$

$e^{-iH_{op}T} = \int e^{iA_c}[d\mathbf{r}(t)]$

$e^{-H_{op}T} = \int e^{-\mathscr{A}_c}[d\mathbf{r}(t)]$

where

$A_c = \int (\frac{1}{2}\dot{\mathbf{r}}^2 \mp V)\,dt$
$\mathscr{A}_c$

$[d\mathbf{r}(t)] \equiv \lim_{\substack{T=N\varepsilon \\ \varepsilon\to 0}} J \prod_n d^3r_n$

$J \propto (1/\varepsilon)^{3/2}$

$\dfrac{N}{T} = \rho \equiv \dfrac{1}{l}$ = fundamental constant

$G(T) \equiv \int e^{iA_d}J \prod_n d^3r_n\,dt_n$

$\mathscr{G}(T) \equiv \int e^{-\mathscr{A}_d}J \prod_n d^3r_n\,dt_n$

where

$A_d = \sum_n \dfrac{(\mathbf{r}_n - \mathbf{r}_{n-1})^2}{2\varepsilon_n}$
$\mathscr{A}_d$

$\qquad \mp \frac{1}{2}\varepsilon_n[V(r_n) + V(r_{n-1})]$

$J \propto (t_n - t_{n-1})^{-3/2}$

where

$$J = \left(\frac{1}{2i\pi\varepsilon}\right)^{3/2}.$$ (4)

The limit $\varepsilon \to 0$ is necessary because in Feynman's formulation the time interval T is divided into N equal intervals of ε each. Thus energy is not conserved if $\varepsilon \neq 0$. For large T, the matrix element $e^{-iH_{op}T}$ is oscillatory. It is more convenient to introduce the analytic continuation $T \to -iT$; in that case the Green's function becomes

$$e^{-H_{op}T} = \int e^{-\mathscr{A}_c}[d\mathbf{r}(t)],$$ (5)

where $\mathscr{A}_c$ is the energy integrated over the path $\mathbf{r}(t)$:

$$\mathscr{A}_c = \int (\tfrac{1}{2}\dot{\mathbf{r}}^2 + V) \, dt.$$ (6)

The new Green's function (5) is identical to that used in statistical physics and is well behaved at large T.

In the discrete version, the continuum Green's function $e^{-iH_{op}T}$ is replaced by $G(T)$ and $e^{-H_{op}T}$ by $\mathscr{G}(T)$, both of which are given in table 3. Because the t_n are now dynamical variables just like $\mathbf{r}_n$, the path integrations are over

$$\prod_n d^3 r_n \, dt_n,$$

where n varies from 1 to N with

$$N/T = \rho \equiv 1/l = \text{fundamental constant}.$$ (7)

For clarity, we now let each t_n vary independently from 0 to T. It is convenient to denote their spacings by a set of positive time intervals $\varepsilon_1, \varepsilon_2,$ ... ε_N, so that $t_1, t_2, t_3, \ldots$ are related to $\varepsilon_1, \varepsilon_1 + \varepsilon_2, \varepsilon_1 + \varepsilon_2 + \varepsilon_3, \ldots$ by a permutation. Each discrete path carries an amplitude e^{iA_d} or $e^{-\mathscr{A}_d}$ given in table 3; the path of stationary phase corresponds to the one described by classical discrete mechanics. In the continuum case it is well known that the amplitudes e^{iA_c} and $e^{-\mathscr{A}_c}$ are related to the transformation matrix elements

$$\prod_t \langle \mathbf{r}(t + dt) | \mathbf{r}(t) \rangle,$$

which describes the fact that the operator $\mathbf{r}_{op}(t)$ at time t does not commute with $\mathbf{r}_{op}(t + dt)$. Here the discrete amplitudes e^{iA_d} and $e^{-\mathscr{A}_d}$ can be viewed as representing the corresponding product of matrix elements

$$\prod_n \langle \mathbf{r}_{n+1}, t_{n+1} | \mathbf{r}_n, t_n \rangle. \tag{8}$$

The discreteness in the new mechanics refers to the discrete number of measurements. Each measurement determines the position of the particle $\mathbf{r}_n$ and the time t_n. Both $\mathbf{r}_n$ and t_n can take on any of the continuous eigenvalues of $(\mathbf{r}_n)_{\mathrm{op}}$ and $(t_n)_{\mathrm{op}}$, which, however, do not commute with $(\mathbf{r}_{n+1})_{\mathrm{op}}$ and $(t_{n+1})_{\mathrm{op}}$, as indicated by (8).

In discrete mechanics there is no Hamiltonian or Lagrangian, but only action. What is the physical meaning of these Green's functions? The answer is provided by the following theorem.

THEOREM For $T \gg l$, if we neglect terms $O(e^{-T/l})$ or $O(e^{-iT/l})$ but keep $O(l/T)$, $O(l^2/T^2)$, etc., then

$$G(T) \sim e^{-i\mathscr{H}T}$$

and   (9)

$$\mathscr{G}(T) \sim e^{-\mathscr{H}T},$$

where

$$\mathscr{H} = \mathscr{H}^\dagger = H_{\mathrm{op}} + \tfrac{1}{8}l^2[\nabla^2 - V, [\nabla^2, V]] + O(l^3) \tag{10}$$

and

$$H_{\mathrm{op}} = -\tfrac{1}{2}\nabla^2 + V$$

is the Hamiltonian in the continuum theory. (The proof is given in reference [2].)

According to (9), $G(T)$ and $\mathscr{G}(T)$ are the Green's functions of a "Schroedinger" equation

$$-\frac{1}{i}\frac{\partial\psi}{\partial T} = \mathscr{H}\psi$$

or its analytic continuation

$$-\frac{\partial\psi}{\partial T} = \mathscr{H}\psi,$$

where the Hermitian $\mathscr{H} = \mathscr{H}^\dagger$ plays the role of an effective Hamiltonian.

Consequently, in this approach the whole physical interpretation of the usual continuum quantum mechanics can be carried over to the discrete version. For example, the eigenvalues of $\mathscr{H}$ are the energy levels of the system and the limit $T \to \infty$ of $e^{-i\mathscr{H}T}$ is the S-matrix, which is unitary.

It is not difficult to carry out high-order corrections in l. For a free particle $V = 0$, the effective Hamiltonian $\mathscr{H}$ is identical to H_{op}; this is then the quantum mechanical version of the classical result: The motion of a free particle is unaltered when the continuum mechanics is replaced by the discrete mechanics.

In the case of a one-dimensional harmonic oscillator $V = \frac{1}{2}\omega^2 x^2$ the continuum spectrum is

$$E_{\mathrm{con}} = \left(n + \tfrac{1}{2}\right)\omega,$$

where $n = 0, 1, 2, \ldots$. The corresponding discrete spectrum is

$$E_{\mathrm{dis}} = \left(1 - \tfrac{1}{4}l^2\omega^2\right)\left(n + \tfrac{1}{2}\right)\omega + O(l^3),$$

where l is the fundamental constant of the theory, given by (7).

Next we consider the example when V is the Coulomb potential experienced by an electron of mass m_e at a distance r from a nucleus of charge Z distributed uniformly within a sphere of radius R:

$$V = \begin{cases} -Z\alpha/r & \text{for } r \geqslant R \\ -\tfrac{1}{2}Z\alpha[3 - r/R^2]/R & \text{for } r \leqslant R, \end{cases} \tag{11}$$

where $\alpha \cong 1/137$ is the fine structure constant. The additional energy shift in the discrete mechanics may be estimated by using the nonrelativistic wave function $\psi(\mathbf{r}) = \langle \mathbf{r} | \ \rangle$. We find for the difference between the discrete and continuum energy levels

$$E_{\mathrm{dis}} - E_{\mathrm{con}} = -\frac{l^2}{4m_e}\langle |(\nabla V)^2| \rangle \cong -\frac{6\pi}{5m_e R}|Z\alpha|\psi(0)|^2. \tag{12}$$

For the 2s orbit in hydrogen

$$|\psi(0)|^2 = (8\pi a^3)^{-1}. \tag{13}$$

where $a = (m_e \alpha)^{-1}$ is the Bohr radius. Setting the agreement between the experimental observation and the theoretical QED calculation of the Lamb shift to an accuracy ε rydberg, we obtain

$$lm_e < \left[\frac{10}{3}(137)^3 \varepsilon m_e R\right]^{1/2}, \tag{14}$$

Table 4

Continuum theory	Discrete theory
$e^{-TH_{op}} = \int e^{-A_c} J \prod_x d\phi(x)$	$\mathscr{G}(T) = \int e^{-A_d} J \prod_{i=1}^{N} d^4 x_i \, d\phi_i$
$\phi(x)$ = dynamical variable	ϕ_i and x_i are both dynamical variables $(i = 1, 2, \ldots, N)$
x = 4-dimensional euclidean coordinate (parameter)	$\dfrac{N}{\text{volume}} = \rho = \left(\dfrac{1}{l}\right)^4$ = fundamental constant
$A_c = \int \left[\tfrac{1}{2}(\partial\phi/\partial x_\mu)^2 - j\phi\right] d^4 x$	$A_d = \sum_{l_{ij}} \tfrac{1}{2}\lambda_{ij}(\phi_i - \phi_j)^2 - \sum_i j_i \phi_i$
Equation of motion:	Equation of motion:
$-\dfrac{\partial^2 \phi}{\partial x_\mu^2} = j(x)$	$\sum_{\substack{j\,\text{linked}\\\text{to}\,i}} \lambda_{ij}(\phi_i - \phi_j) = j_i$
Laplace equation	Kirchhoff law

which, for $R \sim 0.8$ fermi and $\varepsilon \sim 10^{-11}$ [6], gives

$$l < 1.6 \times 10^{-14} \text{ cm.} \tag{15}$$

IV Spin 0 Field

As an example of the relativistic quantum field theory, we first discuss the case of a massless scalar field ϕ interacting with an arbitrary external current j. The comparison between the new discrete theory and the usual continuum formalism is given in table 4.

For simplicity, we give here only the euclidean version. In the continuum theory, the field $\phi(x)$ is the dynamical variable; the space-time coordinate x is just a parameter. Hence, the path integration extends only to $d\phi(x)$. In contrast, the corresponding Green's function in the discrete theory consists of integrations over $d^4 x_i$ as well as $d\phi_i$. Because of the $d^4 x_i$ integration, it is clear that there are both the 4-dimensional rotational symmetry and the translational invariance.

In the discrete formulation, given a 4-dimensional volume

$$\Omega = L^3 T$$

we postulate that there can be at most N measurements; each determines the space-time position x_i of the observation and the value of the field ϕ_i at x_i. The ratio

$$\rho = N/\Omega \equiv (1/l)^4 \tag{16}$$

is a fundamental constant of the theory. For a given set $\{x_i, \phi_i\}$, where $i = 1$, $2, \ldots, N$, the action A_d is identical to that of a random lattice. In order to construct A_d, one must solve the following two problems:

i. Given N points in a volume Ω, it is desirable to couple only nearby points, simulating the local character of the corresponding density of the continuum action. How can we do that?

ii. Assume that an algorithm is given, so that only neighboring pairs of sites, say, i and j, are coupled. Each pair gives a link l_{ij} and contributes to the action a term $(\phi_i - \phi_j)^2$ multiplied by a weight factor λ_{ij}. The sum over all links

$$\frac{1}{2}\sum_{l_{ij}} \lambda_{ij}(\phi_i - \phi_j)^2 \tag{17}$$

replaces the integral

$$\frac{1}{2}\int \left(\frac{\partial\phi}{\partial x_\mu}\right)^2 d^4x \tag{18}$$

in the continuum theory. Because different links l_{ij} have different lengths and orientations, it is reasonable that they should carry different weights λ_{ij}. What would be the best choice for these weight functions?

We observe that if ϕ_i is identified as the "electric potential" and

$$\lambda_{ij}^{-1} = \text{electric resistance between } i \text{ and } j, \tag{19}$$

then the equation of motion

$$\sum_{\substack{j \text{ linked} \\ \text{to } i}} \lambda_{ij}(\phi_i - \phi_j) = j_i \tag{20}$$

is identical to the Kirchhoff law of an electric circuit, with j_i as the external current entering at point i. In this analog problem, we would like to determine these resistances λ_{ij}^{-1} so that the electric potential ϕ_i gives the best discrete approximation to the corresponding continuum solution $\phi(x)$ of the Laplace equation

$$-\frac{\partial^2 \phi}{\partial x_\mu^2} = j(x).$$ (21)

The answers to these two problems are given in the following.

1 Linking Algorithm

Consider a D-dimensional euclidean space. Given an arbitrary distribution of N points (called lattice sites, or simply sites) in a finite volume Ω, our first task is to divide Ω into simplices whose vertices are those sites. To avoid complications due to the boundary, we may assume Ω to be a D-dimensional rectangular volume with the standard periodic boundary conditon. For $D = 2$, the simplices are triangles; for $D = 3$, they are tetrahedra. In general, a D-simplex consists of $D + 1$ vertices. Between any pair of vertices, we draw a straight line which forms a link. Hence, there are altogether

$$C_2^{D+1} = \tfrac{1}{2}(D + 1)D$$

different links per simplex. Regarding the links as sides of triangles, we can form

$$C_3^{D+1} = \frac{1}{3!}(D + 1)D(D - 1)$$

different triangles per simplex. Likewise, since the triangles can be viewed as surfaces of tetrahedra, there are

$$C_4^{D+1} = \frac{1}{4!}(D + 1)D(D - 1)(D - 2)$$

different tetrahedra per simplex. And so on.

For our applications, we wish to draw links only between nearby sites. This is related to the problem of partitioning Ω into nonoverlapping simplices. Let us introduce the following concept of clusters of neighboring sites: Consider an arbitrary group of $D + 1$ random lattice sites. They lie on the surface of a hypersphere in D dimensions, which will be referred to as their circumscribed sphere. If the inside of that sphere is free of all lattice sites, then this group of $D + 1$ sites is called a *cluster* of neighbors. We form a D-simplex for every cluster, using its sites as vertices.

We note that for any $D + 1$ infinitesimal volumes $d^D r_1, d^D r_2, \ldots, d^D r_{D+1}$ in Ω, the probability that each volume element should contain a lattice site is

$$\rho^{D+1} \prod_{i=1}^{D+1} d^D r_i; \tag{22}$$

the probability that there is no lattice point inside their circumsccibed sphere is given by the Poisson formula

$$\exp(-v\rho), \tag{23}$$

where ρ is the site density and v is the volume of the sphere, which is related to its radius R by

$$v = \begin{cases} (2\pi)^{D/2} R^D/D!! & \text{if } D \text{ is even} \\ 2(2\pi)^{(D-1)/2} R^D/D!! & \text{if } D \text{ is odd.} \end{cases}$$

Hence, the probability that $d^D r_1, d^D r_2, \ldots, d^D r_{D+1}$ should form a cluster and therefore be the vertices of a D-simplex is

$$\rho^{D+1} \prod_{i=1}^{D+1} d^D r_i \exp(-v\rho).$$

The factor $\exp(-v\rho)$ ensures that the vertices of the simplex cannot be too far apart.

Let us examine, out of the N given sites, all C_{D+1}^N combinations of possible groupings of $D + 1$ sites. Whenever a group fulfills the above condition of a cluster, a D-simplex is formed. The simplices thus constructed define our basic random lattice, which has the following properties:

THEOREM   (i) There is no overlap between the volumes of any two simplices, and (ii) the volume sum of all simplices is Ω. (The proof of the theorem is given in reference [3].)

In figure 2, we give an example of a 2-dimensional random lattice.

From (22) and (23), we can compute the various kinematic properties of the random lattice. Let

$N_{n/m} \equiv$ average number of n-simplices per m-simplex;

i.e., each m-simplex is on the average shared by $N_{n/m}$ n-simplices. For example, each site (0-simplex) is shared by an average of $N_{1/0}$ links (1-simplices), each link by $N_{2/1}$ triangles (2-simplices), etc. These averages are given in table 5. We note that

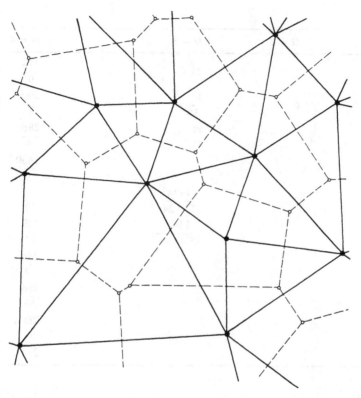

Figure 2
An example of a 2-dimensional random lattice (solid lines) and its dual (dashed lines). Dots
are the lattice sites, and open circles are the corners of the dual cells.

Table 5

	$D = 2$	$D = 3$	$D = 4$
$N_{1/0} =$	6	$2\left(1 + \dfrac{24}{35}\pi^2\right)$	$\dfrac{349}{9}$
$N_{2/0} =$	6	$\dfrac{144}{35}\pi^2$	$\dfrac{590}{3}$
$N_{3/0} =$		$\dfrac{96}{35}\pi^2$	$\dfrac{2860}{9}$
$N_{4/0} =$			$\dfrac{1430}{9}$
$N_{2/1} =$	2	$\dfrac{144\pi^2}{35 + 24\pi^2}$	$\dfrac{177}{17}$
$N_{3/1} =$		$\dfrac{144\pi^2}{35 + 24\pi^2}$	$\dfrac{429}{17}$
$N_{4/1} =$			$\dfrac{286}{17}$
$N_{3/2} =$		2	$\dfrac{286}{59}$
$N_{4/2} =$			$\dfrac{286}{59}$
$N_{4/3}$			2

$$N_{D/(D-1)} = 2,$$

which is the consequence of the fact that in D dimensions, each $(D - 1)$-simplex is shared by two D-simplices. In addition, from the same table we have

$$N_{D/(D-2)} = N_{(D-1)/(D-2)} = \begin{cases} 6 & \text{when } D = 2 \\ 5.2276 & \text{when } D = 3 \\ 4.8475 & \text{when } D = 4, \end{cases}$$

which becomes 4 when $D \to \infty$.

When $D = 4$,

$$N_{1/0} = \frac{340}{9} \cong 37.8 \quad \text{and} \quad N_{2/0} = \frac{590}{3} \cong 197, \tag{24}$$

whereas for a regular "cubic" lattice, $N_{1/0} = 8$ and $N_{2/0} = 24$.

2 Weight Function λ_{ij}

Why can't λ_{ij} be simply a constant, independent of the link l_{ij}? Consider, e.g., a 1-dimensional random distribution of N sites at positions x_n (with $x_1 < x_2 < \ldots < x_{n-1} < x_n < \ldots$). Assign a field variable ϕ_n to each x_n. Most of us will agree that the approximation

$$dx\frac{d^2\phi}{dx^2} \sim \frac{\phi_{n+1} - \phi_n}{x_{n+1} - x_n} - \frac{\phi_n - \phi_{n-1}}{x_n - x_{n-1}} \tag{25}$$

is better than

$$dx\frac{d^2\phi}{dx^2} \sim \lambda(\phi_{n+1} - 2\phi_n + \phi_{n-1}), \tag{26}$$

where λ is a constant. This is because when $\phi(x)$ is a linear function of x, or correspondingly ϕ_n is a linear function of x_n, both $d^2\phi/dx^2$ and the right-hand side of (25) are zero, but not the discrete approximation (26). This obvious criterion can be readily generalized.

In a multidimensional space when $\phi(\mathbf{r})$ is a linear function of the position vector $\mathbf{r}$, the Laplace equation is

$$\frac{\partial^2\phi}{\partial x_\mu^2} = 0,$$

where x_μ is the μ-component of $\mathbf{r}$. We require the weights λ_{ij} in the discrete case to satisfy

$$\sum_{\substack{j \text{ linked} \\ \text{to } i}} \lambda_{ij}(\phi_i - \phi_j) = 0 \tag{27}$$

at all i for an arbitrary distribution of the site-position vectors $\mathbf{r}_1, \mathbf{r}_2, \ldots, \mathbf{r}_n$ when ϕ_i is a linear function of $\mathbf{r}_i$; i.e.,

$$\phi_i = a + \mathbf{b}\cdot\mathbf{r}_i, \tag{28}$$

where a and $\mathbf{b}$ are constants.

To determine λ_{ij}, it is useful to introduce the dual lattice. Let i be an arbitrary site of a D-dimensional random lattice $(i = 1, 2, \ldots, N)$. Take any point p in Ω. We say p belongs to i if i is the nearest site to p. The dual to site i is the volume ω_i consisting of all p belonging to i. Each ω_i is a convex polyhedron (called a cell) in D dimensions. The surface of a cell consists of a number of $(D - 1)$-dimensional faces, formed by points that belong to two or more sites. The intersections of these $(D - 1)$-dimensional faces are $(D - 2)$-dimensional faces, which consist of points each belonging to three or more sites. And so on. The corners (i.e., vertices) of these cells are single points, each belonging to, and only to, $D + 1$ sites. Let c be one of these corners. From our construction we see that c is equidistant from all the $D + 1$ sites to which it belongs. Using c as the center, draw a hypersphere through these $D + 1$ sites. Since no site is nearer to c than the aforementioned $D + 1$ sites, there cannot be any lattice site inside the hypersphere. Hence, these $D + 1$ sites satisfy the definition of a cluster, and form a D-simplex, whose dual is the corner c.

The division of Ω into cells gives the dual of our random lattice. Each link in the random lattice is perpendicular to (but need not intersect) a $(D - 1)$-dimensional face in its dual; each triangle is perpendicular to a $(D - 2)$-dimensional face; etc. This gives a one-to-one correspondence between an n-simplex in the random lattice and a $(D - n)$-dimensional face in its dual. While each D-simplex has a fixed structure (e.g., fixed number of vertices and faces,), this is not so for the cells. For this reason, as we shall discuss, it is often more convenient to formulate the action in terms of simplices, instead of cells. This is particularly true in applications to gauge field theories.

The cells ω_i are called Voronoi polyhedra in the literature [7]. Their construction is a generalization of the Wigner-Seitz procedure for a regular lattice.

In figure 2 we give a two-dimensional example of a random lattice and its dual.

We set the weights λ_{ij} between any pair of sites i and j to be

$$\lambda_{ij} = \begin{cases} 0 & \text{if } i \text{ and } j \text{ not linked} \\ s_{ij}/l_{ij} & \text{if } i \text{ and } j \text{ linked,} \end{cases} \qquad (29)$$

where l_{ij} is the length of the link and s_{ij} is the "volume" of the corresponding $(D - 1)$-dimensional surface in the dual lattice. When $D = 2$, s_{ij} is the border length of the dual cell (see figure 3).

THEOREM   The λ_{ij} given by (29) satisfies

$$\sum_{\substack{j \text{ linked} \\ \text{to } i}} l_{ij}^{\mu} \lambda_{ij} = 0 \qquad (30)$$

at all sites i and

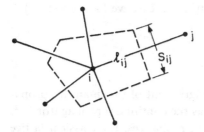

Figure 3
The link l_{ij} and its dual s_{ij}.

$$L^{\mu\nu} \equiv \sum_{i,j} l_{ij}^\mu l_{ij}^\nu \lambda_{ij} = 2\Omega\delta^{\mu\nu}, \qquad (31)$$

where l_{ij}^μ is the μ-component of $\mathbf{l}_{ij} = \mathbf{r}_i - \mathbf{r}_j$ and $\delta^{\mu\nu} = 0$ if $\mu \neq \nu$ and 1 otherwise. (The proof of this theorem is given in reference [5].)

From the identities (30) and (31), we see that when, according to (28), ϕ_i is a linear function of $\mathbf{r}_i$,

$$\sum_j \lambda_{ij}(\phi_i - \phi_j) = \mathbf{b} \cdot \sum_j \lambda_{ij} \mathbf{l}_{ij} = 0$$

at all i, and the sum over all sites i and j

$$\frac{1}{2}\sum_{l_{ij}} \lambda_{ij}(\phi_i - \phi_j)^2 = \frac{1}{4}\sum_{i,j} \lambda_{ij}(\phi_i - \phi_j)^2 = \tfrac{1}{2}\mathbf{b}^2\Omega.$$

The former satisfies condition (27) and (28) for $\sum_j \lambda_{ij}(\phi_i - \phi_j)$ to be a good discrete version of the Laplace operator, and the latter makes the lattice action (17) identical to the continuum value (18) for an arbitrary lattice before the limit $l_{ij} \to 0$, provided that $\phi(\mathbf{r})$ is a linear function of $\mathbf{r}$.

In addition, we can show that if the external current is nonzero only at site 0,

$$j_i = \delta_{i0},$$

then at large distance r_i from 0 the solution ϕ_i becomes the same as that in a continuum; e.g., for $D = 3$ the solution of the Kirchhoff equation satisfies

$$\phi_i \to \frac{1}{4\pi r_i}$$

when $r_i \to \infty$. In the corresponding continuum case we have $\phi = (4\pi r)^{-1}$, which is the solution of

$$-\frac{\partial^2 \phi}{\partial x_\mu^2} = \delta^3(\mathbf{r}).$$

This then ensures that the long wavelength limit of the quantum propagator in the discrete theory is the same as the continuum propagator k^{-2}.

Another pleasant feature is that when $\mathbf{r}_i$ changes, the classical lattice solution ϕ_i and the corresponding action are continuous in $\mathbf{r}_1, \mathbf{r}_2, \ldots, \mathbf{r}_N$ (proved in appendix D of reference [2]). The applicability of the continuity concept and the existence of exact equalities, such as (30) and (31), make it possible to use analytic method in the discrete theory.

V Gauge Theory

As in the conventional treatment [8] of lattice gauge theory, we introduce gauge variables by assigning a group element $U(i,j)$ to a link connecting the points (i.e., sites) i and j in the lattice. For an $SU(N)$ gauge theory, we can set $U(i,j)$ to be an arbitrary $N \times N$ unitary matrix with determinant 1. It is convenient to define both $U(i,j)$ and $U(j,i)$ with $U(j,i) = U(i,j)^{-1}$. Next with each elementary triangle of vertices i, j, and k we associate a group element $U_{ijk} = U(i,j) U(j,k) U(k,i)$. Recall that in our construction of the random lattice we identify a "cluster" of $D + 1$ vertices with the property that the $(D - 1)$-dimensional sphere, on whose surface the cluster of $D + 1$ points lies, has no lattice points in its interior. Each pair of points in such a cluster is joined by a link. An elementary triangle Δ_{ijk} is then formed of any three points i, j, and k in the same cluster and the links between them.

Then discrete action for the lattice is defined using these group variables U_{ijk}:

$$A_d = \frac{1}{g^2} \sum_{\Delta_{ijk}} \kappa_{ijk} f(U_{ijk}), \tag{32}$$

where g is the coupling constant, and the sum extends over all triangles Δ_{ijk}. The coefficients κ_{ijk} are given by

$$\kappa_{ijk} = \tau_{ijk}/\Delta_{ijk}, \tag{33}$$

which is similar to (29), where Δ_{ijk} is the area of the plaquette and τ_{ijk} the $(D - 2)$-dimensional volume of its dual. Just as in (30) and (31) we can prove

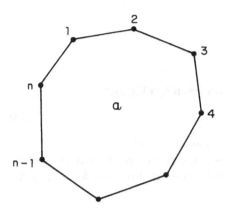

Figure 4
A Wilson loop of area a bounded by links $l_{12}, l_{23}, \cdots, l_{n1}$.

[5] that

$$\sum_k \kappa_{ijk}\Delta_{ijk}^{\mu\nu} = 0$$

for all sites i and j that are linked, and

$$T^{\mu\nu\rho\sigma} \equiv \frac{1}{4}\sum_{\Delta_{ijk}} \kappa_{ijk}\Delta_{ijk}^{\mu\nu}\Delta_{ijk}^{\rho\sigma} = \tfrac{1}{8}(\delta^{\mu\rho}\delta^{\nu\sigma} - \delta^{\mu\sigma}\delta^{\nu\rho})\Omega.$$

The function $f(U)$ obeys

$$f(uUu^{-1}) = f(U)$$

for all group elements u and U; the simplest choice is

$$f(U) = \tfrac{1}{2}\mathrm{tr}(I - U) + \text{c.c.} = \mathrm{Re}[\mathrm{tr}(I - U)],$$

with I = the unit matrix and, for the $SU(N)$ theory, U = an arbitrary $N \times N$ unitary matrix of unit determinant, in accordance with the conventional Wilson formalism.

Next, consider a large loop of area a bounded by links $l_{12}, l_{23}, \ldots$, as shown in figure 4. [The linear dimension of the loop is assumed to be $O(a^{1/2})$ in any direction.] Define

$$\exp(-Ta) \equiv Z^{-1} \int [dr_i][dU(i,j)]e^{-A_d/g^2}W_L, \qquad (34)$$

where

$$Z \equiv \int [dr_i][dU(i,j)]e^{-A_d/g^2}, \tag{35}$$

T is the string tension, A_d is the action given by (32), and

$$W_L = \mathrm{tr}\, U(1,2)\, U(2,3)\ldots, \tag{36}$$

with the product extending over the boundary of α.

In the strong-coupling limit $(g \to \infty)$, one can show that the theory becomes identical to a relativistic string model. For $D = 4$, we find the string tension T is given by

$$T = 2\left(\frac{\pi}{\sqrt{3}}\right)^{1/2} (\rho \ln g^2)^{1/2} \tag{37}$$

with ρ = site density. Furthermore, the thickness d of the string is

$$d^2 = \frac{1}{2\pi T} \ln \alpha. \tag{38}$$

Both (37) and (38) are derived in reference [4]. Thus, unlike a rigid "cubic" lattice, the strong-coupling limit of the random lattice gauge theory is already quite physical, since it carries the phenomenological virtues of a relativistic string model. From (37) we see that quarks are confined in the strong-coupling limit.

As will be shown, the result of the numerical calculation indicates that there is no phase transition in a random lattice nonabelian gauge theory. Consequently, when we change the coupling from strong to weak, quarks should remain confined.

The string tension concerns calculations for a given 2-dimensional figure (Wilson loop). At the next level of complexity, we consider the glueball propagator between two plaquettes, separated by a large time interval T. Let Δ and Δ' be these two plaquettes. Then in the strong-coupling limit we find that these two plaquettes are connected by tetrahedra of the form given in figure 5.

When $T \to \infty$, the corresponding propagator has a leading behavior that can be expressed as a superposition of different angular momentum J states, each having an asymptotic behavior

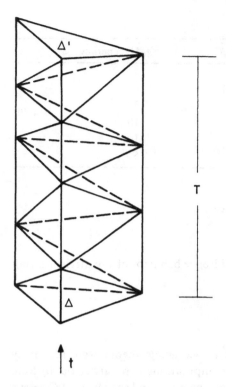

Figure 5
Glueball propagator in the strong-coupling limit.

$$\sim e^{-M_J T}, \tag{39}$$

where M_J is the lowest glueball mass of angular momentum J.

For low angular momentum states, we find that the glueball rotational levels are identical to that of a rigid body of moment of inertia I_1, I_2, and I_3, where

$$I_1 = 0.631\,\rho^{1/4}(\ln g^2)^{5/4}$$

and

$$I_2 = I_3 = 0.341\,\rho^{1/4}(\ln g^2)^{5/4}.$$

Thus the excitation energy $M_J - M_0$ (where M_0 is the ground state mass with $J = 0$) is given by

J	$M_J - M_0$	Degeneracy
1	$\frac{1}{2}(I_1^{-1} + I_2^{-1})$	6
1*	I_2^{-1}	3
2	$2I_1^{-1} + I_2^{-1}$	10
2*	$\frac{1}{2}I_1^{-1} + \frac{5}{2}I_2^{-1}$	10
2**	$3I_2^{-1}$	5

$$(40)$$

For high angular momentum we have

$$M_J \propto \sqrt{J},$$

so that the system exhibits the typical Regge behavior of a relativistic string in rotation.

VI Numerical Results

Numerical programs for a random lattice gauge theory were set up by Friedberg and Ren at Columbia; the computations were carried out by Ren. In figure 6 we give the average plaquette energy u and specific heat C versus $\beta = 1/g^2$ for the $U(1)$ theory. The corresponding plots for an $SU(2)$ theory are given in figure 7. We see that the specific heat has a peak in the $U(1)$ theory, but not in the $SU(2)$ theory. For $U(1)$, the peak becomes steeper when the number of lattice sites increases, suggesting that there is a phase transition. On the other hand, the specific heat curve for $SU(2)$ has no peak, indicating that the passage from strong to weak coupling is a smooth one. Consequently, while both theories are confined in the strong-coupling limit, the weak-coupling limit is consistent with deconfinement in the $U(1)$ theory (QED), but not in a nonabelian gauge theory. Extension to $SU(3)$ and also calculations on Wilson string tension are in progress.

In contrast, we give in figure 8 the recent numerical calculation by N. H. Christ and A. Terrano for the $SU(3)$ gauge theory on a regular lattice. As we can see, there is a sharp peak in the specific heat, suggesting that the transition from strong to weak in a regular lattice is by no means smooth, unlike that in a random lattice.

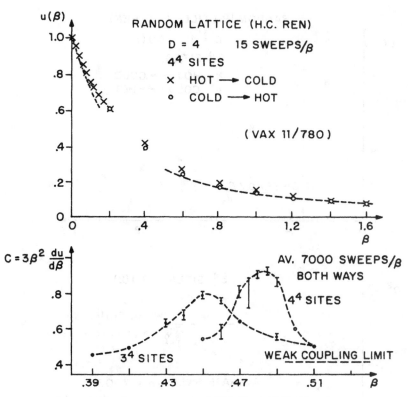

Figure 6
Numerical results for a 4-dimensional $U(1)$ random lattice field theory: (top) average plaquette energy versus $\beta = 1/g^2$; (bottom) specific heat versus β. [Courtesy of H. C. Ren]

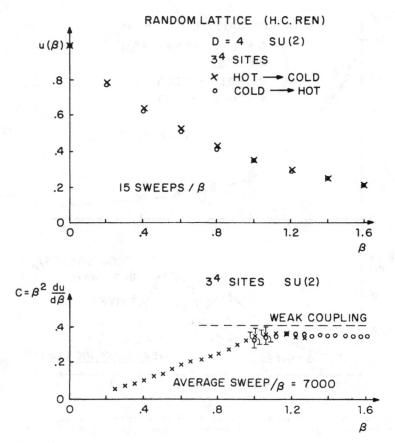

Figure 7
Numerical results for a 4-dimensional $SU(2)$ random lattice field theory: (top) average plaquette energy versus β; (bottom) specific heat versus β. The error bars indicated are applicable to all points. [Courtesy of H. C. Ren]

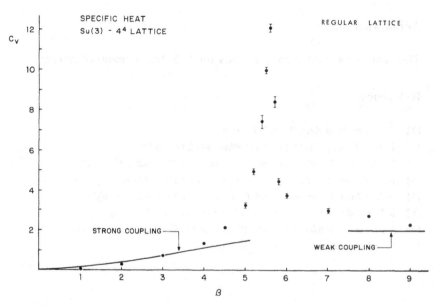

Figure 8
Specific heat versus β for a 4-dimensional $SU(3)$ theory on a "cubic" lattice. [Courtesy of N. H. Christ and A. Terrano]

VII Concluding Remarks

The random lattice guarantees rotational and translational symmetry. At its very minimum, this new approach gives a covariant cutoff parameter, maintains unitarity, removes divergences, and enables one to compute energy levels as a coarse grain average. Thus, it makes possible the quantization of gravity, which is what I am working on at the moment, together with G. Feinberg and R. Friedberg. At its maximum, it may change our concept of space-time in a fundamental way.

For more than three centuries we have been influenced by the precept that fundamental laws of physics should be expressed in terms of differential equations. Difference equations are always regarded as approximations. Here, we try to explore the opposite: Difference equations are more fundamental, and differential equations are regarded as approximations. With today's rapid progress in computing technology, this may well be an opportune period for such a radical departure from the traditional view.

Acknowledgment

This work was supported in part by the U.S. Department of Energy.

References

[1] T. D. Lee, *Phys. Lett.* B122, 217 (1983).

[2] R. Friedberg and T. D. Lee, *Nucl. Phys.* B242, 145 (1984).

[3] N. H. Christ, R. Friedberg, and T. D. Lee, *Nucl. Phys.* B202, 89 (1982).

[4] N. H. Christ, R. Friedberg, and T. D. Lee, *Nucl. Phys.* B210 [FS6], 310 (1982).

[5] N. H. Christ, R. Friedberg, and T. D. Lee, *Nucl. Phys.* B210 [FS6], 337 (1982).

[6] S. R. Lundeen and F. M. Pipkin, *Phys. Rev. Lett.* 46, 232 (1981).

[7] J. M. Ziman, *Models of Disorder* (Cambridge University Press, New York, 1979).

[8] K. Wilson, *Phys. Rev.* D10, 2455 (1974).

Nonperturbative and Topological Methods in Quantum Field Theory

Roman Jackiw

Renormalized perturbation theory, whose development in the West began immediately after the first Shelter Island Conference with Bethe's Lamb shift calculation, gave us in the intervening third of a century the principal tool for extracting physical content from a quantum field theory. Aside from various resummation techniques of the Bethe-Salpeter variety, little effort was expended on the study of nonperturbative effects, even though there can be no doubt that these are necessarily present, since a quantum field theory is, after all, a quantum mechanical system—to be sure one with an infinite number of degrees of freedom, but nevertheless capable of giving rise to familiar quantal phenomena like collective macroscopic excitations and tunneling to name two, about which I shall have more to say. Just as the Born series for a quantum mechanical potential will not directly illuminate such processes, so also the Dyson-Feynman-Schwinger perturbation expansion has little to say about them in field theory. Of course, some individuals made isolated suggestions that there might exist interesting field theoretical structures inaccessible to conventional perturbative analysis— the most important and prescient are Skyrme's conjectures [1]—but the subject did not engender significant research. One can mention two reasons for this lack of interest by the physics community. First, the infinities of field theory had only recently been tamed by the renormalization procedure that is defined perturbatively. Consequently, there was much uncertainty whether one could obtain unambiguous results outside the perturbative framework, or whether such calculations were hopelessly contaminated by the infinities. Second, there were doubts about the relevance of field theoretical dynamics to elementary particle physics, so that close study of any particular field theory was judged only mathematically, but not physically, interesting.

In the last decade, these attitudes have evolved. For better or worse, we have grown accustomed to the infinities and are less intimidated by them, although Dirac and Schwinger counsel against this complacency. Also local field dynamics has vigorously re-established itself as the only viable theoretical framework for elementary-particle physics; first through the successes of current algebra, then through light-cone and quark/parton analyses of deep inelastic processes, and finally through the contemporary synthesis in quantum chromodynamics and quantum flavor dynamics. While perturbation theory remains the preeminent calculational method, now we have also perfected alternative approximate analyses, which have exposed a wealth of nonperturbative effects in quantum field theory. I shall briefly summarize what we have learned, describe the impact on practical physical questions, and then toward the end, I shall indicate some new research directions in this area, of the kind mentioned by Isadore Singer at the beginning of his talk yesterday.*

It has been established that a quantum field theory will in general give rise to particle states and also to processes beyond those that are seen in the Born series. The novel states do not arise by quantizing small fluctuations of a local field present in the Lagrangian, nor are they conventional bound states; rather they occur, when a symmetry is spontaneously broken, as collective, coherent excitations of all the elementary quanta. These are the celebrated soliton states, and are present in any number of dimensions: the kink in one, the Abrikosov-Ginsburg-Landau (Nielsen-Olesen) vortex in two, the 't Hooft-Polyakov monopole in three. While solitons in classical field theories have a long history, it is only recently that we have learned how to quantize them and how to perform approximate yet accurate quantum mechanical calculations [2]. The procedure is an extension of techniques familiar from broken symmetry studies. One finds a finite energy solution to the classical field equations—the soliton—shifts the quantum field by the classical solution, and quantizes the shifted field. In the broken symmetry case, the classical background is constant and is interpreted in the quantum theory as an approximation to the quantum field's vacuum expectation value, while in the soliton case the background is inhomogeneous and gives an approximation to the field form factor in the soliton state. There are technical problems: one needs to preserve translation invariance; infrared divergences from zero Goldstone modes must be cured; quantum numbers that characterize the soliton states must be identified. All these problems have been solved [2], and the following picture emerges: when the coupling constant g is small, the solitons are heavy with mass of order g^{-2} times the characteristic mass scale of the theory. Though more massive than the elementary excitations, the solitons are nevertheless stable, owing to an infinite energy barrier that separates them from ordinary particles. Their stability is further assured by a conserved charge—kink number, vortex number, monopole number—that, however, does not directly arise by Noether's theorem from a Larangian symmetry, but rather characterizes the topological, i.e., large distance, properties of field configurations. The interactions between solitons are strong, $O(g^{-1})$, hence

---

* Editors' note: The paper read to the conference by Singer does not appear in this volume.

difficult to calculate. Those between solitons and ordinary particles are $O(g^0)$; of course conventional particles interact with strength $O(g)$. Therefore, there are three interaction scales. Furthermore, the soliton has both stable and metastable excited states, which are formed either by binding with ordinary particles or by excitation of the soliton's internal degrees of freedom; for example, the charged dyon is an excited state of the monopole [3]. A remarkable property of the solitons is that they may possess unexpected spin and may obey unexpected statistics: although arising in a bosonic theory with integer-spin fields, they can be fermions with half-integer spin. Further novel effects are seen when a Dirac fermion is coupled to a soliton: the most dramatic of these is the emergence of states with fractional fermion number [4].

Turning now to processes, rather than states, the principal result here establishes the occurrence of tunneling in a Yang-Mills theory between energy-degenerate states, which is closely analogous to the tunneling that occurs in a periodic potential, giving rise to band formation [5]. In general, one does not expect such tunneling to happen in a quantum field theory, owing to infinite energy barriers between degenerate states—indeed it is the absence of tunneling that permits spontaneous symmetry breaking. However, in a nonabelian gauge field theory, there are paths in the gauge potential configuration space that avoid the infinite energy barriers and along which tunneling can occur. Semiclassical analysis of tunneling in potential theory is familiarly carried out by identifying these paths as solutions to the classical dynamical equations in imaginary time; similar solutions in Yang-Mills theory are the Belavin, Polyakov, Schwartz, Tyupkin instantons, and they have been used to give a semiclassical description of tunneling in Yang-Mills theory.

All this research has involved particle physicists in a novel activity—solving partial differential nonlinear equations, and this has put us in contact with mathematicians who, as it happens, had come to similar equations on their own. Aside from the satisfaction of participating in a confluence of separate streams of research, we have benefited by learning and using some of the mathematicians' techniques, principally topological analyses of field theory. With these it has been possible to go beyond the non-perturbative but approximate calculations that I have been describing so far and to establish results rigorously.

One very surprising fact that has been found in this way about the Yang-Mills quantum theory is that it is characterized by a hidden parameter, the vacuum angle θ [5]. This comes about in the following way. Yang-Mills theory, like Maxwell theory, is gauge invariant, but its nonabelian gauge transformations, unlike the electromagnetic ones, cannot all be derived by iterating an infinitesmal gauge transformation. Physical results must be gauge invariant, yet quantum mechanical states need not be; according to Wigner, a change of phase is permitted. Of course Gauss's law requires states to be invariant under infinitesmal gauge transformations, and iterating these produces a finite gauge transformation, which by construction is continuously deformable to the identity. All electromagnetic gauge transformations can be reached in this way, but not all the nonabelian ones. Thus we learn that under "large" gauge transformations—as those that cannot be continuously deformed to the identity are called, to distinguish them from the "small" ones, which can be so deformed—physical states of a Yang-Mills theory are only invariant up to a phase, and this phase is the hidden θ parameter of the model. The detailed classification of possible gauge transformations to which I have alluded makes use of the topological fact that the group of all integers is equivalent to Π_3 of the (nonabelian) gauge group; i.e., the mappings of the 3-sphere into the gauge group can be classified by integers. It should be emphasized that the existence of the angle does not rely on instantons; it is an exact quantum mechanical statement, while instantons provide an approximate method for calculating the consequence of the angle. Again the analogy with a periodic potential is a useful one: Bloch-Floquet theory makes the exact statement that the wave function acquires a phase when its argument is shifted by the potential's period. These shifts are analogs of large gauge transformations. The consequent tunneling and band formation can be analyzed in the tight-binding approximation, and this is analogous to instanton calculations.

Another important topological lesson has taught us that the axial vector anomaly [6], which these days is central to many issues in gauge theories, is not a curious by-product of field theoretical infinities, but rather a consequence of the topologically nontrivial possibilities that are present in gauge theories, as described by the Pontryagin index.

One further topological technique has proved useful in physics. Frequently we need to know the spectrum of a linear differential operator—for example, to analyze quantum fluctuations in some background. Zero eigenvalue modes, when they occur in the spectrum, are the most crucial ones for physical application. While one may solve the differential equations explicitly, to determine these important modes, we have now learned that there exist a priori procedures, called "index theorems," that classify the zero modes. The Atiyah-Singer index theorem, first used in the physics of instantons, is an important example [7], while its various extensions appear in other applications. I shall not here have time to discuss the mathematical or physical import of "indices," but Singer's talk yesterday and Witten's tomorrow* address the point adequately.

I now must assess the impact on practical physics that these theoretical discoveries have had. The oldest results—those associated with the axial vector anomaly—seem to be in good shape: anomaly controlled low energy processes like $\pi^0 \to 2\gamma$ proceed at the calculated rate [8]; our prediction that quarks and leptons must balance in number, so that there be no anomalous obstruction to renormalizability, keeps apace with the discovery of new leptons and quarks [9]; the symmetry that is anomalously violated—chiral $U(1)$—does not occur in Nature [10]. However, further, more recent results pose problems: the vacuum angle is PT noninvariant; moreover, its magnitude is not calculable owing to the axial anomaly. While one is intrigued by this unexpected source of CP violation, the stringent bounds on the neutron's electric dipole moment demand that θ be effectively zero; yet there is no principle guaranteeing this—we are facing a puzzle not unlike that of the cosmological constant in gravity theory.

Among solitons, the monopole is the most important for particle physicists, since it should arise in those theories that unify the strong interactions with the electroweak. But we all are disappointed that after fifty years of looking, we still have not found one—maybe I should say that we haven't found more than one [11]. This now poses a problem—the "unified" theory demands a monopole, but it is not seen. There are more problems: It has been suggested that monopoles catalyze proton decay [12]—but we hear from Goldhaber that the proton seems stable; more recently it was startlingly alleged that monopoles in unified theories can obstruct color gauge invariance [13]—but colored states are not supposed to exist. Is there something wrong with the predictions, or are we working with the wrong model, or will experiments eventually come into line? At present, we do not know—so the monopole soliton remains obviously important, perhaps only in a negative way, and it is unclear what its ultimate application will be.

* Editors' note: The paper read to the conference by Witten appears in this volume; Singer's does not.

Thus far, particle physics has not absorbed all the nonperturbative results that have been uncovered; especially the more exotic ones, like boson-fermion conversion or fractional fermion charge, seem far from phenomenological application. But in condensed matter physics, the last of these—fractional charge formation—has found practical application and is used to explain experimental data. Unlike particle physicists, who hope that one Hamiltonian explains all phenomena, condensed matter people have many Hamiltonians, each one describing some different material, and with this great variety one can realize all quantum mechanical possibilities. Certainly there are systems with spontaneous symmetry breaking—the effect was first established within condensed matter physics. Certainly there are solitons—these are the domain walls between different symmetry states of the system. Of course there are fermions moving in the field of the soliton —the electrons of the conduction band. In one such system, polyacetylene, these three facts come together to produce observed phenomena that have been interpreted as soliton-induced charge fractionization [14]. This story is very beautiful, because it can be presented very generally, without reference to details, utilizing only the quantum mechanical properties of fermions interacting with topological solitons. Also, the discussion may be given from many points of view: chemistry, condensed matter physics, particle field theory, mathematics. This vividly demonstrates how different disciplines have come, each in their own way, to the same conclusions about Nature.

Let me sketch briefly a very simple pictorial description of the fractionization phenomenon in polyacetylene. That material is a 1-dimensional polymer with doubly degenerate ground states. (The degeneracy arises for reasons that are called the "Peierls instability" but need not concern us here [14].) Two distinct patterns for bonding electrons to the carbon atoms, which comprise the polyacetylene chain, characterize the two ground states; call them A and B. They are illustrated in (a) and (b) of figure 1. A kink or soliton is a defect in the regular single bond–double bond alteration pattern, and state B with two solitons is depicted in (c) of the figure. Let us compare state B without solitons [i.e., (b)] to that with two solitons [i.e., (c)]. They differ only in the region between solitons, and the one without solitons carries five links in the interval, while the two-soliton state has four links. Inserting two solitons produces a defect of one link. If we now imagine separating the two solitons, so that each acts independently of the other, the

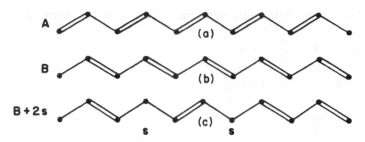

Figure 1
Polyacetylene: (a) vacuum A; (b) vacuum B; (c) vacuum B with two solitons.

quantum numbers of the missing link must be equally shared by each
soliton; i.e., the link number fractionizes into halves. This is the essence of
fermion fractionization, whose further consequences have been experimentally observed [15]. I must stress that the fractionization is without
quantum fluctuation; it concerns eigenvalues and not merely expectation
values [16].

Let me now describe a new direction and new results of topological
investigations in field theory. This current topic concerns one of the oldest
questions of modern physics: quantization of physical quantities. The first
such quantization, the ordinary quantum mechanical one, quantizes dynamical attributes of a system, like energy or angular momentum, and it
involves Planck's constant $\hbar$. In recent times we have seen another type of
quantization, arising classically for topological reasons — the quantization
of soliton number or instanton number. This quantization makes no reference to $\hbar$; finiteness of the classical energy or action requires it. However,
there is a third mechanism for quantization that arises from the conjunction
of topological and quantum mechanical principles. This happens when one
is dealing with a system that produces consistent classical dynamics but
upon being quantized becomes inconsistent, for gauge invariance reasons,
unless parameters (i.e., coupling constants) are quantized. Although field
theorists are only now exploring this phenomenon [17], its first example
was given by Dirac fifty years ago, in his quantization of magnetic charge. I
shall review that classic result, to show how gauge invariance plays a central
role and to set the stage for field-theoretic generalizations.

A particle of mass m and charge e interacts by the Lorentz force law with a Dirac magnetic monopole of strength g, described by vector potential $\mathbf{A}$:

$$m\ddot{\mathbf{r}} = \frac{e}{c}\dot{\mathbf{r}} \times \mathbf{B}, \qquad \mathbf{B} = g\frac{r}{r^3} = \nabla \times \mathbf{A}.$$

This dynamics follows from the Lagrangian

$$L = \frac{1}{2}m\dot{\mathbf{r}}^2 + \frac{e}{c}\dot{\mathbf{r}} \cdot \mathbf{A}(\mathbf{r}).$$

The equation of motion is obviously gauge invariant; however, the Lagrangian is not. Under a static gauge transformation $\mathbf{A} \rightarrow \mathbf{A} + \nabla\theta$, L changes by a total derivative:

$$\mathbf{L} \rightarrow \mathbf{L} + \frac{e}{c}\dot{\mathbf{r}} \cdot \nabla\theta = L + \frac{d}{dt}\left(\frac{e}{c}\theta\right),$$

so the action $I = \int dt\, L$ changes by end-point contributions. But classical mechanics is entirely determined by the dynamical equation of motion, which is gauge invariant, hence obviously consistent; classical mechanics does not care about the action! Quantum mechanics, on the other hand, uses the exponential of the action $\exp(i/\hbar)I$, which must be gauge invariant for consistency. Hence any gauge change of the action must be an integral multiple of $2\pi\hbar$. One can show that for a magnetic monopole the end-point terms in the gauge transformed action give $4\pi(eg/c)$; therefore gauge invariance requires the Dirac quantization condition on the parameters of the problem: $eg/c = (\hbar/2)n$. This derivation relies on the action, which is appropriate to a functional formulation of quantum mechanics like Feynman's, and we see that the monopole is described by a class of multivalued actions labeled by integers. One may also give a Hamiltonian argument. Recall that the angular momentum in a charged particle-monopole system has a radial component of magnitude eg/c. Therefore, conventional quantum mechanics of angular momentum gives the quantization condition by requiring that amount to be $\hbar$ times an integer or half-integer. The mathematical background to the Dirac quantization condition is the fact that $\Pi_1(U(1))$ is equivalent to the group of integers; i.e., the map of the unit circle into the gauge group, here $U(1)$, is classified by integers.

The first field theoretical example that requires quantized parameters is a 3-dimensional gauge theory. In 3 dimension, it is possible to have a term in the Lagrangian that explicitly gives a mass to the gauge potential; nevertheless the equation of motion is gauge covariant [18]. The Lagrangian, written in matrix notation for the gauge fields

$$L = \frac{1}{2g^2} \operatorname{tr} F^{\mu\nu} F_{\mu\nu} - \frac{m}{2g^2} \varepsilon^{\mu\nu\alpha} \operatorname{tr}\left(F_{\mu\nu} A_\alpha - \frac{2}{3} A_\mu A_\nu A_\alpha\right),$$

implies the following equation of motion:

$$\mathscr{D}_\mu F^{\mu\nu} + \frac{m}{2} \varepsilon^{\nu\alpha\beta} F_{\alpha\beta} = 0.$$

Here g is a coupling with dimension of [mass]$^{1/2}$ and $\mathscr{D}$ is the covariant derivative. (Upon examining the linear or abelian theory, one learns that the excitations indeed carry mass m [19].) Under a gauge transformation $A_\mu \to U^{-1} A_\mu U + U^{-1} \partial_\mu U$, the action changes according to

$$I = \int d^3x\, \mathscr{L} \to I + m\frac{8\pi^2}{g^2} W(U),$$

$$W(U) = \frac{1}{24\pi^2} \int d^3x\, \varepsilon^{\alpha\beta\gamma} \operatorname{tr}\left[\partial_\alpha U U^{-1} \partial_\beta U U^{-1} \partial_\gamma U U^{-1}\right].$$

Again because Π_3 of the (nonabelian) gauge group is the group of all integers, $W(U)$ is an integer, and gauge invariance sets a quantization condition $4\pi m/g^2 = \hbar n$ [20]. Thus, 3-dimensional gauge theories are defined by multivalued gauge invariant actions, analogous to the magnetic monopole example. Here too one may give a Hamiltonian derivation. In a Hamiltonian formulation, gauge invariance is achieved by requiring physical states to satisfy Gauss's law, which in our massive theory reads

$$\left(\mathscr{D}_i F^{io} + \frac{m}{2} \varepsilon^{ij} F_{ij}\right)\Big|\text{state}\rangle = 0.$$

But this requirement cannot be met unless m is quantized [21].

A second example, now in 4 dimensions, has been given by Witten [22], who considers an $SU(2)$ gauge theory with N doublets of Weyl fermions in the fundamental representation. Because $SU(2)$ is anomaly-free, this appears to be a consistent theory for any N. Yet again gauge invariance sets a limitation in the quantum theory: N must be even. The argument, in a functional integral formulation, proceeds as follows. We begin with the quantum generating functional

$$\mathscr{F} = \int d\psi \, d\bar{\psi} \, dA_\mu \exp\frac{i}{\hbar} \int d^4x \, \mathscr{L},$$

$$\mathscr{L} = \frac{1}{2g^2} \operatorname{tr} F^{\mu\nu} F_{\mu\nu} + i\hbar \sum_{n=1}^{N} \bar{\psi}_n \gamma^\mu (\partial_\mu + A_\mu) \psi_n,$$

and integrate out the Weyl fermions, leaving an effective nonlocal action for gauge field dynamics

$$I = \int d^4x \frac{1}{2g^2} \operatorname{tr} F^{\mu\nu} F_{\mu\nu} - i\hbar N \ln \det^{1/2} \{\gamma^\mu (\partial_\mu + A_\mu)\},$$

whose gauge transformation properties must still be investigated. (The square root on the determinant occurs because one is integrating four-component Weyl fermions.) Witten finds, as a consequence of $\Pi_4(SU(2))$ being the group of two integers (0, 1) under modulo 2 addition, that a gauge transformation changes the effective action by a discrete amount, $I \to I + \hbar N\pi$; hence the restriction: N must be even. The Hamiltonian version of this argument, given by Goldstone [23], is especially intriguing since it presents the result as a consequence of the triangle anomaly, which plays no role in this Lagrangian discussion. The statement that the $SU(2)$ gauge theory is anomaly-free means that the chiral $SU(2)$ currents, to which the $SU(2)$ gauge fields couple, are conserved. However, the $U(1)$ chiral fermion number current possesses an anomalous divergence, and as usual, the conserved fermion number operator F is not gauge invariant, but changes under a gauge transformation by an integral multiple of N [24]. In the quantum theory there are various transformations one can perform on a physical state—for example, rotations by 2π—whose effect is to multiply the state by $(-1)^F$. Clearly, if F is ambiguous by N, N must be even for consistency.

One further 4-dimensional example has also been given by Witten [25]. This concerns a nonlinear σ-model, which summarizes the low energy theorems of $SU(3) \times SU(3)$ current algebra, or, as one would say today, describes the low energy dynamics of QCD. The most obvious and simplest form for the action is

$$I_\sigma^0 = -\frac{F_\pi^2}{16} \int d^4x \operatorname{tr} \partial_\mu U \partial^\mu U^{-1},$$

where U is an $SU(3)$ matrix and F_π is fitted by experiment to 190 MeV. However, when gauged, I_σ^0 does not reproduce the axial vector anomalies.

Wess and Zumino [26] showed how the action should be modified to include anomalies: to I_σ^0 they add a further term, I_{WZ} (whose form is complicated). An arbitrary coefficient c multiplies the Wess-Zumino addition; however, to reproduce the precise numerical value of the anomaly, as calculated from quark triangle graphs, the coefficient must be N_c—the number of colors. Nevertheless, for the moment we leave c unspecified. Once again the transformation properties of the action $I_\sigma = I_\sigma^0 + cI_{WZ}$ reflect a topological fact: $\Pi_5(SU(3))$ is the group of all integers, and as a consequence, consistency of the quantized theory requires c to be an integer, just as the microscopic quark theory predicts. We do not learn the value of the integer, but remarkably a bosonic effective Lagrangian somehow knows about the underlying fermionic structure [27]!

Finally let me comment on the physical import of these new results. I need not add anything more to all that has been said about magnetic monopoles. About the 3-dimensional gauge theory you may be surprised to hear that it is not merely a theoretician's toy, but also is physically relevant, because physical 4-dimensional theories, when studied at high temperature, are effectively described by theories in 3 dimensions [28]. Therefore, our 3-dimensional gauge theory may provide a phenomenological Lagrangian for 4-dimensional high temperature QCD, and the 3-dimensional quantized mass term may arise from high temperature magnetic screening due to the Pontryagin density in QCD, which is responsible for topological effects in 4 dimensions. While we have no proof of this conjecture, let us point out that both the Pontryagin density and our mass term are PT noninvariant. Moreover, there is a mathematical connection between them, through the formula

$$\text{Pontryagin density} = -\frac{1}{16\pi^2}\text{tr}\,{}^*F^{\mu\nu}F_{\mu\nu} = \partial_\mu X^\mu,$$

$$X^\mu = -\frac{1}{16\pi^2}\varepsilon^{\mu\alpha\beta\gamma}\text{tr}\left(F_{\alpha\beta}A_\gamma - \frac{2}{3}A_\alpha A_\beta A_\gamma\right).$$

The topological current X^μ involves the same combination of gauge fields as our mass term does; indeed the latter's contribution to the action is just the 3-dimensional integral of one component of X^μ. This relation between Pontryagin density and topological current has a well-established place in mathematics under the name "Chern-Simons secondary characteristic class" [29].

Witten's restriction that the number of $SU(2)$ Weyl fermion doublets must be even is surely fascinating, but it does not advance phenomenology since no one is considering models with such fermion content.

Finally we come to the Wess-Zumino term, with its coefficient set necessarily to an integer; we take it to be three—the number of colors. It has recently been shown that by taking the Skyrme nonlinear σ-model [1] (in the Skyrme action, I_σ^0 is supplemented by higher derivative interactions, which give rise to solitons [30]), together with the anomaly-producing Wess-Zumino term

$$I = I_{\text{Skyrme}} + cI_{\text{WZ}} = I_\sigma^0 + \alpha \int d^4x \, \text{tr}\{[\partial_\mu U U^{-1}, \partial_\nu U U^{-1}]^2\} + cI_{\text{WZ}},$$

$$c = N_c = 3,$$

the solitons of the resulting theory—which may still be viewed as a description of low-energy QCD—are fermions with quantum numbers of the low-lying baryons [31]. I spoke of Skyrme at the beginning—he suggested years ago that baryons might be solitons [1]—and his suggestion has new come to fruition in a beautiful synthesis of several topological and nonperturbative ideas: axial vector anomaly, solitons and their unexpected quantum numbers, and quantized Lagrangian parameters.

This is where the subject stands today. While the experimental questions that it has produced remain unsettled, I hope you find, as I do, that it possesses the elegance and generality appropriate to descriptions of fundamental phenomena in Nature [32].

Acknowledgment

This work was supported in part through funds provided by the U.S. Department of Energy under contract DE-AC02-76ER03609.

References

[1]   T. Skyrme, *Proc. Roy. Soc.* A260, 127 (1961).

[2]   For reviews see R. Jackiw, *Rev. Mod. Phys.* 49, 681 (1977); R. Rajaraman, *Solitons and Instantons* (North-Holland, Amsterdam, 1982).

[3]   J. Goldstone and R. Jackiw, in *Gauge Theories and Modern Field Theory*, R. Arnowitt and P. Nath, editors (MIT Press, Cambridge MA, 1976).

[4]   For a review see R. Jackiw, in *Quantum Structure of Space and Time*, M. Duff and C. Isham, editors (Cambridge University Press, Cambridge, 1982).

[5]   For reviews see R. Jackiw, *Rev. Mod. Phys.* 52, 661 (1980); Rajaraman, reference [2].

[6]   For a review see R. Jackiw, in *Lectures on Current Algebra and Its Applications*, S. Treiman, R. Jackiw, and D. Gross (Princeton University Press, Princeton NJ, 1972).

[7]   For reviews see R. Jackiw, C. Nohl, and C. Rebbi, in *Particles and Fields*, D. Boal and A. Kamal, editors (Plenum Press, New York, 1978); T. Eguchi, P. Gilkey, and A. Hanson, *Phys. Rep.* 66, 213 (1980).

[8]   H. Fukuda and Y. Miyamoto, *Prog. Theoret. Phys.* 4, 347 (1949); J. Steinberger, *Phys. Rev.* 76, 1180 (1949); J. Schwinger, *Phys. Rev.* 82, 664 (1951); J. Bell and R. Jackiw, *Nuovo Cimento* A60, 47 (1969); S. Adler, *Phys. Rev.* 177, 2426 (1969); S. Glashow, R. Jackiw, and S. Shei, *Phys. Rev.* 187, 1416 (1969).

[9]   D. Gross and R. Jackiw, *Phys. Rev.* D6, 477 (1972); C. Bouchiat, J. Iliopoulos, and P. Meyer, *Phys. Lett.* B38, 519 (1972).

[10]   The anomalous nonconservation is caused by the tunneling mentioned; see G. 't Hooft, *Phys. Rev. Lett.* 37, 8 (1976) and *Phys. Rev.* D14, 3432 (1976); R. Jackiw and C. Rebbi, *Phys. Rev. Lett.* 37, 172 (1976); C. Callan, R. Dashen, and D. Gross, *Phys. Lett.* B63, 334 (1976).

[11]   B. Cabrera, *Phys. Rev. Lett.* 48, 1378 (1982).

[12]   V. Rubakov, *Zh. Eksp. Teor. Pis'ma Red.* 33, 658 (1981) [*JETP Lett.* 33, 644 (1981)] and *Nucl. Phys.* B203, 311 (1982); C. Callan, *Phys. Rev.* D25, 2141 (1982) and D26, 2058 (1982); V. Rubakov and M. Sevebryakov, *Nucl. Phys.* B218, 240 (1983).

[13]   P. Nelson and A. Manohar, *Phys. Rev. Lett.* 50, 943 (1983); A. Balachandran, G. Marmo, N. Mukunda, J. Nilsson, E. Sudarshan, and F. Zaccaria, *Phys. Rev. Lett.* 50, 1553 (1983).

[14]   W.-P. Su, J. Schrieffer, and A. Heeger, *Phys. Rev. Lett.* 42, 1698 (1979) and *Phys. Rev.* B22, 2099 (1980). For a comparison of this condensed matter approach to fermion fractionization with the earlier field theoretic analysis [R. Jackiw and C. Rebbi, *Phys. Rev.* D13, 3398 (1976)] see R. Jackiw and J. Schrieffer, *Nucl. Phys.* B190 [FS 3], 253 (1981).

[15]   For a summary of experiments see A. Heeger, *Comments Solid State Physics* 10, 53 (1981); for general summaries see S. Kivelson in *Solitons*, S. Trullinger and V. Zakharov, editors (North-Holland, Amsterdam, in press) and W.-P. Su in *Handbook on Conducting Polymers*, T. Skotheim, editor (Dekker, New York, in press).

[16]   S. Kivelson and J. Schrieffer, *Phys. Rev.* B25, 6447 (1982); R. Rajaraman and J. Bell, *Phys. Lett.* B116, 151 (1982); J. Bell and R. Rajaraman, *Nucl. Phys.* B220 [FS 8], 1 (1983); Y. Frishman and B. Horovitz, *Phys. Rev.* B27, 2565 (1983); R. Jackiw, A. Kerman, I. Klebanov, and A. Semenoff, *Nucl. Phys.* B225 [FS 9], 233 (1983).

[17]   S. Deser, R. Jackiw, and S. Templeton, *Phys. Rev. Lett.* 48, 975 (1982) and *Ann. Phys.* (*NY*) 140, 372 (1982).

[18]   R. Jackiw and S. Templeton, *Phys. Rev.* D23, 2291 (1981); J. Schonfeld, *Nucl. Phys.* B185, 157 (1981).

[19]   These models exhibit gauge invariant masses for gauge fields, without the intervention of Higgs fields. Thus they supplement the 2-dimensional Schwinger model (massless spinor QED in two dimensions) as examples of this phenomenologically important effect. It is noteworthy that in both cases the mass arises for topological reasons: the Schwinger model uses the Pontryagin class through its axial vector anomaly; the 3-dimensional models rely on the Chern-Simons characteristic—as explained later. For a review of topological mass generation see R. Jackiw, in *Asymptotic Realms of Physics*, A. Guth, K. Huang, and R. Jaffe, editors (MIT Press, Cambridge, MA, 1983). Also, 3-dimensional gravity allows construction of a similar topological mass term; see reference [17] and S. Deser, in *Quantum Theory of Gravity*, S. Christensen, editor (Adam Hilger, Bristol 1984).

[20]   Reference [17]. For a review see R. Jackiw, in *Gauge Theories of the Eighties*, R. Raitio and J. Lindors, editors, Lecture Notes in Physics 181 (Springer Verlag, Berlin, 1983).

[21]   J. Goldstone and E. Witten (1982) (unpublished). Their argument is reviewed in reference [20].

[22]   E. Witten, *Phys. Lett.* B117, 324 (1982).

[23]   J. Goldstone (1982) (unpublished).

[24]   I. Gerstein and R. Jackiw, *Phys. Rev.* 181, 1955 (1969); Jackiw and Rebbi, reference [10].

[25]   E. Witten, *Nucl. Phys.* B223, 422 and 433 (1984).

[26]   J. Wess and B. Zumino, *Phys. Lett.* B37, 95 (1971).

[27]   All this is paralleled in 2 dimensions: 2-dimensional QCD also has axial vector anomalies, hence an effective nonlinear σ-model must have a Wess-Zumino term to reproduce them. Its coefficient will again be quantized. These points are discussed by A. Polyakov and P. Wiegman, *Phys. Lett.* B131, 121 (1983), in the course of their solution to the quantized, 2-dimensional nonlinear σ-model.

[28]   For a review see D. Gross, R. Pisarski, and L. Yaffe, *Rev. Mod. Phys.* 53, 43 (1981).

[29]   S. Chern, *Complex Manifolds without Potential Theory*, 2nd ed. (Springer Verlag, Berlin, 1979).

[30]   See also L. Faddeev, *Lett. Math. Phys.* 1, 289 (1976).

[31]   A. Balachandran, V. Nair, S. Rajeev, and A. Stern, *Phys. Rev. Lett.* 49, 1124 (1982); Witten, reference [23]. Phenomenological application of these ideas is being attempted—see M. Rho, A. Goldhaber, and G. Brown, *Phys. Rev. Lett.* 51, 747 (1983); J. Goldstone and R. Jaffe, *Phys. Rev. Lett.* 51, 1518 (1983); G. Adkins, C. Nappi, and E. Witten, *Nucl. Phys.* B228, 552 (1983).

[32]   For a view of multivalued actions see R. Jackiw, in *Relativity, Groups and Topology II*, B. DeWitt and R. Stora, editors (North-Holland, Amsterdam, 1984).

Supersymmetry and the Index Theorem

Bruno Zumino

1 Introduction

The main interest in supersymmetry (SUSY) comes from the hope that supersymmetric field theories will provide a deeper unification of the fundamental forces of nature than is afforded by the ordinary gauge principle alone. Work in this direction is still very intensive, but, in spite of numerous very interesting investigations, it seems that a new idea will be needed before the promise is fulfilled [1].

The greatest impact of SUSY so far has been of a more mathematical nature. On the one hand, there are the special renormalization propertiesof SUSY quantum field theories (which could solve the hierarchy problem in GUTs). They are less divergent than generic field theories [2] and in some cases they are even convergent, in a precise sense. This is true of the $N = 4$ SUSY Yang-Mills theory [3, 4] and of a whole class of $N = 2$ SUSY gauge theories [4].

In addition, SUSY has influenced pure mathematics. Ed Witten was probably the first to realize that there are branches of pure mathematics where supersymmetric ideas have been used implicitly for some time, and that using more explicitly the language of SUSY could sharpen the picture and streamline the demonstrations. This applies in particular to the Atiyah-Singer index theorem [5], as explained in a beautiful lecture by I. Singer at this conference.* Conversely, mathematical ideas like that of the index of an elliptic operator, when generalized to quantum field theory, give rise to the concept of the Witten index [6], which proves to be a useful tool in the nonpertubative analysis of spontaneous SUSY breaking. This is a beautiful example of the cross fertilization between mathematics and physics, a continuing phenomenon in the history of science, which is especially evident these days.

Witten's suggestion that the Atiyah-Singer index formula could be understood in terms of a suitable quantum mechanical supersymmetric system has been worked out independently by Alvarez-Gaumé [7] and Friedan and Windey [8], who have used the Feynman path integral techniques (suitably adapted to the presence of fermionic variables). Rigorous mathematical derivations along similar lines have been subsequently given by Getzler [9], who uses a kind of Hamiltonian description, and by Bismut [10], who uses Wiener integrals to give rigorous estimates. (I have not seen this paper. R. Stora has told me of its existence.)

---

*Editors' note: The paper read to the conference by Singer does not appear in this volume.

Clearly, the problem of the index formula has been solved many times, in a more or less rigorous way. Still, none of the derivations seems entirely intuitive and transparent. The mathematical papers are complicated by (surely very legitimate) requirements of rigor, while the papers by physicists are somewhat clouded by the use of path integrals, whose definition is always a little tricky. I shall try to give below a more intuitive picture of the derivation of the index formula, worked out in collaboration with Juan Mañes [11]. We believe that it can be turned into a fully rigorous derivation.

I shall describe here only the case of the Dirac operator on a compact Riemannian manifold of even dimension n. The more general case when the spinors interact also with a Yang-Mills potential can be treated in an analogous way [11]. This more general case (twisted spin complex) is known to contain all classical geometric complexes [12] studied in mathematics: de Rham (Euler), signature (Hirzebruch), Dolbeault. They can all be described in terms of (suitably twisted) spinors. Furthermore, the index theorem for the (twisted) Dirac operator actually implies that theorem for a general elliptic complex, in the sense that there is a homotopy relation. So, the study of the Dirac operator is as general as one may wish.

Our method is clearly closely related to the different techniques used in references [7–10]. For instance, the fact (which we exploit) that the WKB expressions are better and better approximations as the time interval gets smaller and smaller is also the basis for the definition the Feynman path integral [13]. On the other hand, the classical limits of operators and amplitudes (which we use in an essential way) are presumably related to the symbols of operators used in Ref. [9], although we did not succeed in establishing a direct connection.

2 The Dirac Operator in Curved Space and Its Index

We consider a compact Riemannian manifold of even dimension n. The Dirac operator is

$$\sqrt{2}Q = \tfrac{1}{2}\{\gamma^\mu, \pi_\mu\}. \tag{2.1}$$

We have written the right-hand side as an anticommutator so that the operator is hermitean (the reason for the $\sqrt{2}$ will become clear later). Here

$$\gamma^\mu = \gamma^a e_a{}^\mu(x), \tag{2.2}$$

where the (x-independent) gamma matrices γ^a satisfy

$$\{\gamma^a, \gamma^b\} = 2\delta^{ab} \tag{2.3}$$

and $e_a{}^\mu(x)$ is the inverse vielbein field. In absence of Yang-Mills field, the covariant momenta are

$$\pi_\mu = p_\mu + \tfrac{1}{2}\omega_{\mu ab}\sigma^{ab}. \tag{2.4}$$

Here

$$\sigma^{ab} = \frac{1}{4i}[\gamma^a, \gamma^b] \tag{2.5}$$

and $\omega_{\mu ab}$ are the rotation coefficients, which, on a Riemannian manifold, are expressible in terms of the vielbein field and its derivatives by well-known formulas. The canonical momenta satisfy

$$[x^\mu, p_\nu] = i\delta_\nu{}^\mu \tag{2.6}$$

and are given, in the x representation, by

$$p_\nu = \frac{1}{i}\frac{\partial}{\partial x^\nu}. \tag{2.7}$$

The world indices μ, ν, etc., and the orthonormal indices a, b, etc., take the values $1 \ldots n$.

   In order to study the properties of the Dirac operator (Green's functions, eigenvalues, etc.), one considers an associated quantum mechanical system having as bosonic dynamical variables the coordinates x^μ and the momenta p_μ and having as fermionic variables

$$\psi^a = \frac{1}{\sqrt{2}}\gamma^a, \tag{2.8}$$

normalized so that their anticommutator is, like that for Fermi fields,

$$\{\psi^a, \psi^b\} = \delta^{ab}; \tag{2.9}$$

ψ^a commutes with x^μ and p_μ. The Hamiltonian is taken to be

$$H = Q^2 \tag{2.10}$$

and is a nonnegative operator in the Hilbert space on which the dynamical variables operate. The dynamical variables vary in time according to the

Heisenberg equations of motion. For some purposes it is useful to consider an imaginary time (or reciprocal temperature)

$$\beta = it. \tag{2.11}$$

The idea of considering the associated dynamical system is rather old and was exploited by Schwinger [14] to solve problems closely related to the one considered here.

The operator

$$\gamma = \gamma^1 \gamma^2 \cdots \gamma^n \tag{2.12}$$

anticommutes with all γ^a (n is even) and with the operator Q:

$$\{\gamma, Q\} = 0. \tag{2.13}$$

It is the generalization of the 4-dimensional γ^5 and can be used to define positive and negative chirality spinors S^+ and S^-:

$$\gamma S^+ = S^+, \qquad \gamma S^- = -S^-. \tag{2.14}$$

Because of (2.13) the operator Q changes the chirality of a spinor. Since the operator H commutes with γ, its eigenvectors can be assigned a definite chirality. Now H commutes also with Q; therefore from each eigenstate of positive chirality, corresponding to a positive eigenvalue of H, one can obtain an eigenstate of negative chirality by applying Q to it. Eigenstates come in degenerate pairs of opposite chirality, provided their eigenvalue is positive. For the eigenvalue zero this argument to longer works because, if a state satisfies

$$HS = 0, \tag{2.15}$$

it satisfies also

$$QS = 0; \tag{2.16}$$

i.e., it is a zero mode of the Dirac operator. The difference between the number v_+ of zero modes of positive chirality and the number v_- of zero modes of negative chirality is called the index. There is a simple formula for the index,

$$v_+ - v_- = \sum (\pm) e^{-\lambda_i \beta} = \mathrm{Tr}(\gamma^5 e^{-\beta H}), \tag{2.17}$$

where the sum is over all eigenstates, with the plus sign for positive chirality and the minus sign for negative chirality. Remember that $\lambda_i \geq 0$: for positive eigenvalues the terms in the sum cancel in pairs. There remains only the contribution of the eigenvalue zero, and that equals the index. The result is clearly independent of β.

We can attribute to positive chirality states the property of being bosonic and to negative chirality states that of being fermionic (the opposite convention would be equally good). The operators ψ^a (or γ^a) change by one the number of fermions, which we call F. Clearly the operator $(-1)^F$ equals $+1$ on bosonic states and -1 on fermionic states. Therefore

$$(-1)^F = \gamma. \tag{2.18}$$

The operator Q can be interpreted as a conserved SUSY charge. The dynamical system is supersymmetric and the Hamiltonian is given by (2.10) as the square of the SUSY charge. The index of the Dirac operator can be written as

$$v_+ - v_- = \text{Tr}((-1)^F e^{-\beta H}) \tag{2.19}$$

and is now interpreted as the difference between the number of bosonic and the number of fermionic zero energy (vacuum) states: it is Witten's index [6] for the SUSY system. The right-hand side of (2.19) is the *supertrace* of the operator $e^{-\beta H}$.

Let us denote by $\mathscr{S}$ the SUSY transformation generated by Q. One finds easily

$$\mathscr{S}x^\mu = -i[x^\mu, Q] = \psi^\mu, \tag{2.20}$$

$$\mathscr{S}\psi^\mu = i\{\psi^\mu, Q\} = i\pi^\mu, \tag{2.21}$$

where

$$\psi^\mu = \psi^a e_a{}^\mu \tag{2.22}$$

and

$$\pi^\mu = \tfrac{1}{2}\{g^{\mu\nu}, \pi_\nu\}. \tag{2.23}$$

On the other hand, the time derivative of x^μ is given by

$$\dot{x}^\mu = -i[x^\mu, H] = -i[x^\mu, Q^2] \tag{2.24}$$

$$= -i\{[x^\mu, Q], Q\} = \{\psi^\mu, Q\} = \pi^\mu.$$

Therefore (2.20) and (2.21) can be written as

$$\mathscr{S}x^\mu = \psi^\mu, \tag{2.25}$$

$$\mathscr{S}\psi^\mu = i\dot{x}^\mu. \tag{2.26}$$

The operator $\mathscr{S}$ is odd (it exchanges bosonic and fermionic variables) and it satisfies

$$\mathscr{S}^2 = i\frac{\partial}{\partial t}. \tag{2.27}$$

The effect of $\mathscr{S}$ on π^μ is also given by

$$\mathscr{S}\pi^\mu = -i[\pi^\mu, Q], \tag{2.28}$$

but here the right-hand side is a little more complicated. Still (2.27) is valid, because of (2.10).

Our task is to evaluate the index (2.19) or

$$\int dx\, \mathrm{tr}\, \langle x|(-1)^F e^{-\beta H}|x\rangle, \tag{2.29}$$

where tr is the trace over the fermionic variables alone. Since the index is independent of β, it can be evaluated in the limit of small β, or small $t = -i\beta$. On the other hand, it is known that for small t the quantum mechanical transition amplitudes are well described by their classical (WKB) expressions, which in general are just the leading term in an asymptotic expansion for small $\hbar$. Therefore, a t-independent quantity like the index can be evaluated completely in terms of classical quantities such as the Hamilton-Jacobi function. One need only generalize the usual classical phase space concepts to fermionic variables. [In the "classical" limit the ψ^a-anticommute [15] instead of satisfying (2.9).]

3 Phase Space and WKB Quantization for a SUSY System

For a bosonic system with n degrees of freedom the classical Hamilton-Jacobi function is given by

$$S(q_t, q_0; t) = \int_{q_0, 0}^{q_t, t} L\, dt', \tag{3.1}$$

where the integral of the Lagrangian L is over a classical *solution* of the equations of motion joining the configurations $(q_0, 0)$ and (q_t, t). It satisfies

$$p_t = \frac{\partial S}{\partial q_t} \tag{3.2}$$

and

$$p_0 = -\frac{\partial S}{\partial q_0}. \tag{3.3}$$

Here and in the following we omit the indices which number the various degrees of freedom. The WKB form for the quantum amplitude is given by the asymptotic expansion as $\hbar \to 0$,

$$\langle q_t | e^{-(t/\hbar)H} | q_0 \rangle \cong (2\pi i \hbar)^{-n/2} \sqrt{D} \, e^{(i/\hbar)S}, \tag{3.4}$$

where

$$D = (-1)^n \det \frac{\partial^2 S}{\partial q_t \, \partial q_0} \tag{3.5}$$

is the van Vleck determinant [13], which satisfies a continuity equation.

For the generalization to fermionic variables one requires a slight modification of these ideas, which we first describe still for the bosonic case. Let us introduce the modified Hamilton-Jacobi function

$$S'(p_t, q_0; t) = S(q_t(p_t, q_0), q_0; t) - p_t q_t(p_t, q_0), \tag{3.6}$$

which satisfies

$$q_t = -\frac{\partial S'}{\partial p_t} \tag{3.7}$$

and

$$p_0 = -\frac{\partial S'}{\partial q_0}. \tag{3.8}$$

One has the asymptotic expansion

$$\langle p_t | e^{-i(t/\hbar)H} | q_0 \rangle \cong (2\pi \hbar)^{-n/2} \sqrt{D'} \, e^{(i/\hbar)S'}, \tag{3.9}$$

where

$$D' = (-1)^n \det \frac{\partial^2 S'}{\partial p_t \, \partial q_0}. \tag{3.10}$$

The exact left-hand sides of (3.4) and (3.9) are related by a Fourier transformation, while the functions S and S' are related by the Legendre transformation (3.6), (3.2), and (3.7). This is consistent as $\hbar \to 0$, since it is well known that the Legendre transformation arises from the Fourier transformation when one evaluates the integrals by the steepest descent method. The functions S' satisfies the Hamilton-Jacobi partial differential equations

$$\frac{\partial S'}{\partial t} + H\left(p_t, -\frac{\partial S'}{\partial p_t}\right) = 0, \tag{3.11}$$

$$\frac{\partial S'}{\partial t} + H\left(-\frac{\partial S'}{\partial q_0}, q_0\right) = 0 \tag{3.12}$$

and the initial condition

$$S'\bigg|_{t=0} = -p_t q_0, \tag{3.13}$$

which implies that

$$D'\bigg|_{t=0} = 1. \tag{3.14}$$

This contrasts with the singular behavior of S and D for $t \to 0$; for instance,

$$D \sim t^{-n/2}. \tag{3.15}$$

For the case of Fermionic variables, there is an analog of the function S', while S has no analog in general. Let us consider the simple example of the Hamiltonian

$$H = (f/2)(\bar{\eta}\eta - \eta\bar{\eta}) \tag{3.16}$$

where f is a constant and (we take $\hbar = 1$)

$$\{\bar{\eta}, \eta\} = 1, \tag{3.17}$$

$$\{\eta, \eta\} = \{\bar{\eta}, \bar{\eta}\} = 0. \tag{3.18}$$

The equations of motion are

$$i\dot{\eta} = f\eta, \qquad i\dot{\bar{\eta}} = -f\bar{\eta}, \tag{3.19}$$

with solutions

$$\eta_t = e^{-itf}\eta_0, \qquad \bar{\eta}_t = e^{itf}\bar{\eta}_0. \tag{3.20}$$

The "classical" Lagrangian corresponding to (3.16) is

$$L = i\bar{\eta}\dot{\eta} - f\bar{\eta}\eta, \tag{3.21}$$

where now the variables η and $\bar{\eta}$ simply anticommute. The momentum conjugate to η is

$$\frac{\partial L}{\partial \dot{\eta}} = -i\bar{\eta}. \tag{3.22}$$

For the classical motion L vanishes identically. Only the analog of the last term in (3.6) survives in the function

$$S'(\bar{\eta}_t, \eta_0) = -i\bar{\eta}_t\eta_t(\bar{\eta}_t, \eta_0) = -i\bar{\eta}_t e^{-itf}\eta_0. \tag{3.23}$$

By analogy with (3.7) and (3.8) it satisfies

$$\frac{\partial S'}{\partial(-i\bar{\eta}_t)} = e^{-itf}\eta_0 = \eta_t(\bar{\eta}_t, \eta_0) \tag{3.24}$$

and

$$\frac{\partial S'}{\partial \eta_0} = i\bar{\eta}_t e^{-itf} = i\bar{\eta}_0(\bar{\eta}_t, \eta_0). \tag{3.25}$$

By analogy with (3.11) and (3.12), S' satisfies the Hamilton-Jacobi equations

$$\frac{\partial S'}{\partial t} + H\left(\bar{\eta}_t, \frac{\partial S'}{\partial(-i\bar{\eta}_t)}\right) = 0, \tag{3.26}$$

$$\frac{\partial S'}{\partial t} + H\left(-i\frac{\partial S'}{\partial \eta_0}, \eta_0\right) = 0, \tag{3.27}$$

where H here is the classical Hamiltonian corresponding to (3.16):

$$H = f\bar{\eta}\eta. \tag{3.28}$$

Finally, the initial condition is [see (3.13)]

$$S'\Big|_{t=0} = -i\bar{\eta}_t\eta_0. \tag{3.29}$$

In order to construct the analog of (3.9) we define

$$(D')^{-1} = -i\frac{\partial^2 S'}{\partial \bar{\eta}_t \partial \eta_0} = e^{-itf} \tag{3.30}$$

and take

$$\langle \bar{\eta}_t | e^{-itH} | \eta_0 \rangle = \sqrt{D'} \, e^{iS'} = e^{\bar{\eta}_t e^{-itf} \eta_0 + itf/2}. \tag{3.31}$$

Observe the power minus one in the definition (3.30) of D', which is a fermionic determinant [16]. A coordinate representation amplitude can be constructed by a fermionic Fourier transformation [17]

$$\langle \eta_t | e^{-itH} | \eta_0 \rangle = \int_{\bar{\eta}_t} e^{-\bar{\eta}_t \eta_t} \langle \bar{\eta}_t | e^{-itH} | \eta_0 \rangle, \tag{3.32}$$

and one finds that

$$\langle \eta_t | e^{-itH} | \eta_0 \rangle = \eta_0 e^{-itf/2} - \eta_t e^{itf/2}. \tag{3.33}$$

It is easy to verify that this amplitude satisfies the appropriate Schrödinger equation as well as the initial condition for $t = 0$:

$$\langle \eta_t | \eta_0 \rangle = \eta_0 - \eta_t = \delta(\eta_t - \eta_0). \tag{3.34}$$

So we find that, in the case of fermionic variables, one can work also in the coordinate representation. However, in general, the Fourier transformation (3.32) cannot be approximated by a Legendre transformation and the analog of the function S does not exist; only the actual amplitude given by (3.32) exists.

If we set $\eta_0 = \eta_t$ in (3.33) and integrate, we obtain

$$\int_{\eta_t} \langle \eta_t | e^{-itH} | \eta_t \rangle = e^{-itf/2} - e^{itf/2}. \tag{3.35}$$

Now, the Hamiltonian (3.16) has two eigenstates

$$|f/2\rangle, \qquad |-f/2\rangle, \tag{3.36}$$

with the indicated eigenvalues and Fermion number F equal to 1 and 0, respectively. Indeed

$$\eta |-f/2\rangle = 0, \qquad |f/2\rangle = \bar{\eta} |-f/2\rangle,$$
$$F = \bar{\eta}\eta. \tag{3.37}$$

Therefore

$$\mathrm{tr}\left((-1)^F e^{-itH}\right) = e^{-itf/2} - e^{itf/2}, \tag{3.38}$$

which agrees with (3.35). One can also write

$$\text{Tr}((-1)^F e^{-itH}) = \int_{\eta_t} \int_{\bar{\eta}_t} \sqrt{D'}\, e^{iS'}\Big|_{\eta_0=\eta_t} e^{-\eta_t \eta_t}. \tag{3.39}$$

The result (3.38) is not independent of t because the purely fermionic system under consideration is not supersymmetric. If we had added the appropriate bosonic counterpart (particle in a constant magnetic field [13]) the supertrace would have a denominator which exactly cancels the t dependence of (3.38). For a more general system, however, the supertrace calculated by the WKB formulas will not be t independent and will agree with the t independent exact quantum mechanical expression only as $t \to 0$.

The above example motivates the following prescription, which we shall not justify further here (see [11] for a more complete argument). From the classical Lagrangian, or from the Hamiltonian of a SUSY system, compute the function

$$S'(p_t, q_0, \bar{\eta}_t, \eta_0; t), \tag{3.40}$$

and from it the superdeterminant D' (see [16]) of the matrix of the second derivatives

$$\begin{pmatrix} \dfrac{\partial^2 S'}{\partial p_t\, \partial q_0} & \dfrac{\partial^2 S'}{\partial p_t\, \partial \eta_0} \\[2ex] -i\dfrac{\partial^2 S'}{\partial \bar{\eta}_t\, \partial q_0} & -i\dfrac{\partial^2 S'}{\partial \bar{\eta}_t\, \partial \eta_0} \end{pmatrix}. \tag{3.41}$$

In the amplitude

$$(2\pi)^{-n/2}\sqrt{D'}\, e^{iS'} \tag{3.42}$$

set

$$q_0 = q_t, \qquad \eta_0 = \eta_t. \tag{3.43}$$

Finally take the Fourier transformation and the trace; i.e., multiply by

$$(2\pi)^{-n/2} e^{ip_t q_t - \eta_t \eta_t} \tag{3.44}$$

and integrate over q_t, p_t, $\bar{\eta}_t$, and η_t. If one leaves out the q_t integration, one has the index density. The p_t integration can be approximated by the Legendre transformation, so actually one finds oneself in a mixed representation [before the identification (3.43), q_t, q_0 for the bosonic variables, $\bar{\eta}_t$, η_0 for the fermionic variables]. The evaluation of S' is simplified by the fact that it is needed only up to some order in t and only for values of the variables satisfying (3.43).

4 Derivation of the Index Formula

The classical Lagrangian for the Dirac dynamical system discussed in section 2 is

$$L = \mathscr{S}\left(\frac{1}{2i}\dot{x}^{\mu}g_{\mu\nu}(x)\psi^{\nu}\right)$$

$$= \frac{1}{2}\dot{x}^{\mu}g_{\mu\nu}\dot{x}^{\nu} + \frac{i}{2}\psi^{\mu}g_{\mu\nu}D_{t}\psi^{\nu},$$

(4.1)

where the covariant time derivative is given by

$$D_{t}\psi^{\nu} = \dot{\psi}^{\nu} + \dot{x}^{\lambda}\Gamma_{\lambda\rho}{}^{\nu}\psi^{\rho}$$

(4.2)

and

$$\Gamma_{\lambda\rho}{}^{\sigma} = \tfrac{1}{2}g^{\nu\sigma}(\partial_{\lambda}g_{\rho\nu} + \partial_{\rho}g_{\lambda\nu} - \partial_{\nu}g_{\lambda\rho})$$

(4.3)

are the Christoffel connection coefficients associated with the metric

$$g_{\mu\nu} = e_{\mu}{}^{a}e_{\nu}{}^{a}.$$

(4.4)

The equivalence of the two forms in (4.1) is easily verified using (2.25) and (2.26). The fact that L can be obtained by applying to something the operator $\mathscr{S}$ ensures that the system is supersymmetric. Indeed, by (2.27), if we apply $\mathscr{S}$ to L, we obtain a time derivative and the integrated action is invariant, with suitable boundary conditions. In terms of the fermionic variables ψ^{a} [see (2.22)] the Lagrangian takes the form

$$L = \frac{1}{2}\dot{x}^{\mu}g_{\mu\nu}\dot{x}^{\nu} + \frac{i}{2}\psi^{a}D_{t}\psi^{a}$$

(4.5)

with

$$D_{t}\psi^{a} = \dot{\psi}^{a} + \dot{x}^{\lambda}\omega_{\lambda}{}^{a}{}_{b}\psi^{b}.$$

(4.6)

This follows from the well-known relation

$$\Gamma_{\lambda\rho}{}^{\nu} = (\partial_{\lambda}e_{\rho}{}^{a})e_{a}{}^{\nu} + e_{\rho}{}^{b}\omega_{\lambda b}{}^{a}e_{a}{}^{\nu}.$$

(4.7)

The variables ψ^{a}, with x^{μ} and the conjugate momenta

$$p_\mu = \frac{\partial L}{\partial \dot{x}^\mu} = g_{\mu\nu}\dot{x}^\nu + \frac{i}{2}\omega_{\mu ab}\psi^a\psi^b \tag{4.8}$$

are independent and can be taken as a basis for quantization. The equations of motion following from (4.5) are

$$D_t\dot{x}^\mu = \frac{i}{2}\dot{x}^\lambda R_\lambda{}^\mu{}_{,ab}\psi^a\psi^b \tag{4.9}$$

and

$$D_t\psi^a = 0, \tag{4.10}$$

where $R_{\lambda\mu,ab}$ is the Riemann tensor constructed from $\omega_{\mu,ab}$. The classical Hamiltonian corresponding to (4.5) is

$$H = \tfrac{1}{2}\pi_\mu g^{\mu\nu}\pi_\nu, \tag{4.11}$$

where

$$\pi_\mu = p_\mu + \frac{1}{2i}\omega_{\mu ab}\psi^a\psi^b. \tag{4.12}$$

It is easy to see that the expression

$$Q = \psi^\mu\pi_\mu \tag{4.13}$$

is conserved; it corresponds to (2.1) in the classical limit. The Hamiltonian (4.11) can be written as one-half the (super-) Poisson bracket of Q with itself.

In order to apply the ideas of section 3 it is convenient to replace first the real variables ψ^a $(a = 1, 2, \ldots, n)$ with the complex variables

$$\eta_1 = \frac{1}{\sqrt{2}}(\psi_1 + i\psi_2), \qquad \eta_2 = \frac{1}{\sqrt{2}}(\psi_3 + i\psi_4), \ldots \tag{4.14}$$

and their complex conjugates (remember that n is even). To apply the prescription given at the end of section 3 we imagine to have performed the p_t integration, or the corresponding Legendre transformation. It turns out that, after the identification (3.43), the exponential vanishes, so we must evaluate only the determinant D (mixed representation). The solution of the equations of motion (4.9) and (4.10) is determined by the initial data x_0, π_0, and ψ_0. A particular solution is

$$x = x_0, \qquad \dot{x} = \pi = 0, \qquad \psi = \psi_0. \tag{4.15}$$

We are interested in calculating [see (3.7) and (3.8)]

$$\frac{\partial^2 S}{\partial x_t \partial x_0} = \frac{\partial p_t}{\partial x_0} = -\frac{\partial p_0}{\partial x_t} = -\left(\frac{\partial x_t}{\partial p_0}\right)^{-1} = -\left(\frac{\partial x_t}{\partial \pi_0}\right)^{-1} \qquad (4.16)$$

for $\pi_0 = 0$. The simplest way is to use the variational equations about the solution (4.15), which are

$$(\delta x^\mu)^\cdot = g^{\mu\nu}(x_0)\delta\pi_\nu, \qquad (4.17)$$

$$(\delta\pi_\mu)^\cdot = R_\mu{}^\nu\delta\pi_\nu, \qquad (4.18)$$

and

$$(\delta\psi^a)^\cdot + (\delta x^\lambda)^\cdot \omega_\lambda{}^a{}_b(x_0)\psi_0{}^b = 0, \qquad (4.19)$$

where we have introduced the matrix

$$R_\mu{}^\nu = -\frac{i}{2}R_\mu{}^\nu{}_{,ab}(x_0)\psi_0{}^a\psi_0{}^b. \qquad (4.20)$$

The explicit solution is, in matrix notation,

$$(\delta\pi)_t = e^{tR}(\delta\pi)_0, \qquad (4.21)$$

$$(\delta x)_t = (\delta x)_0 + g^{-1}\frac{e^{tR}-1}{R}(\delta\pi), \qquad (4.22)$$

and we do not need the solution of (4.19). Now, (4.22) implies that

$$\left.\frac{\partial x_t}{\partial \pi_0}\right|_{\pi_0=0} = g^{-1}\frac{e^{tR}-1}{R}$$

$$= g^{-1}\frac{e^{tR/2}-e^{-tR/2}}{R}e^{tR/2}. \qquad (4.23)$$

The last exponential factor has determinant one, since the matrix R is antisymmetric. Therefore

$$\det\frac{\partial\pi_0}{\partial x_t} = \det\frac{\partial^2 S}{\partial x_t \partial x_0} = \det g_{\mu\nu}\det\frac{R/2}{\sinh tR/2}, \qquad (4.24)$$

and we find the result for the index

$$(2\pi)^{-n/2}\int_{x_0} dx_0\sqrt{\det g_{\mu\nu}}\int_{\psi_0}\sqrt{\det\frac{R/2}{\sinh tR/2}}. \qquad (4.25)$$

In order to check the independence of t, observe that

$$\sqrt{\det \frac{R/2}{\sinh tR/2}} = \frac{1}{t^{n/2}}\sqrt{\det \frac{tR/2}{\sinh tR/2}}. \tag{4.26}$$

Now R contains two ψ_0. Only the term with $n\ \psi_0$ in the expansion of the square root survives after the ψ_0 integration, and that term carries the power $t^{n/2}$, which cancels the $t^{n/2}$ in the denominator. Therefore

$$\int_\psi \sqrt{\det \frac{R/2}{\sinh tR/2}} = \int_\psi \sqrt{\det \frac{R/2}{\sinh R/2}}. \tag{4.27}$$

Similarly, one can absorb the factors 2π, and (4.25) becomes

$$\int_{x_0} dx_0 \sqrt{\det g_{\mu\nu}} \int_{\psi_0} \sqrt{\det \frac{R/4\pi}{\sinh R/4\pi}}. \tag{4.28}$$

This is our result for the index, in agreement with references [7] and [8].

We now wish to compare (4.28) with the standard mathematical expression for the $\hat{A}$ genus [12]. We observe that, if $C_{abc\ldots}$ is an antisymmetric tensor,

$$\int dx \sqrt{\det g_{\mu\nu}} \int_\psi C_{abc\ldots}\psi^a\psi^b\psi^c\cdots = \int dx \sqrt{\det g_{\mu\nu}}\, C_{abc\ldots}\,\varepsilon^{abc\ldots}$$

$$= \int_M dx^\mu dx^\nu dx^\lambda \cdots e_\mu{}^a e_\nu{}^b e_\lambda{}^c \cdots C_{abc\ldots} \tag{4.29}$$

$$= \int_M dx^\mu dx^\nu dx^\lambda \cdots C_{\mu\nu\lambda\ldots},$$

where the x integration is over the Riemannian manifold M. Therefore we can rewrite (4.28) as

$$\int_M \sqrt{\det \frac{\mathscr{R}/4\pi i}{\sinh \mathscr{R}/4\pi i}}, \tag{4.30}$$

where

$$\mathscr{R} = \mathscr{R}_{ab} = \tfrac{1}{2}R_{\mu\nu,ab}\,dx^\mu dx^\nu \tag{4.31}$$

is now the curvature two-form. Formula (4.30) is the standard mathematical expression for the index of the Dirac operator on a Riemannian manifold, in the absence of a Yang-Mills field [12].

Acknowledgment

This work was supported in part by the Director, Office of Energy Research, Office of High Energy and Nuclear Physics, Division of High Energy Physics, of the U.S. Department of Energy under contract DE-AC03-76SF00098 and in part by the National Science Foundation under research grant PHY81-18547.

References

[1]   Recent reviews are B. Zumino, lecture at the XVIII Solvay Conference, Austin, Texas, Nov. 1982, *Phys. Reports* 104, 87 (1984); L. Hall, Harvard University Preprint HUTP-84/A012. For the method of dimensional reduction in supergravity see the lecture by M. J. Duff at this conference.*

[2]   J. Wess and B. Zumino, *Phys. Lett.* B49, 52 (1974); J. Iliopoulos and B. Zumino, *Nucl. Phys.* B76, 310 (1974); S. Ferrara, J. Iliopoulos and B. Zumino, *Nuc. Phys.* B77, 41 (1974); B. Zumino, *Nucl Phys.* B89, 535 (1975); M. Grisaru, W. Siegel, and M. Rocek, *Nucl. Phys.* B159, 420 (1979).

[3]   S. Mandelstam *Phys. Lett.* B121, 30 (1983), *Nucl. Phys.* B213, 149 (1983).

[4]   See the review by P. C. West delivered at this conference.*

[5]   M. F. Atiyah and I. M. Singer, *Ann Math.* 87, 484, 546 (1968); 93, 119 (1971).

[6]   E. Witten, *Nucl. Phys.* B202, 253 (1982); S. Cecotti and L. Girardello, *Phys. Lett.* B110, 39 (1982); H. Nicolai, "Spontane Symmetriebrechung in Einfach-Supersymmetrischen Theorien," Karlsruhe University Thesis (1975), unpublished.

[7]   L. Alvarez-Gaumé, *Commun. Math. Phys.* 90, 161 (1983); Harvard University Preprint HUTP-83/A035.

[8]   D. Friedan and P. Windey, University of Chicago Preprint EFI 83–65; P. Windey, lectures at the 23rd Cracov School of Theor. Phys., CERN Preprint TH 3758 (1983).

[9]   E. Getzler, *Commun. Math. Phys.* 92, 163 (1983).

[10]   J. M. Bismut, Orsay Preprint (1983), submitted to *Comm. Math. Phys.*

[11]   Juan Mañes and B. Zumino (in preparation).

[12]   See, for instance, T. Eguchi, P. B. Gilkey, and A. J. Hanson, *Phys. Reports* 66, 213 (1980).

[13]   W. Pauli, *Lectures on Physics*, Vol. 6 (MIT Press, Cambridge, MA, 1983), p. 161. [German original *Feldquantisierung* (1950–51).]

[14]   J. Schwinger, *Phys. Rev.* 82, 664 (1951).

[15]   See, e.g., B. Zumino, in *Fundamentals of Quark Models*, I. M. Barbour and A. T. Davies, editors (Scottish Universities Summer School in Physics Publ., 1977), p. 549.

[16]   R. Arnowitt, P. Nath, and B. Zumino, *Phys. Lett.* 56, 81 (1975).

[17]   F. A. Berezin, *The Method of Second Quantization* (Academic Press, New York, 1966).

* Editors' note: The papers read to the conference by Duff and West appear in this volume.

Superunification and the Seven-Sphere

M. J. Duff

We show how $N = 8$ supersymmetry can break spontaneously to $N = 1$ at the Planck scale via a Kaluza-Klein compactification of $d = 11$ supergravity on the squashed seven-sphere. Features unique to Kaluza-Klein supergravity are (i) the massless gravitino of the $N = 1$ phase comes from a *massive* $N = 8$ supermultiplet, (ii) the scalars developing nonzero VEVs also belong to massive $N = 8$ supermultiplets, and (iii) parity remains unbroken when $N = 8$ breaks to $N = 1$.

Next we ask whether the resulting $N = 1$ theory can provide a realistic $SU(3) \times SU(2) \times U(1)$ unification and speculate that it might if some of the gauge bosons and fermions are composite as in the EGMZ model. In contrast with their model, however, we avoid unwanted helicities and problems with their noncompact E_7. Moreover, we suggest a scheme in which the electroweak $SU(2) \times U(1)$ is a subgroup of the $d = 11$ general coordinate group but the strong $SU(3)$ is a subgroup of the $d = 11$ local Lorentz group and are not, therefore, to be combined into a GUT. The special properties of the seven-sphere also suggest a possible solution of the cosmological constant problem involving fermion condensates.

1 Introduction

I am going to describe some work carried out in collaboration with Moustafa Awada, Chris Pope, and Bengt Nilsson [8, 10, 15, 54], but before doing so let me say a few words about extended supersymmetry.

Since its discovery in 1976, supergravity [44] has evolved along rather diverse lines: the $N = 8$ route and the $N = 1$ route. Those who do the maximal $N = 8$ supersymmetry [35, 42] have several motivations. First, it is beautiful. Second, one has the feeling that if Nature cares at all about supersymmetry, it would be crazy to stop at $N = 1$. Third, $N = 8$ supergravity is the only truly unified field theory we have. It is the only known theory in which gravity and all the other particles of lower spin appear in one and the same multiplet.[1] This is not true of $N < 8$ supersymmetry and, in particular, it is not true of $N = 1$. One also expects, therefore, that the $N = 8$ theory will have the best behavior from the point of view of ultraviolet divergences [1].

Those who do $N = 1$ supersymmetry, on the other hand, do so because it is the only supersymmetry likely to be relevant for particle physics phenomenology at present energies. For a recent review of $N = 1$ particle physics, see Zumino [45]. Consequently, the hope has sometimes been expressed

---

1. Thus it realizes Einstein's dream of replacing the "base wood" of matter by the "pure marble" of geometry.

that these two approaches could be linked if $N = 8$ supergravity were to break spontaneously to $N = 1$ at Planckian energies.

We shall see in section 3 that this hope is indeed fulfilled: $N = 8$ supersymmetry breaks spontaneously at the tree level to $N = 1$ at a scale given by gM_p where g is the coupling constant of the Yang-Mills gauge group $SO(8)$ and M_p is the Planck mass. At the same time $SO(8)$ breaks spontaneously to $SO(5) \times SU(2)$, but parity remains unbroken. However, the $N = 8$ theory posessing the above properties is not the ungauged supergravity of Cremmer and Julia [35] nor even the gauged $SO(8)$ theory of deWit and Nicolai [42]. Rather, it is the $N = 8$ theory proposed by Pope and myself [1, 2, 5, 10] obtained by spontaneous compactification of 11-dimensional supergravity [31] on the seven-sphere, S^7.

Some time ago [1, 2] we pointed out that the field equations of $N = 1$ supergravity in $d = 11$ dimensions admit of vacuum solutions corresponding to the product of $d = 4$ anti-de Sitter space-times the seven-sphere; and that since S^7 admits 8 Killing spinors and since its isometry group is $SO(8)$, this gives rise via a Kaluza-Klein mechanism to an effective $d = 4$ theory with $N = 8$ supersymmetry and local $SO(8)$ invariance.

Since then there has been considerable progress in the understanding of S^7 *compactifiction of* $d = 11$ supergravity [3 → 28], and in sections 2 and 3 we give an up to date account paying particular attention in section 3 to the Higgs-Kibble, spontaneous symmetry-breaking interpretation [5] of the five different S^7 solutions that are presently known. This involves new results on the tower of massive $N = 8$ supermultiplets. Whatever one's view of the physical relevance of extra dimensions and supergravity, the Kaluza-Klein S^7 approach to $d = 11$ supergravity discussed in sections 2 and 3 is mathematically sound and, few would dispute, beautiful.

Section 4, however, will be on much shakier ground because here we turn to the other theme of this lecture and ask whether the resulting $N = 1$ theory can provide a realistic unification, bearing in mind that $SO(8) \not\supset SU(3) \times SU(2) \times U(1)$. We shall speculate that it might if some of the gauge bosons and fermions are composite, as in the model of Ellis, Gaillard, Maiani, and Zumino [51]. In contrast with their model, however, we avoid unwanted helicities and problems with their noncompact E_7. Moreover, we suggest a scheme in which the electroweak $SU(2) \times U(1)$ [$\subset SO(8)$] is a subgroup of the $d = 11$ general coordinate group but that strong $SU(3)$ [$\subset SO(7)$] is a subgroup of the $d = 11$ local Lorentz group and are not, therefore, to be combined into a simple group as in conventional Grand

Unified Theories. We also suggest solutions to the cosmological constant problem and the strong CP problem. The all-important, and still unresolved, question concerns realistic fermion representations, and in particular, whether a vectorlike preon theory can give rise to flavor-chiral bound states. Thus although major problems remain, the results are not altogether discouraging.

Finally, in section 5 we examine some rival approaches to S^7 supergravity and discuss the possible interpretation of some very recent results.

2 Kaluza-Klein Supergravity

Back in the 1920s Kaluza and Klein [29] tried to unify gravity with electromagnetism by giving an extra fifth dimension to space-time. Nowadays, of course, we are more ambitious and would like to include the strong and weak interactions as well. The current renaissance in Kaluza-Klein theories is based on the hope that this may be possible with a few more extra dimensions.

We have argued elsewhere [2, 5, 10] that the only viable Kaluza-Klein theory is supergravity and that the only way to do supergravity is via Kaluza-Klein. Some of the reasons are

i. The maximum dimension that one can consistently formulate a supersymmetric field theory is 11 [30]. This $d = 11$ theory of Cremmer, Julia, and Scherk [31] is unique and describes the interaction of gravity g_{MN} with the gravitino Ψ_M and a 3-index gauge field A_{MNP}. These fields have 44, 128, and 84 degrees of freedom, respectively. The continuous symmetries are $d = 11$ general covariance, local $N = 1$ supersymmetry, local $SO(1, 10)$ Lorentz invariance, and abelian gauge invariance of A_{MNP}. There is also a discrete symmetry where one performs on odd number of space or time reflections and sends A_{MNP} to $-A_{MNP}$. (The signature is $- + + + + + + + + + +$.) This theory has only one parameter, the gravitational coupling, and supersymmetry forbids any tampering with the Lagrangian [65]. Either this is the theory of everything or it's wrong: there are no half-measures!

ii. The basic idea of Kaluza-Klein is that what we perceive to be internal symmetries in 4 dimensions are really space-time symmetries in the extra dimensions. Carrying this logic to its ultimate conclusion would demand that *all* symmetries in Nature are space-time symmetries. This extreme Kaluza-Klein philosophy is realized by $d = 11$ supergravity both in the continuous and discrete symmetries.[2] For example, charge conjugation in $d = 4$ is parity in the extra dimensions [32]. (This extreme case would not be true of $N > 1$ supersymmetry, which would require $d < 11$ dimensions.)

iii. Unless one introduces Yang-Mills fields already in the higher dimensional theory, which would contradict the fundamental spirit of Kaluza-Klein discussed in (ii), one cannot obtain light fermions in the $d = 4$ theory by starting with the Dirac Lagrangian. Lichnerowicz's theorem [33] requires that one must start with the Rarita-Schwinger Lagrangian, and this automatically requires supergravity. (For other, less attractive, ways out, see [2].)

iv. Modern approaches to higher dimensional theories demand that they exhibit "spontaneous compactification" [34], i.e., that the field equations admit of stable ground-state solutions of the form ($d = 4$ space-time) × (compact extra dimensions). Compactness ensures that the $d = 4$ mass-spectrum is discrete and also guarantees charge quantization. Higher dimensional theories where the extra dimensions are compact are indistinguishable from four dimensional theories with a very special mass spectrum, and this is why Kaluza-Klein could be realistic despite the science-fiction overtones of extra dimensions. However, not all higher dimensional theories exhibit spontaneous compactification. In $d = 11$ supergravity spontaneous compactification not only works but the dimension of space-time ($d = 4$) is now *output* rather than input.

To see this, consider the $d = 11$ field equations [35] with $\langle \psi_M \rangle = 0$,

$$R_{MN} - \tfrac{1}{2} g_{MN} R = \tfrac{1}{3} [F_{MPQR} F_N{}^{PQR} - \tfrac{1}{8} g_{MN} F^2], \tag{1}$$

$$\nabla_M F^{MPQR} = -\frac{1}{576} \varepsilon^{PQRM_1 \cdots M_4 M_5 \cdots M_8} F_{M_1 \cdots M_4} F_{M_5 \cdots M_8}, \tag{2}$$

where $M, N = 1, \ldots, 11$ and $F_{MNPQ} = 4\partial_{[M} A_{NPQ]}$ is the field strength of the A_{MNP} potential. Denote by x^μ the coordinates of space-time, whose dimension is not yet specified, and by y^m the coordinates of the extra dimensions. Maximal symmetry for the spacetime ground-state requires that we look for solutions of the form

---

2. One might argue that the gauge transformation on A_{MNP}, namely, $\delta A_{MNP} = \partial_{[M} A_{NP]}$, where $\Lambda_{NP} = -\Lambda_{PN}$, is arbitrary, is not a "space-time" symmetry. Recently, however, Bars and Macdowall [87] have shown that the gravity-matter system (g_{MN}, A_{MNP}) may be reinterpreted as pure gravity in first-order form (e_M^A, ω_M^{AB}) with an extra symmetry on the spin-connection ω_M^{AB}. Thus reinterpreted, the gauge invariance is indeed a space-time symmetry of pure gravity!

$$\langle g_{\mu\nu} \rangle = \overset{0}{g}_{\mu\nu}(x), \qquad \langle F_{\mu\nu\rho\sigma} \rangle = \overset{0}{F}_{\mu\nu\rho\sigma}(x),$$
$$\langle g_{mn} \rangle = \overset{0}{g}_{mn}(y), \qquad \langle F_{mnpq} \rangle = \overset{0}{F}_{mnpq}(y), \tag{3}$$

with all components with mixed indices equal to zero. Moreover $\overset{0}{F}_{\mu\nu\rho\sigma}(x)$ must be an invariant tensor. This means that

$$\overset{0}{F}_{\mu\nu\rho\sigma} = 3m\varepsilon_{\mu\nu\rho\sigma}, \tag{4}$$

where m is a constant. Hence, as pointed out by Freund and Rubin [36], we have singled out $d = 4$ for space-time since $F_{\mu\nu\rho\sigma}$ has rank 4. For the moment let us also set $F_{mnpq} = 0$ for simplicity; then substituting (3) and (4) into (1) and (2), we find that (2) is trivially satisfied, while (1) yields the product of a 4-dimensional Einstein space-time with the usual Minkowski signature

$$R_{\mu\nu} = -12m^2 g_{\mu\nu}, \tag{5}$$

which we take to be anti-de Sitter space (AdS) $= SO(3,2)/SO(3,1)$, and a 7-dimensional Einstein space with Euclidean signature

$$R_{mn} = 6m^2 g_{mn} \tag{6}$$

and hence compact since $m^2 > 0$. There are still infinitely many solutions of (6) and, a priori, each could claim to be the ground state. We need some criterion for distinguishing the false vacua from the true vacuum. A necessary condition for the true vacuum is that it be stable, and one reason for stability would be an unbroken supersymmetry. For a supersymmetric vacuum, we require that $\langle \psi_M \rangle$ stay zero under a supersymmetry transformation, i.e., that $\langle \delta\psi_M \rangle = 0$, where

$$\delta\psi_M = D_M\varepsilon - \frac{i}{144}(\hat{\Gamma}^{NPQR}{}_M + 8\hat{\Gamma}^{PQR}\delta^N_M)F_{NPQR}\varepsilon \equiv \bar{D}_M\varepsilon. \tag{7}$$

For solutions of the type discussed, we have

$$\bar{D}_\mu = D_\mu + m\gamma_\mu\gamma_5, \tag{8}$$

$$\bar{D}_m = D_m - \frac{m}{2}\Gamma_m, \tag{9}$$

where we have used the decomposition of the $d = 11$ Γ matrices

$$\hat{\Gamma}_A = (\gamma_\mu \times \mathbb{1}, \gamma_5 \times \Gamma_a). \tag{10}$$

The number of unbroken supersymmetries is thus given by the number of

Table 1
Known solutions with S^7 topology

Solution	Holonomy	Supersymmetry	Gauge group	Reference
I round	1	$N = 8$	$SO(8)$	[1, 2, 10]
II left-squashed	G_2	$N = 1$	$SO(5) \times SU(2)$	[8]
III round + "torsion"	$SO(7)$	$N = 0$	$SO(7)$	[3]
IV right-squashed	G_2	$N = 0$	$SO(5) \times SU(2)$	[15]
V right-squashed + "torsion"	$SO(7)$	$N = 0$	$SO(5) \times SU(2)$	[15]

"Killing spinors" on the $d = 7$ manifold, i.e., the number of solutions to

$$\bar{D}_m \eta = 0 = \left(\partial_m - \frac{1}{4} \omega_m{}^{ab} \Gamma_{ab} - \frac{m}{2} e_m{}^a \Gamma_a \right) \eta, \tag{11}$$

where $\eta(y)$ is an 8-component (commuting) spinor in $d = 7$, and where $\omega_m{}^{ab}$ and $e_m{}^a$ are the spin connection and siebenbein of the ground-state solution to (6). Such Killing spinors satisfy the integrability condition

$$[\bar{D}_m, \bar{D}_n] \eta = -\tfrac{1}{4} C_{mn}{}^{ab} \Gamma_{ab} \eta = 0, \tag{12}$$

where $C_{mn}{}^{ab}$ is the Weyl tensor. The subgroup of spin (7) generated by these linear combinations of the spin (7) generators Γ_{ab} corresponds to the holonomy group $\mathcal{H}$ of the generalized connection of (11). Thus the maximum number of unbroken supersymmetries, N, is equal to the number of spinors left-invariant by $\mathcal{H}$.

The maximally symmetric solution of (6) is given by the "round" $S^7 = SO(8)/SO(7)$ for which $C_{mn}{}^{ab} = 0$, $\mathcal{H} = 1$ and $N = N_{\max} = 8$. However, there also exist other solutions that are topologically seven-spheres that deviate from the maximally symmetric geometry by being "squashed," and solutions for which the $A_{mnp}(y)$ "torsion" field is also nonzero. These vacua have symmetry group $G \subset SO(8)$ and $N < 8$. To date, five S^7 solutions are known and are given in table 1.

This table suggests that these different solutions correspond to different phases of the same effective 4-dimensional theory, solution I being the symmetric phase and solutions II–V being spontaneously broken phases. This brings us to another attractive feature of the Kaluza-Klein theory based on $N = 1$, $d = 11$ supergravity:

v. The unique properties of S^7 provide a Kaluza-Klein origin for the spontaneous breakdown of gauge symmetries, supersymmetries, and the discrete symmetries of C, P, and CP.

The ability to find solutions with squashing [8, 15], is due to the remarkable result (valid for spheres only of dimension $4k + 3$, $k \leqslant 1$) that S^7 admits not one, but two Einstein metrics [37]. The ability to find solutions with "torsion" [3], is related to the equally remarkable result that, apart from group manifolds, S^7 is the only compact space to admit an "absolute parallelism" [38]. More remarkable than both, in the author's opinion, is that in $d = 11$ supergravity these somewhat abstract results from differential geometry are translated into something familiar in the effective 4-dimensional theory: the Higgs-Kibble mechanism [5].

3 Higgs-Kibble Interpretation

A   Vacuum I: The Round S^7

Since the round S^7 admits 8 Killing spinors and since its isometry group is $SO(8)$, the Kaluza-Klein mechanism will give rise to an effective $d = 4$ theory with $N = 8$ supersymmetry and local $SO(8)$ invariance. The group $SO(8)$ posesses the unique property of "triality"; i.e., there are three inequivalent 8-dimensional representations with Dynkin labels $(1, 0, 0, 0)$, $(0, 0, 0, 1)$ and $(0, 0, 1, 0)$ to which we conventionally assign the names vector (8_v), spinor (8_s), and conjugate spinor (8_c), respectively. (We follow the notation of Slansky [39] except that s and v are interchanged.) This triality will be crucial in what follows. In addition to the familiar massless $N = 8$ supermultiplet of spins $\{2, \frac{3}{2}, 1, \frac{1}{2}, 0^+, 0^-\}$ with $SO(8)$ content $\{1, 8_s, 28, 56_s, 35_v, 35_c\}$, this Kaluza-Klein theory also describes an infinite tower of massive, spin 2 supermultiplets with masses quantized in units of m, the inverse radius of S^7. They correspond to irreducible representations of $OSp(4/8)$ but will be reducible under $SO(8)$ [4]. They are obtained by taking the tensor product of the massless multiplet with Dynkin labels $\{(0, 0, 0, 0), (0, 0, 0, 1), (0, 1, 0, 0), (1, 0, 1, 0), (2, 0, 0, 0), (0, 0, 2, 0)\}$ and the representation $(n, 0, 0, 0)$, which correspond to eigenmodes of the scalar Laplacian on S^7, and which yield the massive gravitons [15]. This done, we allow each massless spin 2 to eat a spin 1 and spin 0 to become a massive spin 2; each massless spin $\frac{3}{2}$ to eat a spin $\frac{1}{2}$ to become a massive spin $\frac{3}{2}$; and each massless spin 1 to eat a spin 0 to become a massive spin 1. Of course,

Table 2
Massive $N = 8$ supermultiplets

Spin	$SO(8)$ representation
2	$(n, 0, 0, 0)$
$\frac{3}{2}$	$(n, 0, 0, 1) + (n - 1, 0, 1, 0)$
1	$(n, 1, 0, 0) + (n - 1, 0, 1, 1) + (n - 2, 1, 0, 0)$
$\frac{1}{2}$	$(n + 1, 0, 1, 0) + (n - 1, 1, 1, 0) + (n - 2, 1, 0, 1) + (n - 2, 0, 0, 1)$
0^+	$(n + 2, 0, 0, 0) + (n - 2, 2, 0, 0) + (n - 2, 0, 0, 0)$
0^-	$(n, 0, 2, 0) + (n - 2, 0, 0, 2)$

since space-time is anti-de Sitter space, the quantity "mass' must be defined accordingly (see [57, 41, 59, 89]). For example, massless spin $\frac{3}{2}$ require an apparent mass term, and massless scalars must obey the conformal wave equation. Moreover, members of the same $N = 8$ massive multiplet can have different de Sitter mass. The results are shown in table 2.

For example, the lightest such multiplet has the $SO(8)$ content $\{8_v, 8_c + 56_c, 56_v + 160_v, 160_c + 224_{vc}, 112_v, 224_{cv}\}$ and the next to lightest has the $SO(8)$ content $\{35_v, 56_s + 224_{vs}, 28 + 350 + 567_v, 8_s + 160_s + 672_{vc} + 840_s, 1 + 294_v + 300, 35_s + 840_s'\}$. Note that the s and c type representations occur only for spins $\frac{3}{2}, \frac{1}{2}, 0^-$, while spins 2, 1, and 0^+ have either no subscript or a v subscript. Note also that, although $N = 8$, the maximum spin of these massive supermultiplets is 2 and not 4. This phenomenon of "multiplet shortening" is discusses by Freedman and Nicolai [59].

The infinity of massive gravitinos is due to an infinity of spontaneously broken supersymmetries, as may be seen by Fourier expansion of the $d = 11$ supersymmetry parameter in harmonics on S^7:

$$\varepsilon(x, y) = \sum_k \varepsilon_k(x) Y^k(y). \tag{13}$$

The lowest mode, which is 8-fold degenerate, corresponds to the unbroken $N = 8$ supersymmetry and to the 8 massless gravitinos, whereas the remaining ε_k transorm states of different mass into each other. Similar remarks apply to the other $d = 11$ local symmetries of general covariance, local $SO(1, 10)$, and gauge invariance of A_{MNP}. In this way, one obtains an infinite-dimensional noncompact superalgebra that generalizes the finite-dimensional nonsuperalgebras described by Salam and Strathdee [50].

Note, moreover, that since the massive modes have masses quantized in units of m, the inverse radius of S^7, and since the $SO(8)$ Yang-Mills coupling constant g obeys the Kaluza-Klein relation [27]

$$g^2 = 16\pi G m^2,$$

where G is Newton's constant, the massless sector is in no sense a limiting case of the complete theory. Throughout this work, therefore, we shall always advocate that the massive states be retained.

As a matter of technical interest, however, one may ask whether the theory can be consistently truncated to include only the massless supermultiplet and if so whether this truncation coincides with $N = 8$ gauged $SO(8)$ theory of de Wit and Nicolai [42]. It has been conjectured that this is indeed the case [1, 2, 5, 10]. Verifying this conjecture would require a complete nonlinear analysis of the $d = 4$ equations of motion obtained from $d = 11$, and, to date, only the linearized theory has been obtained in full detail [9, 10]. However, a crucial necessary condition is that the space-time cosmological constant Λ must be related to the $SO(8)$ gauge coupling constant g by the same formula [42],

$$4\pi G\Lambda = -3g^2,$$

as obtains in the $N = 8$ phase of the de Wit-Nicolai theory. Weinberg [43] has shown how the coupling constant may be calculated starting from pure Einstein gravity in higher dimensions. Applied to S^7 of radius m^{-1}, the formula gives $g^2 = 64\pi G m^2$, which, combined with $\Lambda = -12m^2$ from (5), yields

$$16\pi G\Lambda = -3g^2,$$

which disagrees with the de Wit-Nicolai value and would disprove the conjecture. However, it turns out that the presence of the A_{MNP} field in $d = 11$ supergravity leads to a modification of Weinberg's calculation [27]. When this is taken into account it changes the relation between g and m^2 to $g^2 = 16\pi G m^2$ and hence gives precisely the de Wit-Nicolai relation. This reinforces the conjecture and suggests that the "hidden" symmetry of $SO(7)$ discussed in section 4 might be enlarged to the $SU(8)$ of the de Wit-Nicolai theory. With $F_{\mu\nu\rho\sigma}$ given by (4), the criterion for unbroken supersymmetry is given by (11), and on the round S^7 there are 8 solutions, denoted η^I ($I = 1, \ldots, 8$), where the index I corresponds to the 8_s of $SO(8)$. Note, however, that by inserting a minus sign into (4), one obtains a new vacuum solution for which the criterion of unbroken supersymmetry is obtained by sending m to $-m$ in (11). On the round S^7 there are still 8 solutions, denoted $\eta^{I'}$ ($I' = 1, \ldots, 8$), where the index I' now corresponds to the 8_c of $SO(8)$. All of this discussion again goes through provided the s and c labels are interchanged.

B Vacuum II: The Left-Squashed S^7

The squashed S^7 has metric $ds^2 = c^2 e^a \otimes e_a$ ($a = 1, \ldots, 7$), where $c = (2 \times 7^{1/2} m\lambda)^{-1}(1 + 8\lambda^2 - 2\lambda^4)^{1/2}$ is chosen to give $R = 42m^2$ and

$$e^0 = d\mu, \qquad e^i = \tfrac{1}{2}\sin \mu \omega_i, \qquad e^{\hat{\imath}} = \tfrac{1}{2}\lambda(v_i + \cos \mu \omega_i), \qquad (14)$$

where $i = 1, 2, 3$ and $\hat{\imath} = 4, 5, 6 = \hat{1}, \hat{2}, \hat{3}$, where

$$v_i = \sigma_i + \Sigma_i, \qquad \omega_i = \sigma_i - \Sigma_i, \qquad (15)$$

and where σ_i and Σ_i are left-invariant one-forms satisfying the $SU(2)$ algebra

$$d\sigma_1 = -\sigma_2 \wedge \sigma_3, \qquad d\Sigma_1 = -\Sigma_2 \wedge \Sigma_3 \qquad (16)$$

and cyclic permutations. The constant parameter λ describes the degree of distortion. The round S^7 solution of (6) corresponds to $\lambda^2 = 1$, and describes the coset space $SO(8)/SO(7)$ for which $C_{mn}{}^{ab} = 0$, $\mathcal{H} = \mathbb{1}$ and $N = N_{max} = 8$. The left-squashed S^7 solution of (6) corresponds to $\lambda^2 = 1/5$ and describes the Einstein metric on the coset space $Sp(2) \times Sp(1)/ Sp(1) \times Sp(1)$ [15, 20], where $Sp(2) \simeq SO(5)$ and $Sp(1) \simeq SU(2)$, for which $C_{mn}{}^{ab} \neq 0$, $\mathcal{H} = G_2$, and $N_{max} = 1$. Moreover, $N = N_{max} = 1$ because there is indeed one solution of (11), which we denote by the singlet η. In constrast with the $\eta^I(y)$ of the round S^7, the components of η are actually constant in the orthonormal basis of (14). This yields an $N = 1$, $SO(5) \times SU(2)$ vacuum.[3] The massless sector consists of a $2-\tfrac{3}{2}$ in the representation $(10, 1) + (1, 3)$, together, possibly, with some $\tfrac{1}{2}-0$ supermultiplets.

The squashed S^7, however, exhibits the following subtlety. If one reverses the orientation of the orthonormal basis of (14), by sending each e^a to $-e^a$, for example, one obtains a new S^7 solution of (6), hereafter referred to as right-squashed to distinguish it from the left-squashed S^7 discussed previ-

---

3. Geometrically, the squashed seven-sphere corresponds to an S^3 bundle over S^4, i.e., to the $K = 1$ $SU(2)$ Yang-Mills instanton in 4-dimensional euclidean space [8]. Squashing corresponds to changing the length of the S^3 fibres relative to the S^4 base. Thus the symmetries $SO(5)$ and $SU(2)$ are those familiar from instanton physics.

ously. Although we still have $\mathcal{H} = G_2$, there are no longer any solutions of (11), and so this vacuum has $N = 0$. [The same effect could have been obtained by keeping the left-squashed S^7 but inserting a minus sign into (4)]. Because of triality, there are three inequivalent $SO(5) \times SU(2)$ subgroups of $SO(8)$. Left-squashing picks out $[SO(5) \times SU(2)]_c$ under which 8_c goes into $(5, 1) + (1, 3)$, and both 8_s and 8_v go into $(4, 2)$, while right-squashing picks out $[SO(5) \times SU(2)]_s$ under which 8_s goes into $(5, 1) + (1, 3)$ and both 8_c and 8_v go into $(4, 2)$. (The third subgroup, $[SO(5) \times SU(2)]_v$, may also be obtained by squashing S^7 but in an inhomogeneous manner that does not yield an Einstein metric.)

Let us now discuss the Higgs interpretation of the breaking $N = 8/SO(8)$ $\rightarrow N = 1/[SO(5) \times SU(2)]_c$. First we observe that when $SO(8)$ breaks to $[SO(5) \times SU(2)]_c$, the massless $N = 8$ multiplet decomposes as follows: $1 \rightarrow (1, 1)$; $8_s \rightarrow (4, 2)$; $28 \rightarrow (10, 1) + (5, 3) + (1, 3)$; $56_s \rightarrow (16, 2) + (4, 4) + (4, 2)$; $35_v \rightarrow (10, 3) + (5, 1)$; $35_c \rightarrow (14, 1) + (5, 3) + (1, 5) + (1, 1)$. These states by themselves can never form $N = 1$ supermultiplets. In particular, we note that *all* the eight gravitinos must acquire a mass. It follows that the single massless gravitino in the $N = 1$ phase must come from the *massive* sector of the $N = 8$ phase.

To proceed, we note that massless gravitinos in $d = 4$ correspond to $-7m/2$ modes of the Dirac operator $\not{D} \equiv \Gamma^m D_m$. That $\bar{D}_m \eta = 0$ implies $\not{D}\eta = -7m\eta/2$ is clear from (9). By consideration of the integral of $|\bar{D}_m\eta|^2$, the converse is also seen to be true after some straightforward algebra. In reference [15], therefore, we calculated the Dirac operator in the squashed S^7 metric with arbitrary λ, and looked for the eigenmode ψ that, in the case $\lambda^2 = 1/5$, is the solution η of $\bar{D}_m \eta = 0$. The result was that $\psi = \eta$ for all λ^2 and that the corresponding eigenvalue was

$$-\tfrac{3}{2}7^{1/2}m(1 + 2\lambda^2)(1 + 8\lambda^2 - 2\lambda^4)^{-1/2}.$$

Thus the eigenvalue $-7m/2$ mode of $\not{D}$ on the $\lambda^2 = 1/5$ sphere has come from a $-9m/2$ mode on the $\lambda^2 = 1$ sphere. Now the Dirac eigenmodes on the round S^7 are $+[n + (7/2)]m$ and $-[n + (7/2)]m, n \geqslant 0$, the modes being in the $(n, 0, 1, 0)$ and $(n, 0, 0, 1)$ representations, respectively [15, 28]. The singlet massless gravitino on the $\lambda^2 = 1/5$ squashed S^7 has therefore come from the next-to-lowest level of Dirac eigenmodes on the round S^7, which is a 56_c. It is, in fact, the singlet in the decomposition $56_c \rightarrow (10, 1) + (10, 3) + (5, 3) + (1, 1)$ of $SO(8)$ breaking down to the $[SO(5) \times SU(2)_c]$ isometry group of the left-squashed sphere. We refer to this mechanism, whereby states which are apparently very massive $M^2 \sim g^2/G$ in the $N = 8$

Table 3
The Space Invaders Scenario

Round S^7	Left-squashed S^7
$\lambda^2 = 1$	$\lambda^2 = 1/5$
$N = 8$	$N = 1$
$SO(8)$	$[SO(5) \times SU(2)]_c$

phase zoom down from the Planckian sky to become massless in the $N = 1$ phase, as the Space Invaders Scenario (see table 3).

The Higgs effect and super-Higgs effect work as follows. Consider

$$g_{mn}(x, y) = \overset{0}{g}_{mn}(y) + h_{mn}(x, y),$$

where $\overset{0}{g}_{mn}(y)$ is the metric on the round S^7. Fourier expanding $h_{mn}(x, y)$ in harmonics on the round S^7,

$$h_{mn}(x, y) = \sum_k S_k(x) Y^k_{mn}(y), \tag{17}$$

one obtains the 35_v scalars of the $N = 8$ massless sector from the lowest mode together with an infinite tower of massive scalars, $S_k(x)$, where k denotes some representation of $SO(8)$. In the symmetric phase

$$\langle g_{mn}(x, y) \rangle = \overset{0}{g}_{mn}(y) \rightarrow \text{all} \langle S_k(x) \rangle = 0, \tag{18}$$

but deformations of S^7 away from its maximally symmetric geometry correspond to nonzero vacuum expectation values for some of these scalars:

$$\langle g_{mn}(x, y) \rangle \neq \overset{0}{g}_{mn}(y) \rightarrow \text{some} \langle S_k(x) \rangle \neq 0. \tag{19}$$

In order to determine which are responsible for the breaking of $SO(8)$ down to $SO(5) \times SU(2)$, we first note that in the orthonormal basis of (14), the metric deformation h_{ab} that takes the round S^7 into the squashed S^7 is of the form

$$h_{ab} \sim \text{diag}(0, 0, 0, 0, 1, 1, 1). \tag{20}$$

A straightforward calculation in the round S^7 background shows that this is a Killing tensor

$$\nabla_{(l}h_{mn)} = 0. \tag{21}$$

As discussed in [10], there are 336 symmetric rank-two Killing tensors on the round S^7, transforming as a 1, a 35_v, and a 300 under $SO(8)$. The 35_v are associated with the massless scalars (in the $n = 0$ level of table 2) discussed previously, while the 1 and 300 describe massive scalars (in the $n = 2$ level of table 2). The singlet Killing tensor is just a constant multiple of the metric and has constant trace; the trace of the 35_v gives a 35_v of eigenfunctions of the scalar Laplacian on S^7, while the 300 is trace-free. From (20) we see that h_{ab} has constant trace and so must be a linear combination of just the 1 and the 300. The singlet corresponds to an overall scaling of the metric, and so the squashing of (14) corresponds to a nonzero VEV for the 300. As a consistency check we note that under $SO(8) \rightarrow SO(5) \times SU(2)$ we have $300 \rightarrow (1, 1) + (5, 3) + (1, 5) + (10, 3) + (14, 1) + (5, 5) + (35, 1) + (35, 3) + (14, 5)$, and we find as expected the singlet $(1, 1)$ together with the Goldstone bosons $(5, 3)$, which give a mass to the $(5, 3)$ spin 1 gauge bosons in the decomposition $28 \rightarrow (10, 1) + (1, 3) + (5, 3)$.

C Vacuum III: The Round S^7 with "Torsion"

If we substitute $\overset{0}{F}_{\mu\nu\rho\sigma} = 2m\varepsilon_{\mu\nu\rho\sigma}$ into (2) we find

$$\nabla_m \overset{0}{F}{}^{mnpq} = \frac{m}{6}\varepsilon^{npqrstu}\overset{0}{F}_{rstu}, \tag{22}$$

and on the round S^7 nontrivial solutions may be found [3] that also solve (6). (We have rescaled $\overset{0}{F}_{\mu\nu\rho\sigma}$ so as to keep m^{-1} the radius of S^7). Moreover, the $\overset{0}{A}_{mnp}$ field of which $\overset{0}{F}_{mnpq}$ is the curl, $\overset{0}{F}_{mnpq} = 4\partial_{[m}A_{npq]}$, admits the interpretation of a parallelizing torsion, i.e.,

$$R_{mnpq}(\Gamma^s{}_{tu} + S^s{}_{tu}) = 0, \tag{23}$$

where

$$S_{mnp} = S_{[mnp]} = 4m\overset{0}{A}_{mnp}. \tag{24}$$

The group that leaves invariant $\overset{0}{g}_{mn}(y)$ and $\overset{0}{A}_{mnp}(y)$ is the $SO(7)$ subgroup of $SO(8)$ [40, 20, 21, 22, 23, 67], denoted $[SO(7)]_c$, under which the massless $N = 8$ supermultiplet decomposes $1 \to 1$; $8_s \to 8$; $28 \to 7 + 21$; $56_s \to 8 + 48$; $35_v \to 35$; $35_c \to 1 + 7 + 27$. Once again, all 8 gravitinos must acquire a mass and all 8 supersymmetries are broken at a common scale, but this time there is no Invasion of the Killing Spinors and $N = 0$ [6,9].

The Higgs effect and super-Higgs effect now work as follows. Fourier expanding $A_{mnp}(x, y)$ in harmonics on the round S^7,

$$A_{mnp}(x, y) = \sum_k P_k(x) Y^k_{mnp}(y), \tag{25}$$

one obtains the 35_c massless pseudoscalars of the $N = 8$ massless sector from the lowest mode together with an infinite tower of massive pseudo-scalars, $P_k(x)$, where k donotes some representation of $SO(8)$. In the symmetric phase

$$\langle A_{mnp}(x, y) \rangle = 0 \to \text{all} \langle P_k(x) \rangle = 0,$$

but nonzero "torsion" means

$$\langle A_{mnp}(x, y) \rangle = \overset{0}{A}_{mnp}(y) \neq 0 \to \text{some} \langle P_k(x) \rangle \neq 0,$$

i.e., nonzero VEVs for some of these pseudoscalars, and hence a spontaneous breakdown of parity invariance. In order to determine which are responsible for the breaking of $SO(8)$ down to $[SO(7)]_c$, we note that (23) and (24) imply that A_{mnp} is a "Yano Killing tensor," i.e.,

$$\nabla_q A_{mnp} = \partial_{[q} A_{mnp]}. \tag{26}$$

One can show that there are 70 rank-3 totally antisymmetric Yano Killing tensors on the round S^7, transforming as a 35_s and a 35_c under $SO(8)$. The 35_c are associated with the massless pseudoscalars discussed previously (in the $n = 0$ level of table 2), while the 35_s describe massive pseudoscalars (in the $n = 2$ level of table 2). They obey

$$\nabla_m F^{mnpq} = \pm \frac{m}{6} \varepsilon^{npqrstu} F_{rstu}, \tag{27}$$

where the $+$ sign refers to 35_c and the $-$ sign refers to 35_s. From (22), therefore, we see that the nonvanishing "torsion" corresponds to a nonzero

Table 4
Different supersymmetry breaking scales

Round S^7	Right-squashed S^7
$\lambda^2 = 1$	$\lambda^2 = 1/5$
$N = 8$	$N = 0$
$SO(8)$	$[SO(5) \times SU(2)]_s$

GRAVITINO
MASS ↑

$M \sim M_p$ 56_c ———————— (4,4)
                                                        (16,2)
                                                        (1,3)
                                                        (4,2)
                                                        (5,1)

$M = 0$ 8_s

VEV for the 35_c (in the $n = 0$ level of table 2). As a consistency check we note that under $SO(8) \to [SO(7)]_c$, $35_c \to 1 + 7 + 27$, and we find, as expected, the singlet together with the 7 Goldstone bosons, which give a mass to the 7 spin 1 gauge bosons in the decomposition $28 \to 7 + 21$.

D   Vacuum IV: The Right-Squashed S^7

As discussed in subsection B, this vacuum breaks $SO(8)$ to $[SO(5) \times SU(2)]_s$, under which the massless $N = 8$ supermultiplet decomposes $1 \to (1,1)$; $8_s \to (5,1) + (1,3)$; $28 \to (10,1) + (5,3) + (1,3)$; $56_s \to (10,1) + (10,3) + (5,3) + (1,1)$; $35_v \to (10,3) + (5,1)$; $35_c \to (10,3) + (5,1)$. As in vacuum III all supersymmetries are broken but no longer at a common scale; the $(5,1)$ gravitinos will have a different mass from the $(1,3)$ (see table 4).

The Higgs effect works in the same way as vacuum II with the 1 and 300 acquiring the nonzero VEVs. The mass spectrum for spins 2, 1, and 0^+ (which have only a v subscript or no subscript) is insensitive to the handedness of squashing and will be the same for both vacuum II and vacuum IV, but the spectrum for spins $\frac{3}{2}, \frac{1}{2}$, and 0^- (which have s and c subscripts) will be different. In particular, there will be a single massless pseudoscalar on the right-squashed sphere since the relevant mass operator has a single zero eigenvalue mode; see subsection E of this section. This corresponds to a

0^- "invasion" since it belongs to the massive 35_s (in the $n = 2$ level of table 2) of the $N = 8$ phase.

E   Vacuum V: The Right-Squashed S^7 with "Torsion"

Consider the constant 3-form

$$\overset{0}{A}_{abc} = \pm \frac{1}{4} \bar{\eta} \Gamma_{abc} \eta, \qquad \bar{\eta} \eta = 1, \tag{28}$$

where η is the constant Killing spinor of the left-squashed S^7. This satisfies the Einstein equations and

$$\nabla_m \overset{0}{F}{}^{mnpq} = -\frac{m}{6} \varepsilon^{npqrstu} \overset{0}{F}_{rstu}. \tag{29}$$

However, comparison with (22) reveals a crucial sign difference, and thus the left-squashed S^7 *does not* admit solutions of this type. Had we chosen $\overset{0}{F}_{\mu\nu\rho\sigma} = -2m\varepsilon_{\mu\nu\rho\sigma}$ instead of $\overset{0}{F}_{\mu\nu\rho\sigma} = +2m\varepsilon_{\mu\nu\rho\sigma}$, this would be a solution, but as discussed previously, this sign change is equivalent to $\overset{0}{F}_{\mu\nu\rho\sigma} = +2m\varepsilon_{\mu\nu\rho\sigma}$ but with left-squashing replaced by right-squashing. So the right-squashed S^7 does admit solutions with "torsion."

In addition to the nonzero VEVs for the 1 and 300 of scalars (in the $n = 2$ level of table 2), the 35_s of pseudoscalars (also in the $n = 2$ level) have acquired a nonzero VEV, as may be seen from (27). As a consistency check we note that under $SO(8) \to [SO(5) \times SU(2)]_s$, $35_s \to (14, 1) + (5, 3) + (1, 5) + (1, 1)$, and we find the singlet as expected. Since the nonzero torsion is a singlet under the vacuum symmetry of the zero-torsion vacuum, the unbroken gauge group remains $[SO(5) \times SU(2)]_s$ but there is, in addition, a spontaneous breakdown of parity. The pseudoscalar acquiring the nonzero VEV is the single pseudoscalar that was massless in vacuum IV.

The $\overset{0}{A}_{mnp}$ field of the right-squashed sphere has, if anything, an even more interesting geometrical interpretation than that of the round sphere with torsion of vacuum III. First of all, unlike that of the round sphere, it is *constant* in the orthonormal basis of (14) and is, in fact, the quantity appearing in the multiplication table of the *octonions* [12, 13, 17, 18, 21]. Moreover, with $S_{abc} = 4m\overset{0}{A}_{mnp}$,

$$R_{mnpq}(\Gamma^s_{\ tu} + S^s_{\ tu}) = C_{mnpq}(\Gamma^s_{\ tu}), \tag{30}$$

so that S_{abc} is *not* the parallelizing torsion. Rather, it is the *Ricci-flattening* torsion [12], since from (30)

$$R_{mp}(\Gamma^s_{tu} + S^s_{tu}) = 0. \tag{31}$$

We shall return to this property in the next section.

We emphasize once again that it is the right-squashed sphere of vacuum IV and not the left-squashed sphere of Vacuum II that admits the "torsion" leading to vacuum V. References, [12, 20] also discuss squashed seven-spheres with "torsion" but do not distinguish between the $N = 1$ solution of Awada-Duff-Pope [8] and the $N = 0$ solution of Duff-Nilsson-Pope [15].

F Stability

A remaining question is that of stability. One would expect vacua I and II to be absolutely stable owing to the residual supersymmetry. Stability of both against dilations and squashing has recently been established at the classical level by Page [19] (see also [16]). This involves showing that the small scalar perturbations ϕ about the ground state satisfy the Breitenlohner-Freedman [41] criterion for positive energy in AdS [88], namely,

$$-\Box\phi + \alpha\phi = 0, \qquad \alpha > +\frac{3R}{16} = -9m^2. \tag{32}$$

The peculiarities of AdS are such that the squashed sphere is stable against squashing even though, as a function of λ^2 the $\lambda^2 = 1/5$ sphere corresponds to a maximum of the effective potential and $\lambda^2 = 1$ to a minimum [66]. Unfortunately, this does not necessarily imply that the right-squashed sphere is also stable since its bosonic spectrum will differ in the 0^- sector. Indeed, there seems no apparent reason why vacua III, IV, and V should be stable, but the analysis required to determine this has not yet been carried out. See, however, [23]. In the next section we shall focus our attention on vacuum II.

4 Toward a Realistic Theory

We have seen in the previous section that the hope of $N = 8$ supergravity breaking spontaneously to $N = 1$ at Planckian energies is indeed fulfilled; that $SO(8)$ breaks to $SO(5) \times SU(2)$ in the $N = 1$ phase, but that, in this phase at least, parity remains unbroken. All this comes about in a rather unconventional way, however, by a Kaluza-Klein compactification of $d =$

11 supergravity on the squashed seven-sphere. Indeed, spontaneous breaking of $N = 8$ to $0 < N < 8$ by a Higgs mechanism is impossible for conventional supergravity theories in 4 dimensions without a corresponding breakdown in parity. It is impossible altogether for the Cremmer-Julia [35] theory because there is no effective potential for the spin 0 fields. (One may break the supersymmetry from $N = 8$ to $N = 6$, 4, 2, or 0 via the Scherk-Schwarz [46] mechanism. This also relies on extra dimensions, but the similarity to the mechanisms described in the present paper ends there. In particular, the surviving supersymmetries (massless gravitinos) are a subset of the original 8. Their mechanism is "spontaneous" in the sense that the Lagrangian has a symmetry not shared by the vacuum. However, the symmetry breaking is not induced by nonzero VEVs for scalar fields, but rather by assigning a specific dependence of the fields on the extra coordinates that are taken to parametrize a 7-torus T^7. This means that the symmetric phase and the broken phase are described by different Lagrangians with different symmetries, as opposed to the usual Higgs mechanism, where the two phases are described by the same Lagrangian, as in the Weinberg-Salam model, for example.) The gauged $N = 8$ theory of deWit and Nicolai [42], on the other hand, does possess a nontrivial effective potential for which a conventional Higgs effect might take place. However, a surviving supersymmetry would necessarily imply parity violation. This is because the massive gravitinos would have to form spin $\frac{3}{2}$ supermultiplets, but parity demands that massive spin $\frac{3}{2}$ supermultiplets require both massive vectors and massive axial vectors [47]. Since one started only with 28 vectors, this is impossible without a nonzero VEV for some of the pseudoscalers and hence a spontaneous breakdown of parity. The reason that Kaluza-Klein supergravity circumvents this theorem is that the surviving supersymmetry is not one of the original eight. The Space Invaders Scenario described in the previous section provides an inverse super-Higgs effect whereby the massless gravitino of the $N = 1$ phase comes from a massive $N = 8$ supermultiplet, as do the Higgs scalars. The possibility of massive $N = 8$ supermultiplets at the preon level is unique to Kaluza-Klein supergravity, and so the spontaneous breaking of $N = 8$ to $N = 1$ without P violation requires extra dimensions. Our conclusions about the breakdown of $N = 8$ to $N = 1$ thus differ from those reached by Ferrara and van Nieuwenhuizen [48], who claim that the Kaluza-Klein approach only describes a subset of all the breaking patterns discussed in their paper. Since they discussed only the case where the surviving massless gravitino of the

$N = 1$ phase is one of the eight massless gravitinos of the $N = 8$ phase, they omitted the possibility of the Space Invaders Scenario.

Although the hope of $N = 8$ breaking to $N = 1$ at the Planck scale has been fulfilled, it is far from clear that those who wished so fervently for it will be pleased with the result. To begin with, one of the main motivations of those who pursue the $N = 1$ route is to use supersymmetry to solve the gauge hierarchy problem in grand unified theories (GUT) based on simple groups like $SU(5)$, but the $SO(8)$ isometry group of S^7 not only fails to contain $SU(5)$, it does not even contain the $SU(3) \times SU(2) \times U(1)$ of the standard model. The $N = 1$ symmetry of $SO(5) \times SU(2)$ is good enough for the electroweak $SU(2) \times U(1)$, but not the strong $SU(3)$. Second, even leaving aside the $SU(3)$ problem and the question of masses, none of the elementary spin $\frac{1}{2}$ fermions in the $N = 1$ theory appears in the right representations to be identified with any of the known quarks and leptons. The most obvious problem is that of chirality. Now $N = 1$ supersymmetry is the only supersymmetry that can accommodate chiral fermion representations, so the fact that $N = 8$ breaks to $N = 1$ rather than $N = 2$, say, might seem like a bonus. The trouble is, as was pointed out by Witten [49], that no *classical* compactification of $d = 11$ supergravity could give rise to any asymmetry between the right- and left-handed spin $\frac{1}{2}$ in the spectrum of $d = 4$ elementary fermions. Third, in common with all the other vacua of table 1, the $N = 1$ vacuum II yields an enormous Planck-sized cosmological constant for space-time that, unlike $N = 1$ theories not obtained from $N = 8$, cannot be fine-tuned to zero.

In the remainder of this section we pose the question whether, in spite of these difficulties, the $N = 1$ theory we have obtained from $N = 8$ stands a chance of describing a realistic $SU(3) \times SU(2) \times U(1)$ theory. The arguments will necessarily be highly speculative.

We shall argue that such a theory might emerge if, following Ellis, Gaillard, Maiani, and Zumino [51] (EGMZ) and Ellis, Gaillard, and Zumino [52] (EGZ), we postulate that some of the gauge bosons and some of the fermions of the standard model are composites formed from the preons of supergravity. This was based on an earlier idea of Cremmer and Julia [35]. It must be admitted at once that we have no better idea than these authors as to the precise dynamics by which such composites might form. However, we shall argue that the composite model based on the Kaluza-Klein theory described earlier has certain advantages over their composite model based on the 4-dimensional Cremmer-Julia theory.

Table 5
Giving masses to unwanted helicities

$N = 8$ $SU(8)$ supermultiplet[a]

Helicity	$\frac{5}{2}$	2	$\frac{3}{2}$	1	$\frac{1}{2}$	0	$-\frac{1}{2}$	-1	$-\frac{3}{2}$
$SU(8)$ representation	$\bar{8}$	$\overline{36}$	$\overline{168}$	$\overline{378}$	504	420	216	63	$\bar{8}$
		$\overline{28}$	$\overline{56}$	70	56	28	8	1	
$N = 1$ G_2 supermultiplets			1	7	14	27	64	77	
$m = 0$									
$(\frac{5}{2}, 2, 2, \frac{3}{2})$			1	1	0	0	0	0	
$(2, \frac{3}{2}, \frac{3}{2}, 1)$			0	1	1	1	0	0	
$(\frac{3}{2}, 1, 1, \frac{1}{2})$			1	2	1	1	1	0	
$(1, \frac{1}{2}, \frac{1}{2}, 0)$			1	2	0	2	0	1	
$m = 0$									
$(1, \frac{1}{2})$			0	0	1	0	0	0	
$(\frac{1}{2}, 0, 0)$			0	1	2	1	1	0	

a. Assume all spins $\frac{5}{2}$, 2, $\frac{3}{2}$, and 1 get masses except 14 of 1, i.e., adjoint of G_2, and that the surviving spins $\frac{1}{2}$ and 0 remain massless.

Specifically,

i. Whereas EGZ merely supposed that $N = 8$ breaks to $N = 1$ through some unspecified mechanism, in our model this actually happens.

ii. The hoped-for bound states of EGMZ are based on an analogy with 2- and 3-dimensional σ-models, but it seems that gauge bosons are *not* generated dynamically in these models when the global invariance group is *noncompact* [53]. In contrast with the EGMZ model, Kaluza-Klein supergravity does not possess a noncompact $E_{7(7)}$. (The Cremmer-Julia E_7 may be broken by the gauging of the vectors fields as in the de Wit-Nicolai version of $N = 8$. However, one recovers the troublesome E_7 in the limit of zero gauge coupling. The advantage of Kaluza-Klein, as explained in the previous section, is that the zero coupling limit does not exist.)

iii. EGMZ chose the massless $SU(8)$ supermultiplet of bound states from which to construct a realistic GUT, but this supermultiplet contains many unwanted helicity states (see table 5). These unwanted states cannot be made supermassive in a way consistent with an unbroken $SU(3) \times SU(2) \times U(1)$. Since our $N = 1$ theory contains the gauge bosons of $SO(5) \times SU(2) \supset SU(2) \times U(1)$ already at the elementary level, we need look only for $SU(3)$ at the composite level. As shown in table 5, the unwanted helicity states of the EGMZ $N = 8$ $SU(8)$ supermultiplet may be made massive in a way consistent with an unbroken $N = 1$ supersymmetry and G_2 gauge symmetry. G_2 contains $SU(3)$ as a maximal subgroup.

The picture we have in mind differs from that of EGMZ in one other vital respect: it is *not* a GUT theory. Rather, the electroweak $SU(2) \times U(1)$ and the strong $SU(3)$ have different origins and are not to be combined into a simple group.

At this stage, there is one important aspect we should like to stress. The existence of two kinds of Yang-Mills gauge boson, "elementary" and "composite," is an automatic consequence of Kaluza-Klein theories and not to be regarded as an extra ad hoc ingredient in the theory. This is due to the fact, rarely stressed in the literature, that Kaluza-Klein theories unify gravity and Yang-Mills theories in *two* different ways. There are two gravitational bosonic symmetries in the d-dimensional theory: d-dimensional general covariance

$$x^M \to x'^M(x) \tag{33}$$

and d-dimensional local Lorentz invariance

$$\delta e_M{}^A(x) = \alpha^A{}_B(x) e_M{}^B(x),$$
$$\delta \omega_M{}^{AB}(x) = D_M \alpha^{AB}(x), \tag{34}$$

which, in our $d = 11$ case, is a local $SO(1, 10)$. Crudely speaking, the elfbeins $e_M{}^A(x)$ are the gauge bosons of general covariance and the spin connections $\omega_M{}^{AB}(x)$ are the gauge bosons of Lorentz invariance. However, whereas $e_M{}^A$ is an "elementary" field, the $\omega_M{}^{AB}$ are "composite" fields. The spin connection has no kinetic term of its own and may therefore be expressed in terms of the elfbein and its derivatives. The Kaluza-Klein mechanism therefore gives rise to both elementary and composite gauge bosons in $d = 4$. The elementary fields $B_\mu(x)$ come from $e_\mu{}^a(x, y)$ and correspond to a gauge group given by the isometry group of the extra dimensions, $SO(8)$ in the case of the round S^7. The composite gauge fields $A_\mu(x)$ come from $\omega_\mu{}^{ab}(x, y)$ and correspond to the tangent space group of the extra dimensions, $SO(7)$ in the case of $d = 7$. These latter fields have no kinetic energy term of their own and may therefore be expressed in terms of the scalar fields coming from $e_m{}^a(x, y)$ and their derivaties.

Thus, although we do not yet understand how the $SO(5) \times SU(2)$ might

Table 6
How $d = 11$ supergravity might give $SU(3) \times SU(2) \times U(1)$[a]

$N = 1$	$d = 11$ G.R.	$SO(1, 10)$
$\downarrow$	$\downarrow$	$\downarrow$
$N = 8$	$d = 4$ G.R. $\times$ $SO(8)$	$SO(7) \times SO(1, 3)$
$\downarrow$	$\downarrow$	$\downarrow$
$N = 1$	$d = 4$ G.R. $\times$ $SO(5) \times SU(2)$	$G_2 \times SO(1, 3)$
$\downarrow$	$\downarrow$	$\downarrow$
$N = 0$	$d = 4$ G.R. $\times$ $SU(2) \times U(1)$	$SU(3) \times SO(1, 3)$

a. G.R. = general relativity.

break to $SU(2) \times U(1)$ or how the $SO(7)$ might break to $SU(3)$, the picture of table 6 suggests itself as a possible breaking pattern. At any finite order of perturbation theory, of course, a nonvanishing VEV for $e_M{}^A$ means that both general coverance and $SO(1, 10)$ are spontaneously broken. Similarly a nonvanishing VEV for $e_m{}^a$ breaks both $SO(8)$ and $SO(7)$, leaving only a residual $SO(8)$ vacuum symmetry. One must assume that this is an artefact of perturbation theory and that the full $SO(8) \times SO(7)$ (or at least some subgroup like $SU(2) \times U(1) \times SU(3)$) is realized dynamically in the non-perturbative regime.

From the work of Cremmer and Julia [35], we know that when the A_{MNP} field is taken into account, the hidden symmetry may be even bigger than $SO(7)$ and might be enlargeable to $SO(8)$ and then to $SU(8)$. [The $SO(7)$ generators Γ_{ab} are supplemented by Γ_a, which together close on $SO(8)$ and then by $\gamma_5 \Gamma_{abc}$ form the 63 generators of a chiral $SU(8)$.] So the $SU(8)$ supermultiplet of EGMZ and its breaking to G_2 may indeed be relevant for the theory obtained from S^7. The relation between this G_2 and the G_2 holonomy group of the $N = 1$ phase remains somewhat obscure however.

Since we are assuming that the correct tree-level ground state is that of vacuum II with $N = 1$ unbroken, it is unlikely that the breaking to $N = 0$ will occur at any finite order of perturbation theory, and we must again look to a nonperturbative mechanism. One possibility, which also has the advantage of providing the required parity breakdown would be spin $-\frac{1}{2}$ fermion condensates i.e., nonzero VEVs for fermion bilinears like

$$\langle \bar{\chi}(x) \gamma_5 \chi(x) \rangle = \text{constant} \neq 0. \tag{35}$$

As it turns out such condensates are in any case required to solve one of the other outstanding problems: the cosmological constant.

Requiring that space-time have vanishing cosmological constant imposes the following conditions on the curvatures and torsions of $d = 11$ supergravity [7]:

$$\overset{0}{R}_{\mu\nu\rho\sigma}(\omega) = 0, \qquad \overset{0}{R}_{mn}(\hat{\omega}) = 0, \tag{36}$$

where

$$\hat{\omega}_{abc} = \omega_{abc} + S_{abc} - S_{acb} + S_{cab},$$

$$S_{MNP}(y) = \begin{cases} \langle \bar{\psi}_a(x, y) \hat{\Gamma}_b \psi_c(x, y) \rangle & \text{when } MNP = abc \\ 0 & \text{otherwise.} \end{cases} \tag{37}$$

In other words, there must exist a Ricci-flat connection in the extra dimensions, but where the torsion $S_{abc}(y)$ is built of fermionic bilinears. (The so-called torsion of vacua III and V built out of the bosonic A_{mnp} fields will not do the trick.) From (37) we see that the required fermion condensates are precisely of the spin $\frac{1}{2}$ parity violating form (35) since the index of the gravitino fields ψ_M lies only in the extra dimensions and since $\hat{\Gamma}_a = \gamma_5 \Gamma_a$.

One not very interesting solution is to have $S_{abc} = 0$ and $\overset{0}{R}_{mn}(\omega) = 0$, i.e., Ricci-flat solutions like T^7 [35] or $K3 \times T^3$ [54]. If, on the other hand, we require $\overset{0}{R}_{mn}(\omega) \neq 0$, as we must to get nonabelian elementary gauge fields, then for most geometries the problem has no solution. The remarkable property of S^7 topologies is that such a Ricci-flat connection does exist [12]. Indeed, we have already seen in (31) that the squashed S^7 admits a Ricci-flat torsion:

$$S_{abc} = \pm m\bar{\eta}\Gamma_{abc}\eta, \tag{38}$$

where η is the Killing spinor of the $N = 1$ left-squashed sphere. [Since we are not obtaining the torsion from A_{mnp}, it does not matter whether it satisfies (29) rather than (22).] Thus if it were to turn out that in the true vacuum

$$\langle \bar{\psi}_a \hat{\Gamma}_b \psi_c \rangle = \pm m\bar{\eta}\Gamma_{abc}\eta, \tag{39}$$

then the true vacuum would have

$$\Lambda = 0. \tag{40}$$

[Another solution would be to take the parallelizing torsion on the round S^7, but this would be something of an overkill with $R_{mnpq}(\hat{\omega}) = 0$, whereas only $R_{mn}(\hat{\omega}) = 0$ is required for $\Lambda = 0$. We are, in any case, assuming vacuum II at the tree level.] Of course we have succeeded only in swapping the problem of why (40) should be true for why (39) should be true. The points we are making are that (a) in $d = 11$ supergravity the cosmological constant of spacetime is *calculable*; a necessary condition for explaining $\Lambda = 0$. Theories where Λ may be "tuned" can never explain this. (b) The special properties of the seven-sphere allow the possibility that the answer is $\Lambda = 0$. Most geometries do not permit this possibility. (c) If $\bar{\psi}_a \hat{\Gamma}_b \psi_c$ has any geometrical significance, equation (39) seems the most natural one. Note, incidentally, that m^{-1}, the inverse radius of S^7, would now be a *calculable* multiple of the Planck mass. Thus, in common with the model suggested by Candelas and Weinberg [55], one would be able to calculate the Yang-Mills coupling constants as pure numbers.

Another calculable number is the CP violating angle θ. The term in the $d = 11$ supergravity Lagrangian

$$\varepsilon^{M_1 M_2 \cdots M_{11}} F_{M_1 \cdots M_4} F_{M_5 \cdots M_8} A_{M_9 \cdots M_{11}}$$

yields a term in the $d = 4$ Lagrangian of the form [5]

$$f_{IJKL}(P) F_{\mu\nu}{}^{IJ} F_{\rho\sigma}{}^{KL} \varepsilon^{\mu\nu\rho\sigma},$$

where f is a function of the pseudoscalars. When $\langle P \rangle \neq 0$ as in vacua III and V, one can obtain a

$$\theta F_{\mu\nu} F_{\rho\sigma} \varepsilon^{\mu\nu\rho\sigma}$$

term. No such term is present in vacuum II, though nonperturbative effects may introduce one. Note, however, that in the scheme we are advocating that the $F_{\mu\nu}{}^{IJ}$ correspond only to the electroweak bosons and *not* to the gluons. Thus we have a rather drastic solution to the strong CP problem: not only are there no $F\tilde{F}$ terms for gluons; there are no FF terms either! One must then ask whether supersymmetry provides any reason for generating the latter dynamically, but not the former.

We have not addressed, nor at present are we able to address, the all-important question of realistic fermion representations. And although $N = 8$ breaks to the only $N > 0$ supersymmetry permitting chiral fermions, namely, $N = 1$, we are left with the question whether a vectorlike preon theory can give rise to a flavor-chiral bound-state theory. Nor do we have anything to say about families, except that $SO(5) \times SU(2)$ has room for another $SO(3)$.

5 Further Questions

It should be clear that the ideas set out in this paper, especially those of section 4, are at a very preliminary stage, and we do not wish to exclude other possible interpretations. Below we list a few random thoughts for future study.

1. We have argued that it is possible to give masses to the bound states of unwanted helicities by the breaking of $N = 8/SU(8)$ to $N = 1/G_2$. Yet another possibility is that there never were any $N = 8$ bound states, but that bound states form only in the $N = 1$ phase. This picture would avoid having to answer embarrassing questions like, "What does it mean to have massless spin 1 bosons not in the adjoint representation of the gauge group, massless spin $\frac{3}{2}$ not in the vector representation, massless spin 2 that are not singlets, and massless spin $\frac{5}{2}$ of any kind?" However, the amount of information obtainable from group theory alone would be accordingly diminished.

2. If our $d = 11$ gravity Lagrangian had included terms quadratic in the Riemann tensor, then the "composite" gauge bosons coming from the spin connection would then be "elementary" with their own Yang-Mills kinetic energy term. Aside from all the problems of higher derivatives, however, the crucial problem of symmetry restoration discussed in section 4 would remain.

3. The chiral nature of the $SU(8)$ was not exploited in the picture outlined in section 4. Yet this may be the key to the question of chiral fermion representations in Kaluza-Klein. A comprehensive, though pessimistic, review of this crucial question has recently been given by Witten [60] (see also [61, 62, 64]). For the possible significance of $SU(8)$ in the de Wit-Nicolai theory, see [42].

4. There was a certain amount of prejudice involved when in section 2 we looked for space-time vacua with maximal symmetry. Other solutions certainly exist (see, for example, [26]). Ideally, one should consider all possible solutions and exclude the undesirable candidates for the ground state by stability considerations. Similar remarks apply to the alternative $F_{\mu\nu\rho\sigma}$ compactification of $d = 11$ supergravity to a $d = 7$ space-time with 4 compact dimensions [36]. This has not yet been done. Indeed one can show that the ground-state solution of $d = 7$ AdS $\times S^4$ is stable and gives rise to

an effective $d = 7$ theory with local $SO(5)$ invariance and $N = 4$ supersymmetry. There is as yet no mathematical way to exclude this obviously unphysical vacuum state.

5. Indeed, although for reasons explained in this paper, we have focused our attention exclusively on $d = 7$ ground-state solutions topologically equivalent to S^7, there are an embarassingly large number (infinity) of other solutions with other topologies. The ones of obvious physical interest are those that fulfill the Witten [49] criteria of having a gauge group $G \supset SU(3) \times SU(2) \times U(1)$ and that therefore yield the correct gauge bosons already at the elementary level. Witten classified all such spaces in [49], and denoted them M^{pqr}, where p, q, and r are integers, but did not enquire whether they provided solutions to the $d = 11$ field equations. Duff and Toms [2] pointed out the $M^{001} = CP^2 \times S^2 \times S^1$ did not admit an Einstein metric, but that $M^{011} = CP^2 \times S^3$ and $M^{101} = S^5 \times S^2$ did, the latter admitting a spin structure so that fermions can be globally defined. More recently Castellani, D'Auria, and Fre [67] have shown that all M^{pqr}, except M^{001}, admit an Einstein metric and, of particular interest, that the $p = q$ spaces admit an $N = 2$ supersymmetry. These will be stable, whereas the rest are unlikely to be. The beauty of these solutions is that the $SU(3)$, $SU(2)$, and $U(1)$ coupling constants will all be related by the geometry even though this is not a GUT theory [2]. Unfortunately, the problem with these solutions lies in the fermion spectrum. Aside from the absence of chirality anticipated by Witten [49, 60], a harmonic expansion of fermion fields on M^{pqr} spaces reveals [64] that nowhere do the right quark and lepton representations appear either in the zero or nonzero modes. One would then have to argue, as in the case of S^7, that the quarks and leptons appear as bound states, but in this case the whole idea of getting the right gauge bosons at the elementary level seems much less compelling. Thus it seems to us that by abandoning all the unique properties of S^7, giving up the squashing = Higgs interpretation of section 3, and giving up the spontaneous breakdown of $N = 8$ to $N = 1$ (the only supersymmetry compatible with chiral fermions), one has gained very little in return.

There are also solutions that are neither S^7 nor have $SU(3) \times SU(2) \times U(1)$. Some of them do have an unbroken supersymmetry, however. Apart from the T^7 solution of Cremmer and Julia [35], for which $\mathcal{H} = \mathbb{1}$ and $N = 8$, there is the $K3 \times T^3$ solution [54], for which $\mathcal{H} = SU(2)$ and $N = 4$. If one likes hidden symmetries, this one is quite rich with a hidden local $SO(22) \times SO(6) \times U(1)$ with the 134 massless scalars belonging to the coset $[SO(22,6)/SO(22) \times SO(6)] \times [SU(1,1)/U(1)]$. (This observation is due to J. Schwarz and P. G. O. Freund [private communication].)

6. Just as we have conjectured [1, 2, 5, 10] that the symmetric vacuum of the de Wit-Nicolai theory corresponds to the round S^7 solution of $d = 11$ supergravity, so we have also conjectured that other asymmetric extrema of the de Wit-Nicolai effective potential might correspond to other solutions of the $d = 11$ theory that deviate from the maximally symmetric geometry, but that still have the same topology, [5, 10]. Of course, not all solutions with S^7 topology can correspond to de Wit-Nicolai extrema since they may involve a Space Invaders Scenario, as in section 3. One possible equivalence is provided by the $SO(7)$ invariant extremum with $N = 0$ recently found by Warner [22] and the Englert solution of vacuum III, where in both cases only massless pseudoscalars acquire nonzero VEVs. Moreover, de Wit and Nicolai have shown the former extremum to be unstable [23]. As in the case of the round S^7 [27], the equality or otherwise of the cosmological constants would provide an interesting check (see [5, 19, 23, 27]). Unfortunately this comparison is very difficult, not least because of the question of what to hold fixed in going from the symmetric phase to the broken one. Page [19] has suggested that the charge Q associated with A_{MNP} and defined in his paper is the relevant quantity. In this case he finds for the 5 vacua of table 1

$$\Lambda_{\rm I} = -2^2 \cdot 3^{4/3} |Q|^{-1/3},$$

$$\Lambda_{\rm II} = -2^2 \cdot 3^{11/3} \cdot 5^{-5/3} |Q|^{-1/3} = \Lambda_{\rm IV},$$

$$\Lambda_{\rm III} = -2^{1/3} \cdot 3^{1/3} \cdot 5^{4/3} |Q|^{-1/3},$$

$$\Lambda_{\rm V} = -2^{1/3} \cdot 3^{8/3} \cdot 5^{-1/3} |Q|^{-1/3}.$$

It is a strange empirical fact, not discussed by Page, that we find

$$\frac{\Lambda_{\rm II}}{\Lambda_{\rm I}} = \frac{\Lambda_{\rm V}}{\Lambda_{\rm III}},$$

the significance of which still eludes us.

Warner [22] has also found new extrema, which appear to have no $d = 11$ counterpart. In particular, he finds one for which $N = 8/SO(8)$ is broken down to $N = 1/G_2$ with both scalars and pseudoscalars developing nonzero VEVs. This provides a concrete example of the necessity of parity breakdown in any spontaneous breaking of $N = 8$ to $N = 1$ without extra dimensions, as discussed in section 4. The relation between this $N = 1$ and this G_2 and the $N = 1$ and G_2 discussed previously remains mysterious.

At the same time as progress in $d = 11$ supergravity has been evolving, there has been significant development in other aspects of Kaluza-Klein theories. The first concerns quantization and Casimir energies [68, 2, 69, 70, 71, 72, 55, 91], and the second concerns Kaluza-Klein cosmology [74, 85]. It will be interesting to see what emerges when all these ideas are fused together. In particular, it still remains to be seen whether, as was conjectured in [2, 5], the vanishing Casimir energy of supersymmetric vacua valid for theories compactified on a torus [71] (i.e., in Poincaré supergravity) extends also to those compactified on a curved manifold—e.g., a sphere (i.e., in de Sitter supergravity).

There has also been interest in $d = 10$ supergravities as candidate Kaluza-Klein theories [81, 82, 83, 84, 90] because they corresponds to the zero slope limit of the fermionic dual string models [84]. The $N = 1, d = 10$ supergravity is ruled out by anomalies in the divergence of the energy momentum tensor [80, 60] and so therefore is $N = 1$ supergravity coupled to $N = 1$ Yang-Mills (which anyway suffers from inconsistencies of its own [78, 79]). This leaves the $N = 2$ supergravities, types IIA and IIB. In [2] it was suggested that one might try obtaining an $SU(3) \times SU(2) \times U(1)$ theory starting not from $N = 1$ in $d = 11$, as proposed by Witten [49], but from $N = 2$ (type IIA) in $d = 10$, which already has a $U(1)$ symmetry. One would then require compactification of the extra 6 dimensions, all of which may now be curved, to yield $SU(3) \times SU(2)$—e.g., on $CP^2 \times S^2$. This was possible because the A_{MNP} field is also present in this theory. Such solutions have since been found explicitly in [90]. However, as was also pointed out in [2], the fermions are neutral under this $U(1)$ (and cannot therefore even be defined on CP^2, which does not admit a spin structure). Type IIB theories are interesting [84], not least because the potential anomaly in the stress tensor exactly cancels [80]. However, no ground-state solutions with four space-time dimensions are known [84] except those corresponding to T^6 or $K3 \times T^2$ [54].

There are other developments in Kaluza-Klein theories that, so far, have been confined to the original 5-dimensional pure-gravity theory. These include monopole solutions [75, 76, 77] and the emergence of Kac-Moody symmetries [86]. The next step will be to apply these ideas to the $d = 11$

theory. Of particular interest will be the super-Kac–Moody-like algebras relating the massive $N = 8$ supermultiplets of section 3 that arise from Kaluza-Klein compactification on the seven-sphere.

Acknowledgment

In addition to my collaborators M. Awada, C. Pope, and B. Nilsson, I have benefited from conversations with B. deWit, L. Dolan, S. Ferrara, D. Freedman, P. Freund, M. Gaillard, M. Gell-Mann, H. Nicolai, C. Orzalesi, S. Randjbar-Daemi, A. Salam, E. Sezgin, J. Schwarz, P. Townsend, N. Warner, S. Weinberg, E. Witten, and B. Zumino. I am also grateful to J. A. Wheeler and S. Weinberg for their hospitality at the Center for Theoretical Physics and at the Theory Group, University of Texas at Austin, where part of this work was carried out. During this time the research was assisted by National Science Foundation grant PHY 8205717 and by organized research funds of the University of Texas at Austin and the Robert A. Welch Foundation. I would also like to thank the Theory Group at Rockefeller University for its hospitality.

References

[1]   M. J. Duff, in *Supergravity 81*, S. Ferrara and J. G. Taylor, editors (Cambridge University Press, Cambridge, 1982), p. 257.

[2]   M. J. Duff and D. J. Toms, in *Unification of the Fundamental Interactions II*, J. Ellis and S. Ferrara, editors (Plenum, New York, 1982).

[3]   F. Englert, *Phys. Lett.* B119, 339 (1982).

[4]   R. D'Auria and P. Fré, *Phys. Lett.* B121, 141 (1983).

[5]   M. J. Duff, *Nucl. Phys.* B219, 389 (1983), and to appear in the Proceedings of The Marcel Grossman Meeting, Shanghai, August 1982.

[6]   R. D'Auria, P. Fré, and P. van Nieuwenhuizen, *Phys. Lett.* B122, 225 (1983).

[7]   M. J. Duff and C. Orzalesi, *Phys. Lett.* B122, 37 (1983).

[8]   M. A. Awada, M. J. Duff, and C. N. Pope, *Phys. Rev. Lett.* 50, 294 (1983).

[9]   B. Biran, F. Englert, B. de Wit, and H. Nicolai, *Phys. Lett.* B124, 45 (1983).

[10]   M. J. Duff and C. N. Pope, in *Supersymmetry and Supergravity 82*, S. Ferrara, J. G. Taylor, and P. Nieuwenhuizen, editors (World Scientific Publishing, 1983).

[11]   P. G. O. Freund. University of Chicago Preprint EFI 82/83.

[12]   F. Englert, M. Rooman, and P. Spindel, *Phys. Lett.* B127, 47 (1983).

[13]   T. Dereli, M. Panahimoghaddam, A. Sudbery, and R. W. Tucker, *Phys. Lett.* B126, 33 (1983).

[14]   B. Morel, University of Geneva Preprint UGVA-DPT 1983/02-378.

[15]   M. J. Duff, B. E. W. Nilsson, and C. N. Pope, *Phys. Rev. Lett.* 50, 2043 (1983), and errata in 51, 846 (1983).

[16]   Y. Fujii, T. Inami, M. Kato, and N. Ohta, University of Tokyo Preprint UT-Komaba 84–4 (revised version).

[17]   F. Gursey and C. H. Tze, *Phys. Lett.* B127, 191 (1983).

[18]   J. Lukierski and P. Minnaert, *Phys. Lett.* B129, 392 (1983).

[19]   D. N. Page, University of Texas Preprint (to appear).

[20]   F. A. Bais, H. Nicolai, and P. van Nieuwenhuizen, CERN Preprint TH 3577 (1983).

[21]   F. Englert, M. Rooman, and P. Spindel, *Phys. Lett.* B130, 50 (1983).

[22]   N. P. Warner, Caltech Preprints CALT-68-992 and *Phys. Lett.* B128, 169 (1983).

[23]   B. de Wit and H. Nicolai, NIKHEF Preprints H/83-7 and H/83-8 (1983).

[24]   P. Tataru-Mihai, Preprint (1983).

[25]   R. G. Moorhouse and R. C. Warner, University of Glasgow Preprint (1983).

[26]   P. van Baal, F. A. Bais, and P. van Nieuwenhuizen, University of Utrecht Preprint (1983).

[27]   M. J. Duff, C. N. Pope, and N. P. Warner, *Phys. Lett.* B130, 254 (1983).

[28]   B. E. W. Nilsson and C. N. Pope, Imperial College Preprint ICTP/82-83/27.

[29]   Th. Kaluza, *Sitzungsber. preuss. Akad. Wiss* 966 (1921); O. Klein, *Z. Phys.* 37, 895 (1926).

[30]   W. Nahm, *Nucl. Phys.* B135, 149 (1979).

[31]   E. Cremmer, B. Julia, and J. Scherk, *Phys. Lett.* B76, 469 (1978).

[32]   P. G. O. Freund, *Symmetries in Particle Physics*, I. Bars, A. Chodos, and C. Tze, editors (Plenum, New York, 1984).

[33]   A. Lichnerowicz, *C. R. Acad. Sci. Paris, Ser. A-B* 257, 7 (1963).

[34]   E. Cremmer and J. Scherk, *Nucl. Phys.* B108, 409 (1976), B118, 61 (1977).

[35]   E. Cremmer and B. Julia, *Nucl. Phys.* B159, 141 (1979).

[36]   P. G. O. Freund and M. A. Rubin, *Phys. Lett.* B97, 233 (1980).

[37]   G. Jensen, *J. Diff. Geom.* 8, 599 (1973).

[38]   E. Cartan and J. A. Schouten, *Proc. K. Akad. Wet. Amsterdam* 29, 933 (1926).

[39]   R. Slansky, *Phys. Rep.* C79, 1 (1981).

[40]   L. Castellani and N. P. Warner, *Phys. Lett.* B130, 47 (1983).

[41]   P. Breitenlohner and D. Z. Freedman, *Phys. Lett.* B115, 197 (1982).

[42]   B. de Wit and H. Nicolai, *Phys. Lett.* B108, 285 (1982), *Nucl. Phys.* B208, 323 (1983).

[43]   S. Weinberg, *Phys. Lett.* B125, 265 (1983).

[44]   D. Z. Freedman, P. van Nieuwenhuizen, and S. Ferrara, *Phys. Rev.* D13, 3214 (1976); S. Deser and B. Zumino, *Phys. Lett.* B62, 335 (1976).

[45]   B. Zumino, (University of California at Berkeley) Preprints UCB-PTH-83/2 and LBL-15819, and lecture at this conference.*

* Editors' note: The paper read to the conference by Zumino appears in this volume.

[46] J. Scherk and J. H. Schwarz, *Phys. Lett.* B153, 61 (1979), *Nucl. Phys.* B153, 61 (1979).

[47] P. Fayet and S. Ferrara, *Phys. Rep.* 32, 249 (1977).

[48] S. Ferrara and P. van Nieuwenhuizen, *Phys. Lett.* B127, 70 (1983).

[49] E. Witten, *Nucl. Phys.* B186, 412 (1981).

[50] A. Salam and J. Strathdee, *Ann. Phys.* 141, 316 (1982).

[51] J. Ellis, M. Gaillard, L. Maiani, and B. Zumino, in *Unification of the Fundamental Interactions*, J. Ellis, S. Ferrara, and P. van Nieuwenhuizen, editors (Plenum, New York, 1980).

[52] J. Ellis, B. Zumino, and M. K. Gaillard, *Acta Physica Polonica* B13, 253 (1982).

[53] A. C. Davis, A. J. Macfarlane, and J. W. Van Holten, *Phys. Lett.* B125, 151 (1983).

[54] M. J. Duff, B. E. W. Nilsson, and C. N. Pope, *Phys. Lett.* B129, 39 (1983).

[55] P. Candelas and S. Weinberg, University of Texas Preprint UTTG-6-83.

[56] P. G. O. Freund, *Phys. Lett.* B130, 265 (1983).

[57] W. Heidenreich, *Phys. Lett.* B110, 461 (1982).

[58] G. W. Gibbons, C. M. Hull, and N. P. Warner, *Nucl. Phys.* B218, 173 (1983).

[59] D. Z. Freedman and H. Nicolai (private communication).

[60] E. Witten, lecture at this conference.*

[61] G. Chapline and R. Slansky, *Nucl. Phys.* B209, 461 (1982).

[62] C. Wetterich, *Nucl. Phys.* B223, 109 (1983).

[63] S. Randjbar-Daemi, J. Strathdee, and A. Salam, *Nucl. Phys.* B214, 491 (1983).

[64] S. Randjbar-Daemi and J. Strathdee (unpublished).

[65] H. Nicolai, P. K. Townsend, and P. van Nieuwenhuizen, *Lett. Nuovo Cimento* 30, 315 (1980).

[66] R. Coquereaux and A. Jadezyk, CERN Preprint TH. 3483 (1983).

[67] L. Castellani, R. D'Auria, and P. Fré, University of Torino Preprint IFTT 427 (1983).

[68] S. D. Unwin, *Phys. Lett.* B103, 18 (1981).

[69] M. J. Duff and D. J. Toms, in Proceedings of the Second Quantum Gravity Seminar, Moscow, October 1981 (CERN Preprint TH 3248).

[70] T. Appelquist and A. Chodos, *Phys. Rev. Lett.* 50 (1983), *Phys. Rev.* D28, 772 (1983).

[71] D. Pollard, Imperial College Preprint ICTP/82-83/11 (1983).

[72] M. A. Rubin and B. D. Roth, *Phys. Lett.* B127, 55 (1983).

[73] E. Witten, *Commun. Math. Phys.* 80, 81 (1981).

[74] A. Chodos and S. Detweiler, *Phys. Rev.* D21, 2167 (1980).

[75] D. Pollard, *J. Phys.* A16, 565 (1983).

[76] D. Gross and M. J. Perry, *Nucl. Phys.* B226, 29 (1983).

[77] R. Sorkin, *Phys. Rev. Lett.* 51, 87 (1983).

[78] T. Kephart and P. Frampton, *Phys. Rev.* D28, 1010 (1983).

[79] G. Sierra and P. K. Townsend, Ecole Normale Preprint (1983).

[80] E. Witten and L. Alvarez-Gaume, Princeton University Preprint (1983).

* Editors' note: The paper read to the conference by Witten appears in this volume.

[81]   M. J. Duff, P. K. Townsend, and P. van Nieuwenhuizen, *Phys. Lett.* B122, 232 (1983).

[82]   S. Randjbar-Daemi, A. Salam, and J. Strathdee, *Phys. Lett.* B124, 349 (1983).

[83]   D. Z. Freedman, G. W. Gibbons, and P. C. West, *Phys. Lett.* B124, 491 (1983).

[84]   J. Schwarz, lecture at this conference.*

[85]   P. G. O. Freund, *Nucl. Phys.* B209, 146 (1982).

[86]   L. Dolan and M. J. Duff, Rockefeller University Preprint RU83/B/64 (August 1983).

[87]   I. Bars and S. Macdowall, *Phys. Lett.* B129, 182 (1983).

[88]   L. F. Abbott and S. Deser, *Nucl. Phys.* B195, 76 (1982).

[89]   S. Deser and R. Nepomechie, Brandeis University Preprint.

[90]   S. Watamura, University of Tokyo Preprint UT-Komaba 83-6 (May 1983).

[91]   An. Ing and Chen Shi, Academia Sinica Preprint (1982).

*The paper read to the conference by Schwarz appears in this volume.

Supersymmetry and Finiteness

P. C. West

Extended theories of rigid supersymmetry are described and their various formulations given. The renormalization properties of these theories are derived, and it is demonstrated that a class of them is finite to all orders of perturbation theory. It is shown that one can add to those theories that are finite certain soft terms and still preserve finiteness. The prospects for finding a realistic theory of extended rigid supersymmetry are discussed.

1 Introduction

It was first pointed out in 1930 that the method of second quantization, which was introduced to avoid negative energy states, led to infinite probabilities. These infinities resulted from radiative corrections that were calculated in a perturbative framework. This fact was greeted with dismay, and it was somewhat later before it was suggested that these infinities could be absorbed into the parameters of the theory. Soon after the successful calculation of the Lamb shift the rules of quantum field theories were formulated. These rules included the method of renormalization. According to this method a quantum field theory contained a parameter Λ which regulated the theory; that is, it rendered the previously infinite answers finite. The calculation of any process involved the bare parameters, the parameter Λ, and the momentum. The bare parameters must also depend on the parameter Λ in a very precise way. In a renormalizable theory, renormalization conditions are used to relate the bare parameters to the physical parameters in just such a way that the terms that arise order by order in perturbation theory and diverge as Λ goes to infinity cancel exactly against Λ divergent terms in the bare parameters. The calculated process still depends on Λ, but only in the form of powers of $1/\Lambda$. When Λ is taken to infinity these terms disappear and the final physical answer is claimed to be independent of whatever method of regularization one chooses.

This is claimed to be true even if the regulator does not respect the symmetries of the theory. In this case one must be careful to add, by hand, correcting terms that maintain the Ward identities. Should one choose not to add these terms or not be aware of a symmetry that required them, then the quantum theory is unlikely to preserve that symmetry. In this sense the renormalization procedure is ambiguous. This procedure of introducing a parameter only to remove it at a later stage has been considered so inelegant by a few physicists that they have doubted the validity of the renormalization method.

In quantum electrodynamics $g - 2$ for the electron when calculated according to the renormalization method is in agreement with experiment to one part in 10^8. It is difficult, however, to draw any firm conclusions about this impressive agreement since we know it is a number calculated in a series that does not converge and cannot be resummed. The latter is connected [1] with the appearance of Landau ghosts, which are a strong indication that QED is not a consistent quantum field theory. There are no such accurate measurements for the nuclear forces.

Of course one could retain the parameter Λ. In this case experimental results will place limits on how small Λ can be. The predictions of the theory, however, may depend on the type of regulator method used, and it may be considered natural to use a method that respects the symmetries of the theory.

Unlike other theories a radiative correction in any quantum field theory involves effects, in the loop graphs, of particles of arbitrarily high energy. That is, it is a theory that requires a knowledge of the whole of physics up to arbitrary energies. One intuitive method of regulation is to introduce a cutoff Λ in momentum. Believing that one can take Λ to infinity and obtain a consistent theory in, say, QED is to believe that the theory will be consistent despite the fact that the effects of the nuclear forces and gravity, which are undoubtedly present, have been omitted. An alternative approach is to keep Λ finite and be content to settle for an effective theory that is valid up to an energy Λ. In this context, demanding that the effective theory have a perturbative scheme that might give one increasingly accurate estimates of physical quantities up to effects due to energies greater than Λ is likely to be equivalent to demanding that the theory be renormalizable. For it is only in a renormalizable theory that one can absorb the divergent Λ terms in the parameters of the theory, leaving effects only polynomial in $1/\Lambda$, and so have a perturbative scheme that might give one increasingly accurate estimates of physical quantities. Even in the context of a renormalizable theory it is far from clear that one can find such an approximation scheme. We must worry about whether the perturbation expansion converges and whether the renormalized parameters are stable as more and more orders in the perturbation expansion are included; that is, the values of the renormalized parameters should not fluctuate wildly from one order of perturbation theory to the next. Even in an asymptotically free theory there is still uncertainty whether the perturbation expansion in the coupling constant converges [3]. While the latter problem is another way of stating the gauge hierarchy problem [2], it is the "technical" solution [4] of this problem that has suggested that supersymmetry may be relevant in nature.

Soon after the discovery of supersymmetric theories [5] it was noticed [6] that they had remarkable renormalization properties. In the simpliest model of supersymmetry, the Wess-Zumino model, it was found that one needed only one wave function renormalization and not the 3 or 13 renormalization constants, as would be naively expected if supersymmetry was preserved or violated, respectively, at the quantum level [7]. It is this property that is being exploited in $N = 1$ supersymmetric models to solve the gauge hierachy problem mentioned. In $N = 1$ supersymmetric models, if we have two very differing mass scales m and M such that $m \ll M$, then the corrections to m in perturbation theory are $\alpha \ln M$ or $\alpha \ln \Lambda$ if M is taken to be the cutoff. Although $\alpha \ln M$ may still be of order one if M is large enough, it is better than contributions like M^2, which would occur in more general theories that involve scalar fields.

The most remarkable renormalization properties occur in the extended models of rigid supersymmetry. At first attention was focused entirely on the maximally extended $N = 4$ supersymmetric Yang-Mills theory. The β function was shown to vanish for this theory at one [8], two [9], and three loops [10]. Shortly after this latter result an argument for the finiteness of this theory to all order was found [11, 12]. Somewhat later, two more arguments [13–15] for finiteness to all orders were given.

More recently, it was realized that the $N = 2$ supersymmetric theories were finite above one loop and contained a class of theories that were finite to all orders [16]. In section 2, theories of extended supersymmetry are discussed in their component field formulation and their $N = 2$ and $N = 1$ superspace formulations. In section 3 arguments for finiteness are discussed in general and used to show the finiteness properties of these $N = 2$ theories. In section 4 the known ultraviolet properties of supergravity theories are summarized.

In section 5 it is shown that one can add certain soft terms, i.e., terms of dimension 3 or less, that break supersymmetry but respect gauge invariance to the class of $N = 2$ finite theories and still maintain finiteness of these theories.

Obviously in a finite theory there is no need to introduce a regulator parameter, and thus one avoids the intricacies discussed previously involved with the subsequent removal of this parameter. The bare parameters are still related to the "physical" parameters by renormalization conditions;

however, now these relations do not involve a regulator, and so the bare parameters are calculable in terms of "physical" parameters of the theory.

One may regard finite field theories as effective theories and introduce a cutoff Λ in momentum as discussed. In this case we will only find powers of Λ^{-1}, and the "physical" and bare parameters will not contain terms which diverge as Λ goes to infinity. Thus the changes between the "physical" and bare parameters found order by order in perturbation theory will not involve $\alpha \ln \Lambda$ terms, but only Λ^{-1}, terms which are very small if Λ has any reasonable value. In this sense finite quantum field theories are much less sensitive to effects at high energies. The finite theories without soft terms are superconformally invariant even at the quantum level, and this fact will play a role in determining the properties of these theories. It is well known that superconformal quantum scalar theories, were such theories to exist, would be free theories. However, these proofs do not apply to theories with gauge invariance. The addition of the soft terms of section 5 break the superconformal invariance and so avoid the problems of associated Goldstone particles. However, even in this case one may expect the superconformal invariance of the underlying theory to play an important role.

The significance of having a finite quantum field theory is not clear; however, in section 6 a preliminary discussion of how one may attempt to construct a realistic finite theory is given.

2 Theories of Extended Rigid Supersymmetry

Supersymmetry [5] is the only known symmetry that combines in a non-trivial manner the Poincaré group and an internal group G, such as $SU(N)$, in the same structure. Supersymmetry bypasses the Coleman-Mandula theorem [17] by introducing supercharges Q_α^i ($i = 1 \to N$) that obey anti-commutation relations; to be specific, they satisfy

$$\{Q_\alpha^i, Q_\beta^j\} = -2(\gamma^\mu C)_{\alpha\beta} P_\mu \delta^{ij} + C_{\alpha\beta} u^{ij} + (\gamma_5 C)_{\alpha\beta} V^{ij}. \tag{2.1}$$

In this equation $U^{ij} = -U^{ji}$ and $V^{ij} = -V^{ji}$ are extra bosonic generators, called central charges [18], which commute with all the other generators of the supersymmetry group. The remaining relations of the supersymmetry algebra are the commutators of the generators P_μ and $J_{\mu\nu}$ of the Poincaré group, the commutators of T_s the generators of the internal symmetry group G, and the relations

$$[Q_\alpha^i, P_\mu] = 0,$$

$$[Q_\alpha^i, J_{\mu\nu}] = \frac{i}{2}(\sigma_{\mu\nu})_\alpha^\beta Q_\beta^i, \tag{2.2}$$

$$[Q_\alpha^i, T_s] = (\delta_\alpha^\beta (t_r)_j^i + (\gamma_5)_\alpha^\beta (l_r)_j^i) Q_\beta^j. \tag{2.3}$$

These relations ensure that Q_α^i, which is a Majorana spinor, belongs to the $(\frac{1}{2}, 0) + (0, \frac{1}{2})$ representation of the Lorentz group and belongs to the $(t_r)_j^i \delta_\alpha^\beta + (\gamma_5)_\alpha^\beta (l_r)_j^i$ representation of G.

The irreducible representations of extended supersymmetry are found by a simple extension of the method used to find the irreducible representations of the Poincaré group.

This method of induced representations, invented by Wigner [19], boils down to the statement that the properties of the particles in an irreducible representation are determined entirely by their behavior in a given frame of reference (i.e., for a given momentum). For the massless case the generators that transform the "rest frame" states into themselves and have nontrivial action are P_μ, those of G, the helicity operator J, and one supercharge for each value of i, say, Q^i, which raises the value of J (i.e., $[Q^i, J] = \frac{1}{2} Q^i$) and satisfies

$$\{Q^i, Q^j\} = \delta^{ij} \tag{2.4}$$

in the frame $(m, 0, 0, m)$.

Starting with the Clifford vacuum (i.e., $Q^i|\lambda\rangle = 0$) of helicity λ (i.e., $J|\lambda\rangle = \lambda|\lambda\rangle$), we can create the states

$$|\lambda\rangle, \qquad Q^{i*}|\lambda\rangle, \qquad Q^{i*}Q^{j*}|\lambda\rangle, \qquad \dots . \tag{2.5}$$

As a result, an irreducible representation of supersymmetry has the following set of states:

$$|\lambda\rangle, \qquad \left|\lambda - \frac{1}{2}, i\right\rangle, \qquad |\lambda - 1, [ij]\rangle, \qquad \dots, \qquad \left|\lambda - \frac{N}{2}, [ijk\cdots]\right\rangle. \tag{2.6}$$

The series terminates after the state with helicity $\lambda - (N/2)$, and the state with helicity $\lambda - (m/2)$ is in the mth rank antisymmetric representation of $SU(N)$.

For example, if $\lambda = +1$ and $N = 4$, we find the irreducible representation

$$|1\rangle, \qquad |\tfrac{1}{2}, i\rangle, \qquad |0, [ij]\rangle, \qquad |-\tfrac{1}{2}, [ijk]\rangle, \qquad |-1, [ijkl]\rangle, \tag{2.7}$$

Table 1

	N				
Spin	1		2		4
1	—	1	—	1	1
$\frac{1}{2}$	1	1	2	2	4
0	2	—	4	2	6

namely, the spectrum of $N = 4$ Yang-Mills; 1 vector, 4 spin $\frac{1}{2}$ and 6 spin 0. This example is special in that it automatically includes the same number of states with a given helicity, λ, as those with the opposite helicity, $-\lambda$. In general, one must add to the representation of (2.6) the irreducible representation with reversed helicities in order to ensure that the representation admits a CPT symmetry.

Table 1 gives all the massless irreducible representations of supersymmetry that include spins and less. For a detailed discussion of the supersymmetry algebra and its irreducible representations see reference [20].

In working across the table from left to right, these theories are the Wess-Zumino model [21], $N = 1$ Yang-Mills [22], the hypermultiplet [23] (or $N = 2$ matter), $N = 2$ Yang-Mills [24], and finally $N = 4$ Yang-Mills [25]. Clearly, from the previous discussion it follows that if $N > 4$, we will have a theory with spins greater than one. The $N = 2$ theories have $U(2)$ internal symmetry, whereas $N = 4$ Yang-Mills has only an $SU(4)$ symmetry. The latter follows from the commutation relation between the $U(1)$ generator B and the supercharge

$$[B, Q_\alpha^i] = \left(\frac{N - 4}{4}\right)(\gamma_5)_\alpha^\beta Q_\beta^i. \tag{2.8}$$

Clearly if $N = 4$, all states in an irreducible representation must have the same action under B. Since $N = 4$ Yang-Mills has all its states, regardless of helicity, in the same representation, the action of B must be trivial.

These theories of extended rigid supersymmetry are the most symmetric, as well as being consistent theories, that are known, and indeed thought to be possible. Inclusion of a spin $\frac{3}{2}$ requires a spin 2 in order to propagate causally [25], and an interacting spin 2 theory without ghosts contains Einstein's theory of gravity and has the resulting doubts concerning its renormalizability.

The $N = 2$ Yang-Mills theory [24] contains the component fields $(A, B, A_\mu, \lambda_{\alpha i}, \mathbf{X})$ all in the adjoint representation of the gauge group. The spinor $\lambda_{\alpha i}$ satisfies the $SU(2)$ Majorana condition

$$\lambda_{\alpha i} = (i\gamma_5 C)_{\alpha\beta}\varepsilon_{ij}\bar{\lambda}^{\beta j}. \tag{2.9}$$

The field $\mathbf{X}$ is a triplet under $SU(2)$ of auxiliary fields. The hypermultiplet [23] ($N = 2$ matter) contains the component fields (A^{ia}, ψ^a, F^{ia}) in the representation R of G as well as their complex conjugates $(A^*_{ia}, \bar{\psi}_a, F^*_{ia})$ in the representation $\bar{R}$ of G. The field ψ_a is a Dirac spinor and the fields F^{ia} are auxiliary fields belonging to a doublet of $SU(2)$. The most general coupling between these two supermultiplets is given by the following action:

$$A = A^{\text{Y.M.}} + A^{\text{matter}} + A^{\text{interaction}} + A^{\text{mass}},$$

where

$$A^{\text{Y.M.}} = \text{Tr} \int d^4x \left\{ -\frac{1}{4}F_{\mu\nu}^2 - \frac{1}{2}(\mathscr{D}_\mu A)^2 - \frac{1}{2}(\mathscr{D}_\mu B)^2 + \frac{1}{2}\mathbf{X}^2 \right.$$

$$- \frac{i}{2}\bar{\lambda}^i\gamma^\mu\mathscr{D}_\mu\lambda_i + \frac{ig}{2}\bar{\lambda}^i[B + \gamma_5 A, \lambda_i]$$

$$\left. + \frac{g^2}{2}([A, B])^2 \right\},$$

$$A^{\text{matter}} = \int d^4x \left\{ -\mathscr{D}_\mu A^{ia}\mathscr{D}^\mu A^*_{ia} - \frac{i}{2}\psi_a \mathscr{D}\psi^a + |F^{ia}|^2 \right\}.$$

This action is invariant under the following $N = 2$ supersymmetry transformations:

$$\delta A_\mu = -i\bar{\varepsilon}^i\gamma_\mu\lambda_i, \qquad \delta B = \varepsilon^i\gamma_5\lambda_i,$$

$$\delta A = -\bar{\varepsilon}^i\lambda_i,$$

$$\delta\lambda_i = + F_{\mu\nu}\sigma^{\mu\nu}\varepsilon_i - i\gamma^\mu\mathscr{D}_\mu(A - \gamma_5 B)\varepsilon_i \tag{2.10}$$

$$- ig[A, B]\gamma_5\varepsilon_i + 2ig(T^s)_b^a A^b (A^* A^*_j)_a \varepsilon^j$$

$$+ i\tau_i^j\varepsilon_j\bar{X},$$

$$\delta\bar{X}^s = \bar{\varepsilon}^i\tau_i^j[\mathscr{D}\lambda_j^s - g[-A + \gamma_5 B, \lambda_i]^s - g(T^s)_b^a A^*_{ia}\psi^b + g(T^s)_b^a A_i^b \psi_a]$$

and

$$\delta A^{ia} = -i\bar{\varepsilon}^i \psi^a,$$

$$\delta \psi^a = 2i\slashed{D}A^{ia}\varepsilon_i + 2\gamma_5 F^{ia}\varepsilon_i + 2ig(T_s)^a_b A^{ib}(-A_s + \gamma_5 B_s)\varepsilon_i, \tag{2.11}$$

$$\delta F^{ia} = +i\bar{\varepsilon}^i \gamma_5 \slashed{D}\psi^a - 2i\overline{g\varepsilon}^i \gamma_5 \lambda_{js} A^{jb}(T_s)^a_b + ig\bar{\varepsilon}^i(B_s - \gamma_5 A_s)(T^s)^a_b \psi^b.$$

The supersymmetry parameter $\varepsilon_{\alpha i}$ satisfies the same $SU(2)$ Majorana constraint as $\lambda_{\alpha i}$, namely,

$$\varepsilon_{\alpha i} = i(\gamma_5 C)_{\alpha\beta}\varepsilon_{ij}\lambda^{\beta j}. \tag{2.12}$$

The reader may check that these transformations do indeed realize the $N = 2$ supersymmetry algebra. Clearly the above action is manifestly invariant under $U(2)$ and this, with $N = 2$ supersymmetry, explains why, apart from the mass term, the action is determined entirely by the value of the one gauge coupling constant g. For example, the coefficient of the term $-(i/2\bar{\psi}_a(ig)\gamma^\mu(A_\mu)^a_b\psi^b$, which is determined by gauge invariance, is related to the coefficient of the term $-ig\bar{\psi}_a A^{ib}(T_s)^a_b \lambda_{is}$ by $N = 2$ supersymmetry. This is easily seen as A_μ is rotated into $\lambda_{\alpha i}$ and ψ^b is rotated into A^{ib} under $N = 2$ supersymmetry. In fact, there are alternative formulations of both these multiplets which differ in their auxiliary field structure.

We now give the $N = 2$ superspace description of these theories. The $N = 2$ supersymmetric Yang-Mills theory is described by the superfield potential $A_{Bi}(\chi_\mu, \theta^j_A, \bar{\theta}_{Aj})$ such that the field strength

$$F_{AiBj} = \varepsilon_{ij}\varepsilon_{AB}W = D_{Ai}A_{Bj} + D_{Bj}A_{\bar{A}i} + \{A_{Ai}, A_{Bj}\} \tag{2.13}$$

is chiral [69], namely,

$$\mathscr{D}_{\bar{A}i}W = 0,$$

$$\bar{\mathscr{D}}^{\dot{A}i}\bar{\mathscr{D}}^{1j}_{\dot{A}}W = \mathscr{D}^{Ai}\mathscr{D}_{A}jW. \tag{2.14}$$

Here $A, \dot{A}; B\dot{B} = 1, 2$, and we are using two component notation; the action is given by [69]

$$A^{\text{Y.M.}} = \int d^4x \, d^4\theta \, W^2 + \text{h.c.} \tag{2.15}$$

This superspace formulation corresponds to having the triplet of auxiliary field $\mathbf{X}$ considered previously. An alternative possibility, which has only one auxiliary field, is given in reference [26].

The $N = 2$ matter can be described by a superfield [27] $\phi_i(\chi_\mu, \theta_{\alpha j}, z)$ (i, $j = 1, 2$), where z is a bosonic coordinate corresponding to the fact that ϕ_i carries an off-shell central charge [28]. This possibility corresponds to having auxiliary fields F^{ia} and F^*_{ia}. At present there is no known way of using this multiplet to carry out super-Feynman rule calculations. This is due to the appearance of the coordinate z, which is not integrated over. An alternative superfield description is given by the dimension-one superfields V, L^{ij}, L^{ijk} [29]. These superfields do not involve an off-shell centrala charge [28] and so can be used for quantization in superspace. The extension of this superfield description to admit complex representations can be found in reference [16]. However, precisely because these multiplets do not have an off-shell central charge, they do not allow mass terms.

The $N = 2$ multiplets can be decomposed into $N = 1$ multiplets. The $N = 2$ Yang-Mills multiplet is described by a chiral superfield φ (a Wess-Zumino multiplet) in the adjoint representation and an $N = 1$ Yang-Mills multiplet described by a superfield V. The $N = 2$ matter consists of two Wess-Zumino multiplets described by chiral superfields X and Y in the representations R and $\bar{R}$ of G, respectively. The action of equation (2.10) then becomes

$$A = \int d^4x\, d^4\theta\, \{\phi^s(e^{gV})^t_s \phi_t + \bar{x}_{a\sigma}(e^{gV}\sigma)^a_b X^b_\sigma$$

$$+ Y_{a\sigma}(e^{-gV}\sigma)^a_b \bar{Y}^b_\sigma\} + \frac{\mathrm{Tr}}{64g^2} \int d^4x\, d^2\theta(W^A W_A)$$

$$+ g\left(\int d^4x\, d^2\theta\, \phi^s(T^\sigma_s)^a_b X^b_\sigma Y_{a\sigma} + \text{h.c.}\right),$$

where

$$W_A = \bar{D}^2(e^{-gV} D_A e^{gV}),$$

$$(V^\sigma)^a_b = V_s(T^\sigma_s)^a_b, \quad \text{and} \quad (V)_{st} = -f_{rst}V^r. \tag{2.16}$$

The index σ corresponds to the possibility of there being several species of matter multiplets.

As noticed before, the term $\bar{X}VX$ is related to the term $XY\varphi$ by $U(2)$ invariance, and consequently these terms must have the same coupling, i.e., g.

3 Ultraviolet Properties of the Extended Theories of Rigid Supersymmetry

There are three arguments for determining the ultraviolet properties of the extended theories of rigid supersymmetry. One argument [14] relies on expressing the theory in light cone coordinates and performing super-Feynman rule power counting to determine the divergences. This argument has so far only been applied to $N = 4$ Yang-Mills. The other two arguments apply to all extended rigid theories, and in this section we shall discuss both these arguments, first outlining the general strategy they employ, which can be applied to other types of supersymmetric theories, and then giving their application to the extended theories of rigid supersymmetry.

The Anomalies Argument ([11, 12])

This was the first argument to be presented, and the strategy it employs utilizes the fact [30] that in any supersymmetric theory the energy momentum tensor $\theta_{\mu\nu}$, some of the internal currents $j_\mu{}^i{}_j$ and the supercurrent $j_{\mu\alpha i}$ lie in a supermultiplet. This follows from the fact that these currents define charges which must obey the algebra of the supersymmetry group [31]. Consequently any superconformal anomalies which these currents possess must lie in a supermultiplet. Typically this supermultiplet of anomalies will include θ_μ^μ, $(\gamma^\mu j_{\mu i})_\alpha$, and $\partial^\mu j_{\mu j}^i$ for some i, j, corresponding to the breaking of dilatation, special supersymmetry, and some of the internal currents, respectively. Clearly, if some of the relevant internal symmetries are preserved (i.e., $\partial_\mu j_j^{\mu i} = 0$ for some i, j), then the anomaly multiplet, if it is irreducible, will vanish, and consequently $\theta_\mu^\mu = 0$. However, θ_μ^μ is proportional to an operator $(F_\mu F^{\mu\nu} + \cdots)$ times the β function, and so the β function must vanish. From this result one can argue in specific formalisms such as the background field method for the finiteness of the theory being considered.

As an example of how this argument works, let us consider $N = 4$ Yang-Mills and assume that the quantum theory preserves supersymmetry, $SU(4)$ invariance, and that when viewed as an $N = 1$ supersymmetric theory the anomalies lie in a chiral supermultiplet, i.e.,

$$\left(\theta_\mu^\mu, (\gamma^\mu j_\mu)_\alpha, \partial_\mu j^{\mu(5)}, C, P\right). \tag{3.1}$$

In this multiplet C and P are objects of dimension 3 that have no known interpretation in terms of symmetries. Now from section 2 we saw that the theory does not admit an additional $U(1)$ symmetry, and hence if $SU(4)$ is preserved, all the chiral currents are preserved. Consequently, in any $N = 1$ decomposition the R current will be preserved, i.e., $\partial_\mu j^{\mu(5)} = 0$. This implies that $\theta^\mu_\mu = 0$, which in turn implies that $\beta(g) = 0$.

A much stronger argument, which assumes an $O(4)$ invariance, supersymmetry, and that the anomalies lie in a multiplet of $N = 2$ supersymmetry, can be found in reference [11].

Let us now apply the argument to show that the $N = 2$ theories of rigid supersymmetry are always finite above one loop and for a special class of these theories are finite to all orders [16].

There exists a superfield formulation of the $N = 2$ rigid theories that is manifestly invariant under $N = 2$ supersymmetry and $U(2)$ internal symmetry. This is the formulation discussed in section 2 which employs the superfield W for $N = 2$ Yang-Mills [24] and the complex superfields V, L^{ij}, L^{ijk} for $N = 2$ matter [29]. Let us regulate this theory by introducing higher derivatives. Since the status of the method of dimensional reduction is unclear [32], this is the only regularization method which we can be sure preserves $N = 2$ supersymmetry and the $U(2)$ invariance of these theories. Then, as is well known, all graphs will be regulated except the primitive one-loop graphs [33, 34]. Now, in general, introducing higher derivatives into a theory alters the infinity structure at one loop i.e., the β function of the theory. However, for the $N = 2$ theories expressed in the formalism given, the additional massive states introduced by the higher derivatives must belong to the $N = 2$ supermultiplets with the content 1 spin 1, 4 spin $\frac{1}{2}$, and 5 spin 0. This follows from the fact that this is the only available massive multiplet which does not possess a central charge. Consequently if the β function at one loop for the theory with higher derivatives is given by [35, 36]

$$\beta(g) = \frac{g^3}{96\pi^2} \sum_\lambda (-1)^{2\lambda} C_\lambda (1 - 12\lambda^2), \tag{3.2}$$

where the sum is over all helicity states, then the β function is the same as in the theory without higher derivative. However, it remains for us to check that this formula for the β function is indeed the correct one.

This property of the $N = 2$ rigid theories would not hold for a general $N = 1$ supersymmetric theory or for the original formulation of the hypermultiplet.

Hence if the theory is finite at one loop, the theory when higher derivative regulated will be finite at one loop and so be rendered finite to all orders by higher derivatives.

As such we can be sure that the first requirement in the anomalies argument is true, namely, that the currents of the quantum $N = 2$ rigid theories belong to a supermultiplet and that the $U(2)$ symmetry is preserved.

Let us now consider what is the possible multiplet that the anomalies of the $N = 2$ supercurrent multiplet could lie in. The $N = 2$ supercurrent [37] is a dimension 2 object J and the anomaly equation is of the form

$$D^{ij} J = \text{anomaly}, \tag{3.3}$$

where $D^{ij} = D^{(i} D^{j)}$.

However, the right-hand side of equation (3.3) must be constructed from the gauge invariant superfields W, L^{ij}, L^{ijkl}, V, and covariant derivatives of these fields, i.e.,

$$D^{ij} J = \bar{D}^{ij} \bar{W}^2$$
$$+ \text{term involving } L^{ij}, L^{ijkl}, \text{ and } V. \tag{3.4}$$

Now the term involving $\bar{W}$ represents an anomaly which contains θ^μ_μ and the divergence of one of the chiral $U(2)$ currents in the same multiplet. The terms involving L^{ij}, L^{ijkl}, and V need not be of this form, but the reader may verify that these particular terms may be absorbed in the definition of J and so removed from the left-hand side of the equation.

In this case the anomaly just given contains only the first term, and this is of the required form in that it contains one of $U(2)$ chiral currents as well as the trace of the energy momentum tensor. Consequently, as discussed, we can apply the logic of the anomalies argument to find that $\beta(g) = 0$ for theories that are finite at one loop.

If the theories have one-loop infinities, we can expect that the $U(2)$ symmetry is preserved except at one loop, and so in this case, as the anomalies discussion given still applies above one loop, the β function will vanish above one loop. A more detailed discussion of this argument will be given shortly [38].

Nonrenormalization Argument

This argument relies on a generalization to extended supersymmetry of the well-known nonrenormalization theorems of $N = 1$ supersymmetry. The general theorem can be stated as follows: In the superspace background field formalism of an N extended supersymmetric theory any contribution, *above one loop*, to the effective action must be

a. an integral over *all* superspace and
b. a gauge invariant function of the superpotentials and the matter fields.

For rigid theories this means that any local contribution above one loop to the effective action must be of the form

$$\int d^4x \, d^{4N}\theta \, \mathscr{L}(A_{\alpha i}, \phi, D_{\alpha j}\phi, \ldots), \tag{3.5}$$

where $A_{\alpha i}$ are the superspace Yang-Mills potentials and ϕ are the matter fields.

The curious one-loop exception arises due to the gauge fixing procedure in the extended superfield formalism. Having fixed the gauge for the Yang-Mills superfield, one finds that the corresponding ghosts have a gauge invariance. Fixing this gauge invariance in the background field formalism, one finds that the ghost for ghost has a gauge invariance. This procedure goes on indefinitely until one fixes the gauge for the ghosts in a way which explicitly involves the background potentials. Hence the nonrenormalization theorem does not apply to graphs which contain this final set of ghosts. Fortunately, however, these ghosts only couple to background fields, and so the violation of the theorem only applies to one-loop graphs.

For $N = 1$ supersymmetry one does not have this problem, and so the nonrenormalization theorem applies to all orders. As an example, let us consider the Wess-Zumino model described by the chiral superfield φ (dim $\varphi = +1$), whose action is

$$A = \int d^4x \, d^4\theta \, |\varphi|^2 + \left\{ \int d^4x \, d^2\theta (m\varphi^2 + \lambda\varphi^3) + \text{h.c.} \right\}. \tag{3.6}$$

Under a renormalization we have

$$\varphi \to Z^{1/2}\varphi, \qquad m \to Z_m m, \qquad \lambda \to Z_\lambda \lambda, \tag{3.7}$$

and consequently we might expect counterterms of the form

$$\int d^4x \, d^4\theta \, (Z - 1)|\varphi|^2 + \left\{ \int d^4x \, d^2\theta \, \{(Z^2 Z_m - 1)\varphi^2 m \right.$$

$$\left. + (Z^{3/2}Z_\lambda - 1)\lambda\varphi^3\} + \text{h.c.} \right\}. \tag{3.8}$$

However, the $N = 1$ nonrenormalization theorem [39] tells us that the last two counterterms cannot arise as they are subspace integrals, and consequently

$$Z^2 Z_m = 1 \qquad \text{and} \qquad Z_\lambda Z^{3/2} = 1. \tag{3.9}$$

In other words, there is in effect only one wave function renormalization [7] in the Wess-Zumino model, rather than the three infinities that one might naively expect.

The $N = 1$ Yang-Mills theory has component field content $(A_\mu; \lambda_\alpha; D)$. It is described in superspace by a potential which is subject to superspace constraints. The action for this theory is

$$\int d^4 x \, d^2 \theta \, W_B W^B + \text{h.c.}, \tag{3.10}$$

where

$$W_B = \bar{D}^2 A_B = \bar{D}^2 (e^{-gV} D_B e^{gV}),$$

and so one might expect under $A_B \rightarrow Z^{1/2} A_B$ the counterterm

$$(Z - 1) \int d^4 x \, d^2 \theta \, W_B W^B + \text{h.c.} \tag{3.11}$$

This term, however, is not forbidden by the nonrenormalization theorem since it can be rewritten as a full superspace integral, namely,

$$(Z - 1) \int d^4 x \, d^4 \theta \, (A_B \bar{D}^2 A^B). \tag{3.12}$$

In the background field formalism we must have the relation $Z_g Z^{1/2} = 1$, in order that the terms in field strength be renormalized in the same way [40], and so $N = 1$ Yang-Mills theory only has one infinity. From the beginning of this section, this discussion has been drawn from reference [13]; it has been discussed in more detail for the specific case of $N = 2$ superfields in reference [15]. The strategy [13] behind the nonrenormalization argument is to apply the nonrenormalization theorem given previously to extended supersymmetry theories and to examine whether any counterterms are possible. One immediately sees that the measure

$$\int d^4 x \, d^{4N} \theta \tag{3.13}$$

has the dimension $-4 + 2N$, which is positive for $N > 3$, and so the dimension of the integrand of any counterterm will be negative, and so such counterterms are unlikely to exist.

Let us now apply [16] the previous chain of arguments to the $N = 2$ rigid theories of supersymmetry. In this case the contributions to the effective action, above one loop, must be a function of the $N = 2$ Yang-Mills potential $A_{\beta j}$ of dimension $\frac{1}{2}$ and the $N = 2$ matter fields L^{ij}, L^{ijk}, and V which are all of dimension 1. They must also be an integral over $\int d^4 x \, d^8 \theta$ that has dimension 0.

Clearly there are no terms of this structure, i.e., of the form

$$\int d^4 x \, d^8 \theta \, \{A_{\alpha i} D_{\beta j} \cdots D_{\gamma k} A_{\delta l} + V D_{\beta j} \cdots D_{\gamma k} V \cdots \}, \tag{3.14}$$

and so above one loop an arbitrary $N = 2$ theory must be finite [16].

Having reviewed the two arguments for the ultraviolet behavior, we notice that they both contain a one-loop exception clause. This results in the anomalies argument because higher derivative regularization does not regulate one loop, and in the nonrenormalization argument because of the need to terminate the infinite sequence of ghosts for ghosts. However, in both arguments we expect that an arbitrary rigid $N = 2$ supersymmetric theory will be finite above one loop.

We therefore consider these theories at one loop. The β function for an arbitrary $N = 2$ theory with m_σ $N = 2$ matter multiplets in the representation $R_\sigma + \bar{R}_\sigma$ of the group G is [16]

$$\beta(g) = \frac{g^3}{16\pi^2} \cdot 2 \cdot \left(\sum_\sigma m_\sigma T(R_\sigma) - C_2(G) \right). \tag{3.15}$$

Clearly the β function vanishes if

$$\sum_\sigma m_\sigma T(R_\sigma) = C_2(G). \tag{3.16}$$

There are many solutions to this equation, one example being if $G = SU(N)$ with m hypermultiplets in the fundamental representation, in which case $T(R) = 1/2$, $C_2(G) = N$, and we find that

$$(m/2) - N = 0. \tag{3.17}$$

Because an $N = 2$ theory only has one coupling constant, it is finite if $\beta(g) = 0$. This is easily seen from the following argument. In terms of the

Figure 1

$N = 1$ superfield formulation of equation (2.16) the model has wave function renormalizations Z_X, Z_Y, Z_φ, and Z_V corresponding to the superfields X, Y, φ, and V, respectively. The only other renormalization is for the coupling constant g. Now the interaction term

$$g \int d^2\theta \, X Y \varphi$$

is not renormalized due to the $N = 1$ nonrenormalization theorem, and consequently

$$Z_g Z_X^{1/2} Z_Y^{1/2} Z_\varphi^{1/2} = 1. \tag{3.18}$$

From the $O(2)$ invariance of $N = 2$ supersymmetry we find that

$$Z_X = Z_Y \quad \text{and} \quad Z_\varphi = Z_V, \tag{3.19}$$

and consequently $Z_g Z_V Z_X^2 = 1$. In the background field gauge, however, $Z_g Z_V = 1$, and so

$$Z_X = 1. \tag{3.20}$$

If $\beta(g) = 0$, then $Z_g = 1$, and so $Z_V = 1$ and the theory is finite. Clearly in the $N = 2$ superfield formalism the $O(2)$ symmetry is manifest, and a similar argument implies finiteness.

It is easy to check this argument at one loop by calculating the $\bar\varphi\varphi$ and $\bar X X$ propagator at one loop. The one-loop $\bar X X$ propagator graphs are given in figure 1, and the reader may verify that these graphs sum to zero due to the negative sign on the VV propagator. The $\bar\varphi\varphi$ propagator is given at one loop by the graphs in figure 2. It is a simple matter to derive the formula for the one-loop β function of equation (3.15) from these graphs.

An explicit check on these results has been made at two loops, where it has been shown that the β function evaluated from the relevant two-loop Feynman graphs receives no two-loop contribution [41].

To summarize, an arbitrary rigid $N = 2$ supersymmetric theory is always

Figure 2

finite above one loop and will be finite to all orders if and only if [16]

$$\sum_\sigma m_\sigma T(R_\sigma) = C_2(G). \tag{3.21}$$

This statement requires some further clarification when one realizes that the β function is in general subtraction scheme dependent. This is not the case when the β function is zero, but for a general theory only the first two terms in the β function are scheme independent [42]. Consequently, for the $N = 2$ theories that are not finite, that is, only possess a β function that vanishes above one loop, we can expect a change of subtraction scheme to induce terms in the β function at three loops and above. This is consistent with either argument for the finiteness of these theories; for if at two loops we add a finite piece to the coupling constant renormalization, Z_g, say, then we shall find that we will get infinities at three-loop order coming from the insertion of this finite term into one-loop graphs, as these one-loop graphs even with insertions are not predicted to be finite by either argument. These infinities will then contribute to the β function at three loops.

A further point concerns the formalism used to evaluate the infinities. No matter what formalism is used, we shall always find, provided we use a supersymmetric regulator method, that the β function is zero. Since there are difficulties with the scheme of dimensional reduction [32], it is conceivable that this will involve the use of higher derivatives in any case. However, in most formalisms other than the background field $N = 2$ superfield formalism mentioned, we shall find that the various renormalization constants are infinite, although one may be able to render them finite by a judicious choice of gauge. Such infinities would occur in a component field calculation where even the gauge fixing term is not supersymmetric and the wave function renormalizations of fields in the same supermultiplet need not be the same.

A more recent argument, based on the Adler-Bardeen theorems, has also been used to derive the renormalization properties of the rigid theories of supersymmetry (see reference [71]).

4 Supergravity Theories and Finiteness

We shall begin by reviewing the problem of ultraviolet infinities in super-gravity theories. In supergravity the component fields $(e_\mu^a, \psi_{\mu a i}, \ldots)$ are contained in the superspace vielbein E_M^π (π a curved index and M a tangent space index). In $N = 1$ supergravity both types of indices take values from one to eight. The introduction of a spin connection Ω_π^{ab} allows the definition of a covariant derivative

$$\mathcal{D}_M = E_M^\pi(\partial_\pi - \tfrac{1}{2}\Omega_\pi^{ab}\mathbb{M}_{ab}), \tag{4.1}$$

where $\mathbb{M}_{ab}$ are the generators of the Lorentz group which acts on flat spinor indices and flat vector indices with the same parameter. The torsion and curvature tensors are defined in the usual way:

$$\{\mathcal{D}_M, \mathcal{D}_N] = -T_{MN}^R\mathcal{D}_R - \tfrac{1}{2}R_{MN}^{cd}\mathbb{M}^{cd}. \tag{4.2}$$

In general, the superspace formalism just given has several problems, the most obvious problem being that it contains fields of too high a spin. To remedy this defect one must place constraints on the torsions and curva-tures [43]. A derivation and discussion of these constraints can be found in reference [44].

In $N = 1$ supergravity one finds [45] that all the remaining curva-tures and torsions can be expressed in terms of three superfields R, $W_{(ABC)}$, and G_{AB}, where R and G_{AB} obey the relations

$$\bar{D}_A R = 0, \qquad \bar{D}_A W_{CDE} = 0, \qquad D^A G_{A\dot{A}} = \bar{D}_{\dot{A}} R, \tag{4.3}$$

$$D^A W_{ABC} = -\frac{i}{2}(D_{B\dot{A}}G_C^{\dot{A}} + D_{C\dot{A}}G_B^{\dot{A}}).$$

The superfields R and $G_{A\dot{B}}$ are proportional at $\theta = 0$ to the auxiliary fields [46] M, N, and b_μ of $N = 1$ supergravity, which are zero by their equation of motion, and so on-shell $R = G_{A\dot{B}} = 0$. As such the on-shell counterterms of $N = 1$ supergravity must be constructed from W_{ABC}, which has a "geo-metric" dimension of $\tfrac{3}{2}$. From the $N = 1$ nonrenormalization theorem it must be an integral over all of superspace, and so the counterterm at l loops must be

$$\kappa^{2(L-1)}\int d^4x\, d^4\theta\, E\mathcal{L}(W_{ABC}, \mathcal{D}_B W_{CDE}, \ldots). \tag{4.4}$$

The first possible such terms on grounds of dimensional analysis occurs at three loops [47],

$$(\kappa^2)^2 \int d^4x \, d^4\theta |W|^2 E, \tag{4.5}$$

where $W = W_{ABC} W^{ABC}$. Clearly, there are an infinite series of possible counterterms whose coefficients must vanish in order for the theory to be renormalizable.

The situation is somewhat similar for the extended supergravity theories up to and including $N = 4$ supergravity. The only on-shell superfields are W_{AB}, W_A, and W for $N = 2, 3$, and 4 supergravity [48]. The $\theta = 0$ components corresponding to the lowest dimension fields that occur in the corresponding superconformal theories. These superfields are chiral as well as satisfying

$$D_i^A \overline{W}_{AB} = 0, \qquad D_i^A \overline{W}_A = 0. \tag{4.6}$$

The three-loop counterterms for $N = 2, 3$, and 4 supergravity are then

$$(\kappa^2)^2 \int d^4x \, d^8\theta |W_{AB} W^{AB}|^2 E,$$

$$(\kappa^2)^2 \int d^4x \, d^{12}\theta |W_A W^A|^2 E, \tag{4.7}$$

$$\kappa^4 \int d^4x \, d^{16}\theta E |WW|^2,$$

respectively.

For $N > 4$ the situation is somewhat different. The on-shell fields still allow counterterms, but these objects are not full integrals over the whole of superspace. For example, in $N = 8$ supergravity the on-shell field is of dimension 0 and of the form W_{ijkl} which satisfies the self-duality condition [49]

$$W_{ijkl} = \frac{1}{4!} \varepsilon_{ijklmnpq} \overline{W}^{mnpq} \tag{4.8}$$

as well as

$$\overline{D}_{\dot{A}(i} W_{jklm)} = 0, \qquad D_\alpha^i W_{jklm} = \tfrac{1}{5} \delta^i_{[j} D_\alpha^n W_{klm]n}. \tag{4.9}$$

$$E_N^\pi \partial_\pi$$  $$E_M^\pi \partial_\pi$$

Figure 3

One may then ask what is the maximally extended superspace formalism that we may be reasonably sure exists and can be used to describe $N = 8$ supergravity? This is an $N = 4$ description obtained by dividing $N = 8$ supergravity into five $N = 4$ multiplets. One multiplet consists of $N = 4$ supergravity plus six $N = 4$ Yang-Mills multiplets, while the other four multiplets are four $N = 4$ multiplets that each contains one spin $\frac{3}{2}$ particle as its highest spin. Then $N = 4$ superspace formalism of the former multiplet is given in reference [52], and that for the latter multiplet is thought to exist. The former multiplet is described by a prepotential V of dimension -6. Consequently, on dimensional grounds the $N = 4$ action for this multiplet is of the form

$$\int d^{16}\theta \, d^4 x \, (VD^{16}V + KV^2 D^{30}V + \cdots). \tag{4.11}$$

In other words the propagator goes like $(p)^{-8}$ and the vertices like p^{15}. Examining the three-loop graph in figure 3, we find it is of the form

$$\int d^{12}p \, \frac{1}{p^{24}} \frac{1}{p^{64}} p^{60} p^{18} p^{-2d}, \quad \text{where } d = \begin{cases} 1 & \text{if } M = \mu \\ \frac{1}{2} & \text{if } M = A, \end{cases} \tag{4.12}$$

which is logarithmically divergent [53]. In this calculation we require a D^{16} for each loop and the $E_N^\pi \partial_{\pi K} VV$ vertices have a factor $p^9 p^{-d}$. The corresponding three-loop counterterm is [50]

$$K^4 \int d^4 x \, D^{[i_1 \cdots i_4],[j_1 \cdots j_4]} \bar{D}^{[k_1 \cdots k_4],[l_1 \cdots l_4]}$$

$$\tag{4.10}$$

$$\cdot (W_{i_1 \cdots i_4} W_{j_1 \cdots j_4} W_{k_1 \cdots k_4} W_{l_1 \cdots l_4}).$$

Now according to our previous discussion and as was pointed out in reference [13], this counterterm is not a full superspace integral and so is forbidden, provided an $N = 8$ superspace description of the theory exists and can be used to quantize the theory within the framework background formalism. Unfortunately, it is far from sure that such a superspace description does exist. The only known auxiliary field formalism [51] (i.e., super-space description) of $N = 8$ supergravity possess an off-shell central charge with the resulting constraints. Although, contrary to some claims, this is a local formulation, no way is known to solve these constraints and quantize with fields involving the central charge. Exactly the same arguments apply to the extended theories for $4 < N \leqslant 8$. Each of them possesses a three-loop counterterm which is a superspace subintegral and consequently would not arise if the maximal possible superspace formulation of the theories existed.

Unfortunately, for $1 \leqslant N \leqslant 8$ there exist counterterms at higher orders. There is, for example, a counterterm for $N = 8$ supergravity at eight loops given by [70]

$$K^{14} \int d^4x \, d^{32}\theta \, E\big(D^{Ai} W_{ijkl} D_{\dot{A}m} \overline{W}^{mjkl}\big)^2.$$

This is a full superspace integral.

This discussion shows that $N = 4$ supersymmetry by itself is not suffi-cient to establish that the three-loop counterterm of $N = 8$ supergravity must vanish. Of course, to do the calculation properly one must take into account all the $N = 4$ multiplets, and it is possible that they could conspire so as to yield a vanishing coefficient for the three-loop counterterm.

The problem has essentially the same origin as for gravity itself, namely supergravity has a dimensional coupling constant, κ. Unlike the case for the rigid theories of supersymmetry, the existence of this parameter allows one to balance the dimension of any number of superfields in the integrand of a counterterm, or in any anomaly equation, and so avoid the logic of either the nonrenormalization or anomaly arguments for finiteness.

We finally examine the ultraviolet properties of the superconformal theories. The $N = 4$ and 2 superconformal theories are described, respec-tively, by the dimension 1 and 0 superfields W and W_{AB} discussed previ-ously. Like all conformal theories, they do not contain a dimensional parameter and their actions are

$$\int d^8\theta \, d^4x \, W^2 \quad \text{and} \quad \int d^4\theta \, d^4x \, W_{AB} W^{AB}, \tag{4.13}$$

respectively. The counterterms in the corresponding quantum theories must be integrals over the whole of superspace as well as being gauge invariant functions of E_A^M. For the $N = 4$ theory $\int d^{16}\theta \, d^4x$ has dimension -4, and so there are no counterterms possible. The $N = 2$ theory could at first sight have the counterterm

$$\int d^4 x\, d^8 \theta\, E. \tag{4.14}$$

However, this term in fact vanishes [54] due to the necessary constraints, and in any case it would not be superconformally invariant. Hence above one loop we can conclude that $N = 2$ and 4 conformal supergravity are finite. At one loop $N = 4$ has no infinities [55], whereas $N = 2$ does have a one-loop infinity [55]. The existence of a finite superconformal theory was noted previously in reference [56].

5 The Addition of Explicit Symmetry Breaking Terms That Preserve Finiteness

One might wonder whether one could add terms to the class of finite $N = 2$ theories, discussed in section 3, and still maintain the finiteness of the theory. The addition of a soft term, that is, a term of dimension 3 or less, will cause the extended supersymmetry and internal symmetry of the theory to be broken explicitly. We shall demand, however, that the soft terms preserve the gauge invariance of the theory. The insertion of a soft term in a graph will always reduce the overall degree of divergence of the graph. However, because the symmetries of the theory have been broken, it is possible to find infinite counterterms of dimension 3 or less which were previously forbidden by the symmetry of the original theory. Should this latter possibility occur, then one may try to add a judicious choice of soft terms in such a way that they generate no new counterterms.

The first terms which were found to maintain finiteness in the context of $N = 4$ Yang-Mills were $N = 1$ supersymmetric mass terms [57]. A general analysis of the soft terms which preserved finiteness for $N = 4$ Yang-Mills was given in references [58] and [59]. The analysis of references [58] and [59] worked in a light cone and $N = 1$ superfield formulation of $N = 4$ Yang-Mills theory, respectively, while an analysis at the level of component fields which found some soft terms that preserved finiteness was given in reference [60]. The addition of $(A^2 - B^2)$-type terms to $N = 4$ Yang-Mills was contained in the general analysis of reference [58] as well as in reference [60]. This particular term was also found later in reference [61].

What soft terms could be added to theories in the class of finite $N = 2$ theories and still maintain finiteness was analyzed in reference [63]. The

Figure 4

strategy we adopt in order to carry out this analysis is as follows. We add to a finite $N = 2$ theory all gauge invariant terms of dimension 3 or less. We then calculate the one-loop infinities induced by these terms and find the necessary and sufficient conditions for finiteness at one loop. It is then shown that any terms which maintain finiteness at one loop will maintain finiteness to all orders.

The one-loop analysis is carried out by expressing the theory in terms of $N = 1$ superfields, as was done in section 2, and then using $N = 1$ super-Feynman rules in conjunction with the spurion technique [63]. The advantage of the spurion technique is that one can work with $N = 1$ superfields even though the terms we are adding violate $N = 1$ supersymmetry. As an example, let us add a term of the form $A^2 - B^2$, where A and B are the scalar and pseudoscalar fields of either X, Y, or φ; that is, we must add, for φ, say,

$$\text{Tr} \int d^2\theta \, d^4x \, L\varphi^2 + \text{h.c.} \sim \text{Tr} \int d^4x \, l(A^2 - B^2), \tag{5.1}$$

where $L = \theta^2 l$. The super-Feynman rules of the theory with this additional term are then the same as the original theory, except that we have the new vertex given in figure 4. The reader may easily verify that there are no one-loop infinite graphs that contain a vertex L, and so at one loop this $A^2 - B^2$ addition preserves the finiteness of the theory.

Consider now adding a term of the form $A^2 + B^2$; in other words, we add

$$\int d^4\theta \, d^4x \, \{ U_1 \, \bar{\varphi}^s (e^{gV})_s^t \varphi_t + U_{2\sigma} \bar{X}_{a\sigma}(e^{gV_\sigma})_b^a X_\sigma^b$$

$$+ \, U_{3\sigma} Y_{a\sigma}(e^{-gV_\sigma})_b^a \bar{Y}_\sigma^b \} \sim \int d^4x \, u(A^2 + B^2), \tag{5.2}$$

where $U_1 = \theta^2 \bar{\theta}_{u_1}^2$, etc. This addition generates the new vertices shown in figure 5. The only infinite one-loop graphs are given in figure 6. These

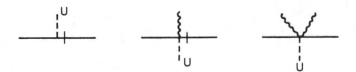

Figure 5

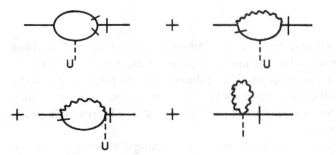

Figure 6

graphs lead to the following infinite contribution to the effective action:

$$
-g^2 \ln \Lambda \int d^4 x \, d^4 \theta \left\{ \varphi^s \varphi_s (C_2(G) U_1 + \sum_\sigma T(R_\sigma)(U_{2\sigma} + U_{3\sigma}) \right.
$$

$$
\left. + (X_\sigma X_\sigma C_2(R^\sigma)(U_1 + U_{2\sigma} + U_{3\sigma}) + (X \leftrightarrow Y) \right\}.
$$

(5.3)

This expression vanishes if

$$
U_1 + U_{2\sigma} + U_{3\sigma} = 0 \qquad \text{for all } \sigma
$$

(5.4)

and

$$
U_1 C_2(G) + \sum_\sigma T(R_\sigma)(U_{2\sigma} + U_{3\sigma}) = 0.
$$

(5.5)

Using the finiteness condition $C_2(G) = \sum_\sigma T(R_\sigma)$, we find that equation (5.4) implies equation (5.5).

Using dimensional analysis, or equivalently super-Feynman rule power counting, we can analyze what divergences a given soft term may give rise to. For example, if we add the $A^2 - B^2$ insertion of equation (5.1), we find that the spurion L has dimension $+1$, and consequently any infinite counterterm induced being an integral overall superspace will be of the form

Table 2

Insertion	A²−B²	A²+B²	$\bar{\chi}\chi$	$\bar{\lambda}\lambda$	A(A²+B²)	A³−3AB²
	$A^2 - B^2$	$A^2 + B^2$	$\bar{\chi}\chi$	$\bar{\lambda}\lambda$	$A(A^2 + B^2)$	$A^3 - 3AB^2$
$A^2 - B^2$						
$A^2 + B^2$			✓			
$\bar{\chi}\chi$	✓	✓			✓	
$\bar{\lambda}\lambda$	✓	✓				✓
$A(A^2 + B^2)$	✓	✓		✓		
$A^3 - 3AB^2$	✓	✓				✓

$$\ln \Lambda \operatorname{Tr} \int d^4 x \, d^4\theta \, L(\varphi \text{ or } X \text{ or } Y). \tag{5.6}$$

However, for semisimple groups [theories with $U(1)$ factors do not belong to the finite class of $N = 2$ theories], this term vanishes, and so there are no possible infinite counterterms.

For the $A^2 + B^2$ insertion of equation (5.2), the spurion U has dimension 0, and so only the following type of counterterm is possible:

$$\ln \Lambda \int d^4\theta \, d^4 x \, U(\bar{\varphi}\varphi + \bar{X}X + \bar{Y}Y). \tag{5.7}$$

In other words, this term only gives rise to a self-infinity, in agreement with the one-loop result given previously.

The addition of other insertions which are of dimension 3 gives rise to more possible counterterms, and the reader is referred to reference [62] for a detailed discussion. The following is a schematic description of this analysis. Denoting the physical field component content of any of the chiral fields X_σ^σ, $Y_{a\sigma}$, φ by A, B, χ, and the spinor in the Yang-Mills multiplet by λ, the one-loop infinities are given in table 2. A check mark in this table indicates the appearance of an infinity. Consider adding a $\bar{\chi}\chi$ insertion; it gives rise to an infinity of the form $A(A^2 + B^2)$. The only way this infinity can be canceled is by adding an appropriate $A(A^2 + B^2)$ soft term and arranging its coefficient in such a way that the $A(A^2 + B^2)$ infinities cancel. Once this has been carried out, it is found that the $A^2 - B^2$ infinities cancel automatically. The remaining $A^2 + B^2$ infinities do not cancel, but they can be canceled by adding an appropriate $A^2 + B^2$ soft term. The resulting soft insertions that produce no infinities are of the form

$$m(A^2 + B^2) + m\bar{\chi}\chi + mA(A^2 + B^2). \tag{5.8}$$

Examination of the coefficients reveals that this term is none other than an $N = 1$ supersymmetric mass term and so can be rewritten in the form

$$m \int d^2\theta \, d^4x \, \mathrm{Tr}\, \varphi^2 \tag{5.9}$$

for the case of φ and similar terms for X_σ and Y_σ.

An alternative set of soft insertions that also induce no infinities is found by adding a mass term for the gaugino $\bar{\lambda}\lambda$. The resulting $A^3 - 3AB^2$ infinity can only be canceled by adding a term of the same form, i.e., $A^3 - 3AB^2$, with a choosen coefficient. Again the $A^2 - B^2$ infinity cancels automatically and the remaining $A^2 + B^2$ infinity can be canceled by adding an appropriate $A^2 + B^2$ term. The resulting combination of terms is of the generic form

$$m\bar{\lambda}\lambda + m(A^2 + B^2) + m(A^3 - 3AB^2). \tag{5.10}$$

Although this is not an $N = 1$ mass term, it is like an $N = 1$ mass term in the sense that it is related by $O(2)$ invariance to the mass term of equation (5.9), and so one would expect such a term to be present.

To summarize the previous discussion, we can obtain a finite theory if for every spinor mass term we add a unique term cubic in the spin zero fields and also add an $A^2 + B^2$ term such that one mass squared relation is satisfied for every species σ of hypermultiplet. For the case of $N = 4$ Yang-Mills we have only one type of hypermultiplet, and the one relation is of the form

$$S\,\mathrm{Tr}\, m^2 = \sum_j m_j^2 (-1)^{2j+1} = 0. \tag{5.11}$$

For $N = 2$ theories, however, the above equation need not hold.

The allowed types of terms that preserve finiteness can be expressed in terms of the following basis:

i. $N = 1$ supersymmetric masses,
ii. any $A^2 - B^2$ masses,
iii. the $N = 1$ like mass term of equation (5.10), and
iv. $A^2 + B^2$ masses that satisfy the relation of equation (5.4).

Having discovered the necessary and sufficient conditions such that the soft terms preserve finiteness at one loop, we now show that these conditions are sufficient to ensure that this finiteness persists to all orders. Consider first the addition of a term of $A^2 - B^2$ type. The analysis leading to equation (5.6) tells us that the theory has no L-dependent infinities at any order of perturbation theory including one loop. We now use an inductive argument of Weinberg [64]. Let us assume that all the graphs up to n loops are finite; then the $(n + 1)$-loop infinities arise either from subdivergences or in the overall superficial divergence of the $(n + 1)$-loop graph. The former type of infinity is absent by assumption, and the latter type of infinity is absent since there are no L-dependent infinities. Since there are no one-loop infinities, by induction there are no infinities to all orders.

The addition of $N = 1$ supersymmetric masses also preserves finiteness to all orders. A simple way to see this result is to observe that no m-dependent infinities can arise because of the $N = 1$ nonrenormalization argument. Finiteness to all orders follows by applying the same argument as for the $A^2 - B^2$ terms considered.

The $N = 1$ like masses of equation (5.10) are finite to all orders as a result of the fact that they are connected to the $N = 1$ mass for the field $\bar{\chi}\chi$ by an $O(2)$ rotation.

This leaves only the $A^2 + B^2$ masses which preserved one-loop finiteness if and only if

$$U_1 + U_{2\sigma} + U_{3\sigma} = 0 \qquad \text{for all } \sigma. \tag{5.12}$$

Adding the term of equation (5.2) to the theory leads to graphs with this insertion which can be found from the graphs of the original theory in the following way.

For every vertex with a vector line leaving in the original graphs there is now the possibility of adding a U insertion:

Similarly for every chiral propagator in the graphs of the original theory there is now the possibility of a U insertion, i.e.,

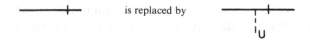

Use of the $N = 1$ super-Feynman rules shows that if the former type of replacement is made in a graph g which in the original theory had an infinity of the form $A(g)$, then it now has an infinity of the form $UA(g)$. The latter type of replacement results in a change in the value of the graph from $A(g)$ to $-UA(g)$. Consequently, given a graph g of the original theory with I_1, $I_{2\sigma}$, and $I_{3\sigma}$ propagators of the $\varphi\bar{\varphi}$, $\bar{X}_\sigma X_\sigma$, and $\bar{Y}_\sigma Y_\sigma$ types, respectively, and N_1, $N_{2\sigma}$, and $N_{3\sigma}$ vertices with a V line leaving φ, X_σ, and Y_σ lines, respectively, we find the effect of the replacements is to produce an infinity of the form

$$N(g) = \left\{ -U_1 I_1 - \sum_\sigma (U_{2\sigma} I_{2\sigma} + U_{3\sigma} I_{3\sigma}) \right.$$
$$\left. + U_1 N_1 + \sum_\sigma (U_{2\sigma} N_{2\sigma} + U_{3\sigma} N_{3\sigma}) \right\} A(g). \tag{5.13}$$

Use of equation (5.4) and relations of the type $I_1 + E_1 = \sum_\sigma V_\sigma + N_1$, where E_1 are the number of external φ lines and V_σ are the number of $X_\sigma Y_\sigma \varphi$ vertices, enables us to rewrite this equation in the form

$$N(g) = \left\{ U_1 E_1 + \sum_\sigma U_{2\sigma} E_{2\sigma} + U_{3\sigma} E_{3\sigma} \right\} A(g).$$

Summing over all graphs g with fixed numbers of external lines, we find that the infinity induced by the insertion is

$$\sum_g N(g) = \left\{ U_1 E_1 + \sum_\sigma (U_{2\sigma} E_{2\sigma} + U_{3\sigma} E_{3\sigma}) \right\} \sum_g A(g).$$

However, the finiteness of the original theory means that the $N = 1$ super-Feynman graphs conspire so that

$$\sum_g A(g) = 0.$$

Consequently the theory is also finite with the $A^2 + B^2$ insertion, provided equation (5.4) holds.

This rather schematic discussion has glossed over the possibility of having more than one insertion in a graph as well as the need to eliminate auxiliary fields in order to channel infinities. A detailed discussion of these points can be found in reference [62].

The analysis only covers the generic type of term that can be added and does not include terms that are allowed by the existence of particular numerically invariant tensors. Such a term is given by

$$\int d^4x \, d^2\theta \, S X_a^1 X_b^2 X_c^3 \varepsilon^{abc} + \text{h.c.,}$$

where $S = \theta^2 s$ and X_a^i belongs to the fundamental representation of $SU(3)$. In fact this term does also preserve the finiteness of the theory, provided we add certain $A^2 + B^2$ mass terms.

6 Realistic Finite Theories

It is interesting to examine what is the field content of those $N = 2$ supersymmetric theories that obey equation (3.21) and so are finite. For $SU(5)$ we must have $p + 3q + 7r = 10$, where p, q, and r are the number of hypermultiplets in the $5 + \bar{5}$, $10 + \overline{10}$, and $15 + \overline{15}$ representations, respectively. For $SO(10)$ one can have p and q hypermultiplets in the $10 + 10$ and $16 + \overline{16}$ representations, respectively, provided $p + 2q = 8$. The group E_6 will result in a finite theory if we use four hypermultiplets in the $27 + \overline{27}$ representation, while the group E_7 requires three hypermultiplets in the $56 + 56$ representation. We observed that for the grand unified groups $SO(10)$, E_6, E_7, and E_8, the theories that are finite contain the known fermions. In fact for these groups these fermions occur in three or four generations plus their minor generations.

As has been stressed, however, an $N = 2$ theory contains mirror fermions in the sense that given any right-handed spinor the theory must also possess a left-handed spinor in the same representation of the gauge group. This fact is an obvious consequence of the fact that the generators of the gauge group and supersymmetry commute and the counting of states discussed in section 2.

At this point it is useful to recall why one likes fermions to have the chiral fermionic structure given in the standard model. The reasons are as follows:

a. They are the only charged fermions observed with masses up to about 10–20 GeV.
b. They explain why the known fermions are almost massless in the sense that their masses are related to the low energy breaking of $SU(3) \times SU(2) \times U(1)$ to $SU(3) \times U(1)$.
c. The chiral anomalies for the known chiral fermions vanish. In the presence of mirror fermions the anomalies would cancel automatically and

on the face of it there would be no reason for the anomalies of the fermions with a small mass to cancel [65].

Whereas it is possible that there may be some other explanations for the almost masslessness of the observed fermions and the vanishing of their anomalies, it is essential, in order to find realistic $N = 2$ theories, to discover a mechanism that splits the known fermions from the mirror fermions in mass, giving the latter masses of order 10–20 GeV or larger. This mass splitting breaks $SU(2) \times U(1)$, and so it must be related to the scale of weak interactions. As such, the mirror fermions are unlikely to have masses much greater than 100 GeV. It would be interesting to examine how big these $SU(2) \times U(1)$ violating masses could be without poluting the predictions of the standard model. There are however, several other problems to be overcome in obtaining a realistic finite theory of $N = 2$ supersymmetry.
Apart from

i. making minor fermions sufficiently massive, one must also
ii. break the grand unified group down to $SU(3) \times SU(2) \times U(1)$ and then to $SU(3) \times U(1)$,
iii. give the observed fermions their observed masses, and
iv. break supersymmetry.

This list of problems applies not only to finite $N = 2$ theories, but to $N = 2$ supersymmetric theories in general. In realistic $N = 1$ supersymmetric models there is usually a hidden sector which breaks supersymmetry. Whether this hidden sector is connected with local supersymmetry supergravity or is only invariant under rigid supersymmetry, its net effect is to add soft supersymmetry breaking terms to a lower energy supersymmetric model. One may therefore simply add soft terms to an $N = 2$ theory which may or may not be finite and attempt to solve the problems listed. Should one be successful one may then inquire as to the origin of these terms. Bearing this in mind, we shall now discuss how one may resolve these problems.

Taking the last problem first, we note that a finite $N = 2$ supersymmetric theory cannot contain a $U(1)$ factor and so cannot break supersymmetry and the tree level. Hence, in any *perturbative* analysis, as for $N = 1$ supersymmetric theories, we shall not break supersymmetry unless it is broken at the tree level [66]. Fortunately, we can use the results of section 5 and add soft terms that still maintain finiteness. It is then easy to achieve (iv) as they break supersymmetry explicitly and give scalars masses which are greater than those of the observed fermions. If the theory is not finite, we can have a $U(1)$ part of the gauge group, and in this case it is possible, according to reference [67], to break supersymmetry. Little work has been done on this mechanism. If no $U(1)$ groups are present, supersymmetry breaking at the tree level is impossible and the previous discussion applies.

One can also use the soft terms to construct potentials that will spontaneously break the grand unified group. What is more difficult is to do this and keep the observed fermions massless until the final breaking to $SU(3) \times U(1)$. For example, because of the term $\varphi X_\sigma Y_\sigma$, there are problems in achieving this by using the field φ. An expectation value for φ will create the mass terms $\langle \varphi \rangle X_\sigma Y_\sigma$, and if the breaking is not from $SU(2) \times U(1)$ to $U(1)$, then this will give large equal masses to the mirror fermions and observed fermions in X and Y. If $\langle \varphi \rangle$ is the final breaking to $SU(3) \times U(1)$ and if the mirror fermions are all assigned to Y fields and the observed fermions in X, then we again find that both types of fermions will get the same mass.

One could attempt to cancel such masses by adding an mXY soft term. However, as $\mathrm{Tr}\,\varphi = 0$, it is impossible to cancel all such masses. Of course we could use the scalar fields in X and Y to break the gauge symmetry; however, this may be difficult to achieve in a finite theory where the number of X and Y is limited.

In the standard model and its grand unified extensions the observed fermion masses originate from replacing the Higgs field by its vacuum expectation value in Yukawa terms. The smallness of these masses compared with the weak scale is translated into the smallness of the Yukawa couplings, which vary in magnitude from 10^{-6} to about $10^{-2}\,g$. In $N = 2$ models, however, the only Yukawa terms in the Lagrangian, apart from terms involving gauginos, are contained in the term $m\varphi X_\sigma Y_\sigma$. Apart from the difficulties mentioned, we observe that these terms would give fermion masses independent of the generation and controlled in magnitude by the gauge coupling g and the $SU(2) \times U(1)$ vacuum breaking expectation value. One can, however, hope to generate fermion masses radiatively. In this case one must be sure to break any tree level symmetries that would usually forbid such terms by adding appropriate soft terms—for example, terms cubic in spin zero fields. Even then, one must worry about achieving the correct magnitude for these masses.

Before investigating prospects for solving all the above problems in a theory that is finite, i.e., an $N = 2$ theory with $C_2(G) = \sum_\sigma T(R^\sigma)$, and appropriate soft terms that preserve finiteness [62], it is worthwhile to attempt to solve these problems within the context of an $N = 2$ theory plus soft terms regardless of finiteness. However, before discussing this work [68], we note that even in a finite theory it is still possible to use the renormalization group equations to study the evolution of the coupling of the theory. Suppose we start with a finite theory which has some of its symmetries broken at a high energy. From the low energy viewpoint the theory, due to the effective integration out of the massive modes, will not appear finite, and so we can study the evolution of the couplings of the low energy theory in the usual way. Toy examples of this mechanism are easy to find simply by adding soft terms with large masses. It would be interesting to study the properties of such evolutions in these very special theories.

Let us begin with an $N = 2$ supersymmetric theory with gauge group $SU(3) \times SU(2) \times U(1)$ [68]. This theory, apart from the usual gauge bosons and their gauginos, will contain a chiral superfield in the adjoint representation. The usual particles $(e, v)_L, e_R, (u, d)_L$, etc., and two Higgs fields H and G may be assigned, without loss of generality, to the chiral superfield X. The mirrors of all these particles are contained in the Y chiral superfields. It is not difficult to find soft terms that give vacuum expectation values to H and G and break the gauge group to $SU(3) \times U(1)$. We can also use soft terms that break supersymmetry and give the scalar particles substantial masses. It is much more difficult to give the fermions appropriate masses. Clearly, they cannot come from the $\varphi X Y$ term, and the vacuum expectation value of φ must be rather small if not zero. They must arise from radiative corrections; however, a mirror mass term $YY\langle H \rangle$ is forbidden from occurring by the symmetry $Y \rightarrow -Y, X \rightarrow -X$. This can only be broken by soft terms of the form $[Y^3]_A$, which do give rise to radiative graphs that generate $YY\langle H \rangle$ terms. Unfortunately, the addition of the soft term $[Y^3]_A$ leads to the potential becoming unbounded below [68].

A much better model is obtained if one starts with the gauge group $SU(3) \times SU(4) \times U(1)$. By adding soft terms, one can find a potential that gives the φ field an off diagonal vacuum expectation value of the form

$$\langle \varphi \rangle = \left(\begin{array}{c|c} 0 & \\ \hline 0 & 0 \end{array} \right),$$

breaking the symmetry to $SU(3) \times U(1)$. The enormous advantage of this strategy is that it enables us to use the $g\varphi X Y$ term to give the mirror fermions a tree level mass. The $SU(2)_L \times SU(2)_R$ subgroup of $SU(4)$ is embedded in the manner

$$\left(\begin{array}{c|c} SU(2)_L & \\ \hline & SU(2)_R \end{array}\right),$$

and so we are allowed to assign fermions and their mirrors in a more subtle way than for the $SU(3) \times SU(2) \times U(1)$ model. For example, for the first generation quarks $q = \binom{u}{d}$ and their mirrors $\tilde{q} = \binom{\tilde{u}}{\tilde{d}}$ we let

$$X = \begin{pmatrix} q \\ \tilde{q} \end{pmatrix}_L \quad \text{and} \quad Y = \begin{pmatrix} \tilde{q} \\ q \end{pmatrix}_R$$

and similarly for the other generations and the electrons and neutrinos. Clearly, the mirror particles acquire tree level mass of order of the weak interaction scale. The mass of the observed fermions can be obtained from the tree level as well as from radiative diagrams that include soft terms. For a discussion of this work see reference [68].

References

[1]   G. Parisi, *Phys. Lett.* B76, 65 (1978); P. Olesen, *Phys. Lett.* B73, 327 (1977).

[2]   E. Gildener and S. Weinberg, *Phys. Rev.* D13, 3333 (1976); M. Veltman, University of Michigan preprint; L. Maiani, *Proc. of the Ecole d'Ete de Physique des Particules, Gif-sur-Yvette, 1979*, p. 3.

[3]   G. 't Hooft, *Phys. Lett.* B109, 474 (1982).

[4]   E. Witten, *Nucl. Phys.* B186, 513 (1981); S. Dimopoulos and H. Georgi, *Nucl. Phys.* B193, 150 (1981); N. Sakai, *Zeit für Physik* C11, 153 (1982).

[5]   Y. A. Gelfand and E. S. Likhtman, *J.E.T.P. Lett.* 13, 323 (1971); D. V. Volkov and V. P. Akulov. *Pis'ma Zh. Eksp. Teor. Fiz.* 16, 621 (1972), *Phys. Lett.* B46, 109 (1973); J. Wess and B. Zumino, *Nucl. Phys.* B70, 139 (1974).

[6]   J. Wess and B. Zumino, *Phys. Lett.* B49, 52 (1974).

[7]   J. Wess and B. Zumino, *Phys. Lett.* B49, 52 (1974); J. Iliopoulos and B. Zumino, *Nucl. Phys.* B76, 310 (1974); S. Ferrara, J. Iliopoulos, and B. Zumino, *Nucl. Phys.* B77, 41 (1974).

[8]   S. Ferrara and B. Zumino, *Nucl. Phys.* B79, 413 (1974).

[9]   D. R. T. Jones, *Phys. Lett.* B72, 199 (1977); E. Poggio and H. Pendleton, *Phys. Lett.* B72, 200 (1977).

[10]   O. Tarasov, A. Vladimirov, and A. Yu, *Phys. Lett.* B93, 429 (1980); M. T. Grisaru, M. Rocek, and W. Siegel, *Phys. Rev. Lett.* 45, 1063, (1980); W. E. Caswell and D. Zanon, *Nucl. Phys.* B182, 125 (1981).

[11]   M. Sohnius and P. West, *Phys. Lett.* B100, 45 (1981).

[12]   S. Ferrara and B. Zumino (unpublished).

[13]   M. Grisaru and W. Siegel, *Nucl. Phys.* B201, 292 (1982).

[14] S. Mandelstam, Proceedings of the XXI International Conference on High Energy Physics, Paris 1982.

[15] K. Stelle, Proceedings of the Paris High Energy Conference 1982. P. Howe, P. K. Townsend, and K. Stelle, Imperial College Preprint.

[16] P. Howe, K. Stelle, and P. West, *Phys. Lett.* B124, 55 (1983).

[17] S. Coleman and J. Mandula, *Phys. Rev.* 159, 1251 (1967).

[18] R. Haag, J. Lopuszanski, and M. Sohnius, *Nucl. Phys.* B88 61 (1975).

[19] E. P. Wigner, *Ann. Math.* 40, 149 (1939).

[20] A. Salam and J. Strathdee, *Nucl. Phys.* B80, 499 (1974). For a review of irreducible representations see D. Z. Freedman, in *Recent Developments in Gravitation*, M. Levy and S. Deser, editors (Cangesse, 1978); S. Ferrara *Supergravity '81*, S. Ferrara and J. Taylor, editors (Cambridge University Press, Cambridge, 1981); P. West, ibid.; P. West, *Introduction to Supersymmetry and Supergravity* (in preparation).

[21] J. Wess and B. Zumino, *Nucl. Phys.* B70, 39 (1974).

[22] S. Ferrara and B. Zumino, *Nucl. Phys.* B79, 413 (1974); A. Salam and J. Strathdee, *Phys. Lett* B15, 1353 (1974).

[23] P. Fayet, *Nucl. Phys.* B113, 135 (1976).

[24] A. Salam and J. Strathdee, *Phys. Lett.* B51, 33 (1974); P. Fayet, *Nucl. Phys.* B113 135 (1976); F. Gliozzi, J. Scherk, and D. Olive, *Nucl. Phys.* B122, 253 (1977); L. Brink, J. Schwarz, and J. Scherk, *Nucl. Phys.* B121, 77 (1977).

[25] S. Deser and B. Zumino, *Phys. Lett.* B62, 335 (1976).

[26] M. Sohnius, K. Stelle, and P. West, *Nucl. Phys.* B173, 127 (1980).

[27] M. Sohnius, *Nucl. Phys.* B165, 483 (1980).

[28] M. Sohnius, K. Stelle, and P. West, *Phys. Lett* B92, 123 (1980).

[29] P. Howe, K. Stelle, and P. Townsend, *Nucl. Phys.* B214, 519 (1983).

[30] S. Ferrara and B. Zumino, *Nucl. Phys.* B87, 207 (1975).

[31] For a discussion of the relation of currents and the superalgebras, see M. Sohnius and P. West (to appear).

[32] W. Siegel, *Phys. Lett.* B84, 193 (1979); L. V. Ardeev, *Phys. Lett.* B117, 317 (1982).

[33] A. A. Slavnov. *Teor. Mat. Fiz.* 13, 1064 (1972); B. W. Lee and J. Zinn-Justin, *Phys. Rev.* D5, 3121 (1972).

[34] S. Ferrara and O. Piquet, *Nucl. Phys.* B93, 261 (1975).

[35] D. J. Gross and F. Wilczek, *Phys. Rev.* D8, 3633 (1973).

[36] T. Curtright, *Phys. Lett.* B102, 17 (1981).

[37] M. Sohnius, *Phys. Lett.* B81, 8 (1979).

[38] P. Howe and P. West (in preparation).

[39] M. T. Grisaru. M. Rocek, and W. Siegel, *Nucl. Phys.* B159 42 (1979).

[40] G 't Hooft, in Proc. XII Winter School of Theoretical Physics in Karpacz; B. S. deWit, in *Quantum Gravity*, 2nd ed., C. J. Isham, R. Penrose, and D. W. Sciama, editors (Oxford University Press, Oxford); L. Abbot, *Nucl. Phys.* B185 189 (1981).

[41] P. Howe and P. West, *Nucl. Phys.* B242, 364 (1984).

[42]   See, for example, D. J. Gross, in *Methods in Field Theory*, R. Balian, editor (North-Holland).

[43]   J. Wess and B. Zumino, *Phys. Lett.* B74, 151 (1978) B79, 394 (1978).

[44]   S. Gates, K. Stelle, and P. West, *Nuclear Phys.* B169, 347 (1980).

[45]   R. Grimm, J. Wess, and B. Zumino, *Nucl. Phys.* B152, 255 (1979).

[46]   K. Stelle and P. West, *Phys. Lett.* B74, 330 (1978); S. Ferrara and P. van Nieuwenhuizen, *Phys. Lett.* B74, 333 (1978).

[47]   S. Deser, J. Kay, and K. Stelle, *Phys. Rev. Lett.* 38, 527 (1977); S. Ferrara and P. von Nieuwenhuizen, *Phys. Lett.* B78, 573 (1978).

[48]   P. Howe, *Nucl. Phys.* B199, 309 (1982), and references therein.

[49]   L. Brink and P. Howe, *Phys. Lett.* B88, 268 (1979).

[50]   R. E. Kallosh. *Phys. Lett.* B99, 122 (1981); P. Howe, K. Stelle, and P. Townsend, *Nucl. Phys.* B191, 445 (1981).

[51]   E. Cremmer, S. Ferrara, K. Stelle, and P. West, *Phys. Lett.* B94, 349 (1980).

[52]   P. Howe, H. Nicholai, and A. van Proeyen, *Phys. Lett.* B112, 446 (1982).

[53]   P. Howe and P. West (unpublished discussion).

[54]   E. Sokatchev, in *Superspace and Supergravity* S. W. Hawking and M. Rocek, editors (Cambridge University Press, Cambridge).

[55]   E. Fradkin and A. Tseytlin, *Phys. Lett.* B110, 117 (1981).

[56]   P. West, Proceeding of La Jolla conference held Jan. 1983 (published by A.P.S.)

[57]   A. Parkes and P. West, *Phys. Lett.* B122, 365 (1983).

[58]   A. Parkes and P. West, *Nucl. Phys.* B222, 269 (1983).

[59]   A. Namazie, A. Salam, and J. Strathdee, *Phys. Rev. Lett.* (to appear).

[60]   J. J. Van der Bij and Y. P. Yao, *Phys. Lett.* 125B (1983) 171.

[61]   S. Rajpoot, J. Taylor, M. Zaimi, *Phys. Lett.* B127 (1983).

[62]   A. Parkes and P. West, *Phys. Lett.* B127, 353 (1983).

[63]   L. Girrardello and M. Grisaru, *Nucl. Phys.* B194, 65 (1982).

[64]   S. Weinberg, *Phys. Rev.* D8, 3497 (1973).

[65]   E. Witten, his paper in this volume for a discussion of this point in relation to dimensional reduction.

[66]   P. West, *Nucl. Phys.* B106, 219 (1976); D. Capper and M. Ramon Medrano, *J. Phys.* G2, 269 (1976); E. Witten, *Nucl. Phys.* B188, 52 (1981).

[67]   P. Fayet, *Nucl. Phys.* B149, 137 (1979).

[68]   P. Aguila, B. Grinstein, L. Hall, G. Ross, and P. West (forthcoming *Nucl. Phys.* B).

[69]   R. Grimm, M. F. Sohnius, and J. Wess, *Nucl. Phys.* B133, 275 (1978).

[70]   P. Howe and U. Lindstrom, *Nucl. Phys.* B181, 487 (1981).

[71]   M. Grisaru and P. West, *Nucl. Phys.* B (to appear).

Einstein Gravitation as a Long-Wavelength Effective Field Theory

Stephen L. Adler

A possible resolution of the difficulties in quantizing general relativity is provided by the suggestion that Einstein gravitation is not a fundamental field theory, but rather is a long-wavelength effective field theory, arising as a scale-symmetry-breaking effect in a renormalizable fundamental theory. In unified theories of this type, Newton's constant will be calculable in terms of fundamental particle masses. The history and current status of these ideas is reviewed.

In the conventional picture of the fundamental forces of physics, as recently reviewed in Weinberg [7], gravitation appears on a quite different footing from the weak, strong, and electromagnetic interactions of the matter fields. The total dynamics, in the usual formulation, is governed by an action functional

$$S = \int d^4x \sqrt{-g} (\mathscr{L}_{\text{matter}} + \mathscr{L}_{\text{gravitation}}). \tag{1a}$$

Here $\mathscr{L}_{\text{matter}}$ is a renormalizable Lagrangian density, containing only dimensionless coupling constants, and $\mathscr{L}_{\text{gravitation}}$ is the Einstein-Hilbert gravitational Lagrangian

$$\mathscr{L}_{\text{gravitation}} = \frac{1}{16\pi G} R, \tag{1b}$$

with R the scalar curvature. Since the coupling constant $(16\pi G)^{-1}$ appearing in the gravitational action has the dimensionality $(\text{mass})^2$, quantization of the gravitational part of (1) leads to a nonrenormalizable field theory. Furthermore, in the conventional view there is no mechanism for relating the gravitational mass scale set by $G^{-1/2}$ to the unification mass of the matter fields. Gravitation thus appears as a phenomenon quite outside the usual framework of theoretical ideas on which elementary particle theory is based.

This statement of the problem of "quantizing gravitation" assumes, however, that the Einstein-Hilbert action is the fundamental quantum action for gravitation. Since all gravitational experiments done to date involve very long wavelengths ($\lambda \gtrsim 10$ cm), there is in fact no experimental evidence for this assumption. Thus, before proceeding to study quantum gravity, we must first address the queston, Is the Einstein theory a fundamental theory, or is it a long-wavelength effective field theory?

A familiar example of a long-wavelength effective field theory is provided by the Fermi theory of weak interactions, as extended by investigations in particle physics over the last fifteen years. For energies well below

100 GeV (i.e., for wavelengths much longer than 10^{-16} cm), the weak interactions are described by the current-current effective action

$$S_{\text{eff}}[\{\text{fermions}\}] = \int d^4x \left(\mathscr{L}_{\text{eff}}^{\text{charged}} + \mathscr{L}_{\text{eff}}^{\text{neutral}} \right),$$

$$\mathscr{L}_{\text{eff}}^{\text{charged}} = \frac{G_F}{\sqrt{2}} (j_{\text{ch}}^{\lambda} + J_{\text{ch}}^{\lambda})(j_{\text{ch}\,\lambda}^{\dagger} + J_{\text{ch}\,\lambda}^{\dagger}),$$

$$j_{\text{ch}}^{\lambda} = \bar{e}\gamma^{\lambda}(1 - \gamma_5)\nu_e + \mu, \tau \text{ terms},$$

$$J_{\text{ch}}^{\lambda} = \bar{u}\gamma^{\lambda}(1 - \gamma_5)(d\cos\theta_C + s\sin\theta_C) + c, t, b \text{ terms},$$

$$\vdots$$

(2)

with G_F the dimensional constant

$$G_F \approx 10^{-5}/m_{\text{proton}}^2. \tag{3}$$

As expected for a theory with a dimensional coupling constant, the Fermi theory is nonrenormalizable, and repeated attempts to quantize the weak interactions starting from the Fermi theory as the fundamental quantum action have met with frustration. It is now known that the Fermi theory is only a long-wavelength effective theory for the weak interactions. The fundamental quantum theory for the weak (and electromagnetic) interactions is the renormalizable gauge theory of Glashow, Salam, and Weinberg (GSW), in which the weak interactions are mediated by the exchange of massive intermediate vector bosons, which obtain their masses from a symmetry-breaking mechanism involving Higgs scalar bosons. When all fermion energies are much lower than the intermediate boson masses, the Fermi theory is recovered from the GSW theory as a low-energy, long-wavelength effective theory for the fermions. In the language of functional integrals, the relation between the Fermi effective theory and the GSW fundamental theory is given by

$$e^{iS_{\text{eff}}[\{\text{fermions}\}]} = \int d\{\text{bosons}\} e^{iS_{\text{fundamental}}[\{\text{fermions}\},\{\text{bosons}\}]}, \tag{4}$$

and from (4) one readily finds an experimentally verified formula relating the Fermi constant to the parameters of the fundamental theory,

$$\frac{G_F}{\sqrt{2}} = \frac{e^2}{8 M_W^2 \sin^2\theta_W}, \tag{5}$$

where e = electric charge, M_W = charged intermediate boson mass, and $\theta_W = SU(2) - U(1)$ mixing angle.

Returning now to gravity, let us assume that the strategy which has worked so successfully for the weak interactions should also be applied to the problem of quantizing gravitation. Thus, we shall assume that the fundamental gravitational action is the renormalizable and classically scale-invariant action

$$S_{\text{fundamental}} = \int d^4x \sqrt{-g}(\alpha C_{\mu\nu\lambda\sigma} C^{\mu\nu\lambda\sigma} + \beta R^2), \tag{6}$$

with $C_{\mu\nu\lambda\sigma}$ the Weyl tensor (the traceless part of the Riemann curvature tensor $R_{\mu\nu\lambda\sigma}$). Quantum corrections break the scale symmetry of (6), and as a result induce an Einstein-Hilbert effective action in the low-energy, long-wavelength limit; this effective action governs observed gravitational phenomena (just as the Fermi effective theory describes low-energy β-decay physics) but is not the fundamental quantum field theory action. The "induced gravitation" approach just sketched has been actively studied over the last few years, as reviewed in Adler [3], and is the viewpoint adopted in the remainder of our discussion.

Continuing to develop the analogy between the Fermi and the Einstein-Hilbert Lagrangians, let us ask what characteristic mass appears in the coupling constant for each effective action. In the microscopic units where $\hbar = c = 1$, the action S is dimensionless, the integration measure d^4x has dimension $(\text{mass})^{-4}$, and thus the Lagrangian density $\mathcal{L}$ has dimension $(\text{mass})^4$. From this fact, and the dimensionalities of the fields appearing in the effective actions, we infer the dimensionalities of the coupling constants already quoted:

1. for Fermi theory,

$$S_{\text{eff}} = \int d^4x \underbrace{\frac{G_F}{\sqrt{2}} \underbrace{\bar{\psi} \ldots \psi \bar{\psi} \ldots \psi}_{\text{dimension (mass)}^6}}_{\text{dimension (mass)}^4}$$

$$\tag{7a}$$

$$\Rightarrow G_F \text{ has dimension } (\text{mass})^{-2}$$

(ψ = fermion field);

2. for Einstein-Hilbert theory,

$$S_{\text{eff}} = \int d^4x \sqrt{-g} \underbrace{\frac{1}{16\pi G} \underbrace{R}_{\text{dimension (mass)}^2}}_{\text{dimension (mass)}^4} \tag{7b}$$

$$\Rightarrow \frac{1}{16\pi G} \text{ has dimension (mass)}^2.$$

Comparing (7a) and (7b), we see that there is an important difference between the behavior of the coupling constant in the two cases. Since the Fermi coupling G_F has the dimensionality of mass to a *negative* power, the dominant contributions to G_F come from the *smallest* mass intermediate states with the relevant quantum numbers, which are the intermediate bosons with mass $M_W \approx 80$ GeV. In contrast, the inverse Newton's constant G^{-1} has the dimensionality of mass to a *positive* power, and hence the dominant contributions to G^{-1} will come from the *largest* characteristic mass scale appearing in physics. This is presumably the Planck mass $M_{\text{Planck}} \sim 10^{19}$ GeV, where the gravitational interactions of elementary particles become of comparable importance to their electroweak and strong interactions.

Just as G_F can be calculated in terms of the more fundamental parameters of the GWS gauge theory, in induced gravity theories one expects Newton's constant G to be calculable in terms of fundamental particle masses. To see how this can come about, we draw on the fact that when $\alpha = g^{-2} > 0$, $\beta = g'^{-2} > 0$, the action of (6) leads to an *asymptotically free* quantum theory, just as the current candidates for unified matter theories are also asymptotically free quantum theories. The term "asymptotically free" refers to the behavior of the coupling constant, which, when radiative corrections are included, is changed from a true constant to a running function of the dominant dimensional variable, assumed for simplicity in the following discussion to be an energy E. In other words, in an asymptotically free theory, quantum effects lead to the replacement

$$g^2 \to g^2_{\text{running}} = \frac{g^2(\mu)}{1 + bg^2(\mu)\log(E/\mu)}, \tag{8}$$

with μ an arbitrary reference mass and with $b > 0$ a constant characteristic of the theory. As the energy E approaches infinity, (8) implies

$$g^2_{\text{running}} \xrightarrow[E \to \infty]{} 0, \tag{9}$$

leading to the vanishing of forces, and hence to free field theory behavior, in the asymptotic high-energy limit. On the other hand, the physics described by (8) becomes strongly interacting at low energies, as is readily seen by rewriting (8) in the form

$$g^2_{\text{running}} = \frac{1}{b \log(E/\mathcal{M})}. \tag{10}$$

Here

$$\mathcal{M} = \mu e^{-1/bg^2(\mu)} \tag{11}$$

is a mass parameter, which can be shown to be μ independent, and which characterizes the theory in the sense that $E \sim \mathcal{M}$ defines the strong coupling regime. We see that as a result of including radiative corrections, a one-parameter family of classical theories characterized by their values of the dimensionless coupling g^2 has been replaced by a one-parameter family of quantum theories characterized by their values of the dimension (mass)^1 scale mass $\mathcal{M}$. As a result of this phenomenon, called *dimensional transmutation*, all dimensional physical parameters in an asymptotically free theory are calculable in terms of $\mathcal{M}$, much as all radiative effects in the familiar, nonasymptotically free, case of quantum electrodynamics are calculable in terms of the fine structure constant α.

Let us now suppose that the gravitational theory of (6) and the unified matter theories, both of which are asymptotically free, can be further unified into an asymptotically free theory with a single classical coupling constant g. Then as a result of dimensional transmutation, g is replaced as a parameter in the quantized theory by a mass parameter $\mathcal{M}$, which presumably should be identified with M_{Planck}. All particle masses and G^{-1} will then be calculable in terms of $\mathcal{M}$, or eliminating $\mathcal{M}$, the ratio

$$G^{-1}/(\text{any particle mass})^2 \tag{12}$$

will be calculable. This scenario naturally accommodates the fact that G^{-1} is of order M^2_{Planck}, but does not explain either why the cosmological constant is very small or why elementary particle masses are small, on the scale of the Planck mass. These unexplained features are a problem in *all* unified models to date, and presumably will eventually be explained by specific kinematical or dynamical features of the ultimate unifying theory.

Having sketched the principal qualitative features of induced gravity theories, let me now survey in a somewhat more technical way their history and current status.

1. The suggestion that the Einstein-Hilbert theory is an effective field theory was first made by Sakharov [6], who proposed that the Einstein-Hilbert action arises from the quantum fluctuations of quantized matter fields in a curved background manifold. (For a survey of this and related early work, see ter Haar, Chudnovsky, and Chudnovsky [5] and Adler [3].)

2. Subsequently, a number of authors (for full references, see Adler [3]) studied Higgs-type models in a curved background manifold, with the action

$$S = \int d^4x \sqrt{-g} [\tfrac{1}{2}\varepsilon\phi^2 R - V(\phi) + \cdots]. \tag{13a}$$

In (13), $V(\phi)$ is a symmetry-breaking double well potential with a global minimum at $\phi^2 = \bar{\phi}^2$, so that when quantized around the stable vacuum, (13a) yields an Einstein-Hilbert action, with the gravitational constant given by

$$\frac{1}{16\pi G} = \tfrac{1}{2}\varepsilon\bar{\phi}^2. \tag{13b}$$

In models of this type, the renormalized coupling ε is necessarily an independent parameter, and so G^{-1} is not calculable in terms of fundamental mass parameters.

3. Further progress stemmed from the observation [1] that in *scalar-free* theories with dynamical breaking of scale invariance, an Einstein-Hilbert action is induced with a calculable Newton's constant G^{-1}. Since a pure nonabelian gauge theory satisfies these calculability criteria, the simplest model for induced gravity is thus a nonabelian gauge theory quantized on a classical background manifold, for which the analog of (4) reads

$$e^{iS_{\text{eff}}[g_{\mu\nu}]} = \int d[A_\lambda] e^{iS_{\text{gauge theory}}[A_\lambda, g_{\mu\nu}]}. \tag{14}$$

4. In the background metric model, the coefficients of the various terms in the long-wavelength expansion of S_{eff},

$$S_{\text{eff}}[g_{\mu\nu}] = \int d^4x \sqrt{-g}\left[\frac{1}{16\pi G_{\text{ind}}}(R - 2\Lambda_{\text{ind}})\right.$$

$$\left. + \alpha C_{\mu\nu\lambda\sigma}C^{\mu\nu\lambda\sigma} + \beta R^2 + \cdots\right],$$

$$(15)$$

can be extracted by taking successive metric variations. Acting on the left- and right-hand sides of (14) with $g_{\mu\nu}\delta/\delta g_{\mu\nu}$ and specializing to flat space-time gives the usual formula for the induced cosmological constant,

$$-\frac{1}{2\pi}\frac{\Lambda_{\text{ind}}}{G_{\text{ind}}} = \langle T_\mu{}^\mu\rangle_0, \tag{16}$$

where $T_\mu{}^\mu$ is the trace of the gauge theory stress-energy tensor and where $\langle\ \rangle_0$ denotes the flat space-time vacuum expectation value. Acting with $(g_{\mu\nu}\delta/\delta g_{\mu\nu})^2$ gives a formula [2, 8] for the induced gravitational constant,

$$\frac{1}{16\pi G_{\text{ind}}} = \frac{-i}{96}\int d^4x\, x^2\langle\mathcal{T}(\tilde{T}(x)\tilde{T}(0))\rangle_0,$$

$$\tilde{T}(x) = T(x) - \langle T(x)\rangle_0, \tag{17}$$

and when higher derivative terms are retained, also a formula [9] for the coefficient of the induced R^2 term,

$$\beta = \frac{-i}{13{,}824}\int d^4x\,(x^2)^2\langle\mathcal{T}(\tilde{T}(x)\tilde{T}(0))\rangle_0. \tag{18}$$

For a general pure nonabelian gauge theory, the coefficients $\Lambda_{\text{ind}}/G_{\text{ind}}$, G_{ind}^{-1} and β given by (16)–(18) are all calculable in terms of the gauge theory scale mass $\mathcal{M}$.

5. To go beyond the background metric model, one must take into account the fact that in a realistic theory, gravity (i.e., the metric $g_{\mu\nu}$) is also quantized. To do this, we make the decomposition

$$g_{\mu\nu} = \bar{g}_{\mu\nu} + h_{\mu\nu}, \tag{19}$$

where $\bar{g}_{\mu\nu}$ is the average background metric and $h_{\mu\nu}$ is a quantum fluctuation. The functional integral formalism for background field quantization then implies that $\bar{g}_{\mu\nu}$ is self-consistently determined by a classical variational principle,

$$\delta S_{\text{eff}}[\{\bar{\phi}_{\text{matter}}\}, \bar{g}_{\mu\nu}] = 0, \tag{20}$$

with $\bar{\phi}_{\text{matter}}$ the average matter fields. In the long-wavelength limit, (20) yields the classical Einstein equations. The self-consistent structure of the calculation is reflected in the fact that the functional form of S_{eff} is itself determined by the quantum fluctuations $h_{\mu\nu}$ around the mean value $\bar{g}_{\mu\nu}$. Acting with $(\bar{g}_{\mu\nu}\delta/\delta\bar{g}_{\mu\nu})^2$ as described, one obtains a formal functional integral expression for the induced gravitational constant G_{ind}^{-1}, including quantum gravity effects [3].

6. Finally, let us return to the basic question: What is the fundamental gravitational action? In the absence of additional nonmetric degrees of freedom (which may well be present!), the renormalizable candidates are

$$\alpha C_{\mu\nu\lambda\sigma}C^{\mu\nu\lambda\sigma} + \beta R^2, \tag{21a}$$

which is scale invariant, and

$$\alpha C_{\mu\nu\lambda\sigma}C^{\mu\nu\lambda\sigma}, \tag{21b}$$

which is conformally invariant. If the Lagrangian of (21b) leads to a finite induced R^2 term [as in the background metric model calculation of (18)], then it is a viable candidate for the fundamental action, while if the induced R^2 term arising from (21b) is divergent, then an R^2 counter term is needed, as in (21a). The classic objection to the fourth-order Lagrangians of (21) is that, in a small fluctuation analysis, they have an energy spectrum which is unbounded from below. In this connection there is a very interesting new global result [4]: *For the Lagrangian given by* (21b), *as well as for that given by* (21a) *with* $\alpha\beta > 0$, *all exact classical solutions have zero energy.* Hence these Lagrangians may well lead in a natural way to satisfactory quantum field theories (without recourse to such unappealing devices as negative metric quantization and special integration contour prescriptions). Thus at this point, the whole subject of the quantization of fourth-order gravitational theories has been reopened, and is an exciting direction for future research in quantum gravitation.

Acknowledgment

This work was supported by the U.S. Department of Energy under grant DE-AC02-76ER02220.

References

[1]   S. L. Adler, *Phys. Rev. Lett.* 44, 1567–1569 (1980).

[2]   S. L. Adler, *Phys. Lett.* B95, 241–243 (1980).

[3]   S. L. Adler, *Rev. Mod. Phys.* 54, 729–766 (1982).

[4]   D. G. Boulware, G. T. Horowitz, and A. Strominger, *Phys. Rev. Lett.* 50, 1726–1729 (1983).

[5]   D. ter Haar, D. V. Chudnovsky, and G. V. Chudnovsky, editors, *A. D. Sakharov: Collected Scientific Works* (Marcel Dekker, New York and Basel, 1982).

[6]   A. D. Sakharov, *Dok. Akad. Nauk. SSSR* 177, 70–71 (1967). [*Sov. Phys. Dokl.* 12, 1040–1041 (1968).]

[7]   S. Weinberg, *Rev. Mod. Phys.* 52, 515–523. (1980).

[8]   A. Zee, *Phys. Rev.* D23, 858–866 (1981).

[9]   A. Zee, *Phys. Lett.* B109, 183–186 (1982).

The Inflationary Universe

Alan H. Guth

I Introduction

Over the past five years, I and a number of other people who have traditionally regarded themselves as particle theorists have begun to dabble in the early universe. Many of our colleagues think that we were motivated mainly by jealousy of Carl Sagan, but I hope to convince you that there is also some serious motivation, which stems directly from recent advances in particle physics itself.

In recent years, experimental particle physics has tended to confirm the notion that the standard $SU(3) \times SU(2) \times U(1)$ model is correct, and that it in fact accounts for essentially all the physics that we have seen. In hopes of discovering new physical laws, many particle theorists have turned to speculations on what happens beyond the standard model, and much of this speculation has centered on grand unified theories (GUTs) [1]. Grand unified theories make a very good prediction for $\sin^2 \theta_W$ (θ_W = Weinberg angle), giving the idea a certain amount of plausibility. But the most dramatic predictions of grand unified theories occur at the extraordinary energy scale of 10^{14} GeV. If we were to try to build a 10^{14}-GeV accelerator with present technology, we could in principle do it, more or less. It would be a linear accelerator with a length of about one light-year. Unfortunately, it seems quite unlikely that such an accelerator will be funded (at least during the Reagan administration). So, if we want to explore the dramatic, 10^{14}-GeV consequences of grand unified theories, we have to turn to the only laboratory to which we have any access at all that has ever reached these energies. And that turns out to be the universe itself, in its very infancy. According to standard cosmology, the universe had a temperature with $kT = 10^{14}$ GeV at about 10^{-35} sec after the big bang. So that is the real motivation for particle theoriests to study the very early universe.

However, I do not think that we are so single-minded that the early universe has become nothing more than a proving ground for grand unified theories. We were brought to the study of cosmology through developments in particle physics, but I think most of us have discovered (or rediscovered) that the origin of the universe is certainly one of the most fascinating questions in science. Fortunately for us, the study of this question has become inextricably linked with our own area of specialization.

What I mainly want to discuss is a scenario called the new inflationary universe. This is a scenario in which the universe supercools by many orders of magnitude below the critical temperature of a phase transition predicted by grand unified theories, and in the process it expands exponentially by many orders of magnitude [2]—hence the name "inflationary." The word "new" refers to a modification of my original proposal, which was suggested independently by Linde [3] and by Albrecht and Steinhardt [4]. They suggested a new mechanism by which the phase transition could take place, solving some crucial problems [2, 5, 6] that were created by my original proposal.

The inflationary model is very attractive because it can solve several very fundamental cosmological problems—which I shall discuss later. If correct, it would also mean that grand unified theory mechanisms are responsible for the production of essentially all the matter, energy, and entropy in the observed universe. So the questions raised are very exciting—let me now get on with some of the details [7].

Since I am a particle theorist, I shall set $\hbar = c = k = 1$, and I shall take the GeV as the fundamental unit for everything. Since I am not a general relativist, I shall not set Newton's constant G equal to one. Instead I shall set $G \equiv 1/M_P^2$, where $M_P = 1.22 \times 10^{19}$ GeV is the Planck mass. Note that 1 GeV $= 1.16 \times 10^{13}$ K $= 1.78 \times 10^{-24}$ g, and that 1 GeV$^{-1} = 1.97 \times 10^{-14}$ cm $= 6.58 \times 10^{-25}$ sec. I have been attempting to communicate with astronomers and I have learned that 1 megaparsec (Mpc) is their way of denoting 1.56×10^{38} GeV^{-1}.

II Grand Unified Theories

Since so much of this work depends on grand unified theories, I shall begin by summarizing the consequences of these theories that are most important to cosmology.

First, grand unified theories predict the existence of a phase transition (or possibly several phase transitions) with a critical temperature $T_c \approx 10^{14}$ GeV. To understand this phenomenon, note that the spontaneous symmetry breaking is accomplished by a scalar Higgs field Φ, which acquires a nonzero vacuum expectation value. The phase transition occurs because this nonzero expectation value is destroyed by thermal fluctuations at high temperatures [9], just as thermal fluctuations destroy the expectation value of the magnetic spin of a ferromagnet at temperatures above the Curie point.

The second important consequence is the existence of magnetic monopoles in the particle spectrum. Any grand unified theory in which a simple group is broken eventually to $SU(3) \times U(1)$ (i.e., QCD and electro-magnetism) will necessarily [10] contain magnetic monopoles of the 't Hooft-Polyakov type [11]. These monopoles are topologically stable knots in the Higgs field expectation value. (In this context one should recall that the Higgs field Φ is generally a multicomponent field, so the topology can be complicated.) The monopole masses are generically of order M_X/α, where M_X is the mass of the superheavy vector bosons and α is the fine structure constant of the grand unified gauge interactions. In the minimal $SU(5)$ theory proposed by Georgi and Glashow [12], $M_X \approx 4 \times 10^{14}$ GeV and $\alpha \approx 1/45$, so the monopole mass is of order 10^{16} GeV.

The third consequence is the nonconservation of baryon number. This means that the proton has a finite lifetime, estimated in the simplest models to lie in the range of 10^{30}–10^{33} years. It also means that the net baryon number of the universe need not be taken as an initial condition, but can instead be generated dynamically [13]. Possibly even more important, it means that there may be no conserved quantity that distinguishes the observed universe from the vacuum. (This last point will be discussed in section V.)

III The Standard Scenario

Now let me discuss what may be called the standard scenario of the very early universe [14]. The universe is assumed at the outset to be homogeneous and isotropic, and hence can be described in comoving coordinates by a Robertson-Walker metric. Whether it is open, closed, or flat, the curvature is negligible at very early times. Thus, the metric takes the simple form

$$ds^2 = -dt^2 + R^2(t)\,d\mathbf{x}^2. \tag{3.1}$$

The expansion of the universe is described by the scale factor $R(t)$, which is governed by the Einstein field equation

$$\left(\frac{\dot{R}}{R}\right)^2 = \frac{8\pi}{3}G\rho, \tag{3.2}$$

where ρ is the energy density and the dot denotes differentiation with respect to the time t. This equation would have an additional term if the curvature were not negligible. The quantity $H \equiv \dot{R}/R$ is the Hubble "constant" (which of course varies as the universe evolves). The mass density is dominated by the thermal radiation of effectively massless (i.e., $M \ll T$) particles at temperature T:

$$\rho = cT^4. \tag{3.3}$$

The constant c depends on the number of effectively massless particle species; for the minimal $SU(5)$ GUT at the highest temperatures, c is about 50. The expansion is taken to be adiabatic, which implies (as long as the number of effectively massless particle species remains unchanged)

$$RT = \text{constant}. \tag{3.4}$$

Putting these relations together, one finds

$$T^2 = \left(\frac{3}{32\pi c}\right)^{1/2} \frac{M_P}{t} \approx \frac{M_P}{40t} \tag{3.5}$$

and

$$R \propto \sqrt{t}. \tag{3.6}$$

To develop some feeling for the relevant numbers, consider $T = 10^{14}$ GeV $\approx T_c$. The equations then give $t \approx 1.9 \times 10^{-35}$ sec and $\rho \approx 1.2 \times 10^{75}$ g cm^{-3}. (For comparison, recall that in these units the density of an atomic nucleus is about 10^{15}.) Objects that are separated by 10^{10} light-years today were separated by only about 6 cm at this early time.

Finally, there must be a phase transition (or possibly several) when the temperature falls to T_c. In the standard scenario it is assumed that this phase transition occurs quickly, without any significant supercooling. The effects on the evolution of $R(t)$ are then negligible.

IV Problems of the Standard Scenario

The standard scenario suffers from four problems, which I shall now describe. It is the existence of these problems that motivates the inflationary scenario.

The first is the monopole problem—that is, the standard scenario leads to a tremendous overproduction [15–17] of magnetic monopoles. To see how this comes about, recall that the monopoles are in fact topologically stable knots in the Higgs field expectation value. If the Higgs field has a correlation length ξ, then one would expect a density of monopoles given roughly by the Kibble relation [19]

$$n_M \approx 1/\xi^3. \tag{4.1}$$

When the universe cools below the critical temperature $T_c \approx 10^{14}$ GeV of the GUT phase transition, it becomes thermodynamically favorable for the Higgs field to align uniformly over large distances. However, it takes time for these correlations to be established. Note that the horizon length, defined as the distance that a light pulse could have traveled since the initial singularity, is given in the radiation-dominated era by $l_H = 2t$, where t is the age of the universe. Thus, causality alone implies [20] that

$$\xi \lesssim 2t. \tag{4.2}$$

Equations (4.1) and (4.2) can be used to obtain an approximate lower bound on n_M immediately after the phase transition. One can also calculate the entropy density s, and one finds

$$n_M/s \gtrsim 10^{-13}. \tag{4.3}$$

At these densities the rate of monopole-antimonopole annihilation can be shown to be insignificant [15, 16, 21]. Entropy is essentially conserved in this scenario, so the ratio n_M/s should be about the same today. One then finds that today one would have

$$\Omega \equiv \rho/\rho_c \gtrsim 3 \times 10^{11}, \tag{4.4}$$

where $\rho_c = 3H^2/8\pi G$ is the critical mass density [which gives a precisely flat $(k = 0)$ universe]. Such a huge value of Ω is clearly impossible. It would imply, for example, that the current age of the universe would be $\lesssim 30,000$ years. (I am told that an age of this order is considered acceptable in some circles—but I have checked with my friends, and they are all confident that such an age can be ruled out.) Thus, some mechanism must be found to suppress the production of magnetic monopoles.

The second problem of the standard scenario is known as the horizon problem; it was first pointed out by Rindler [22] in 1956. The observational basis for this problem is the uniformity of the cosmic background radiation, which is known to be isotropic in temperature to at least one part in 10^4. This fact is particularly difficult to understand when one considers the existence of the horizon length, the maximum distance that light could have traveled since the beginning of time. Consider two microwave antennas pointed in opposite directions. Each is receiving radiation that is believed to have been emitted (or "decoupled") when the hot plasma of the early universe converted to neutral atoms—when T was about 4,000 K at $t \approx 10^5$ years. At the time of emission, these two sources were separated from each other by over 90 horizon lengths [23]. The problem is to understand how two regions over 90 horizon lengths apart came to be at the same temperature at the same time. Within the standard scenario this large scale homogeneity cannot be explained; rather, it must be assumed as an initial condition.

The third problem of the standard scenario was pointed out by Dicke and Peebles [24] in 1979, and is known as the flatness problem. The basis of the problem is the fact that today the ratio between the actual mass density and the critical mass density (i.e., $\Omega \equiv \rho/\rho_c$) is conservatively known to lie in the range

$$0.01 < \Omega < 10. \tag{4.5}$$

No one is surprised by how narrowly this range brackets $\Omega = 1$. However, within the evolution of the standard model, the value $\Omega = 1$ is an *unstable* equilibrium point. Thus, to be near $\Omega = 1$ today, the universe must have been very near to $\Omega = 1$ in the past. When T was 1 MeV (at $t \approx 1$ sec), Ω had to equal one to an accuracy of one part in 10^{15}. At the time of the GUT phase transition, when T was about 10^{14} GeV, Ω had to be equal to one to within one part in 10^{49}! In the standard model this precise fine tuning must be assumed (without explanation) to be a property of the initial conditions.

The fourth problem is the difficulty of understanding the formation of structure in the universe. While the universe appears homogeneous when one averages over lengths of a few times 10^8 light-years, on smaller scales there is a complicated clustering of matter into galaxies, clusters of galaxies, superclusters of clusters, etc. In order to account for the evolution of this structure, the standard scenario relies on the assumption of an ad hoc spectrum of inhomogeneities in the initial conditions. The fact that this spectrum is unexplained is a drawback in itself, but the situation becomes even more puzzling when the starting time is taken as early as $t \approx 10^{-35}$ sec. Gravitational instabilities [25] cause the inhomogeneities $\delta\rho/\rho$ to grow linearly with t, as long as the wavelength of the inhomogeneity exceeds the Hubble length $H^{-1} = 2t$. Thus, an early starting time requires a peculiarly small initial inhomogeneity. For example, a galactic scale inhomogeneity has a wavelength that implies linear growth until $t \approx 10^9$ sec; at that time one requires $\delta\rho/\rho \approx 10^{-4}$ to account for subsequent galactic evolution.

Thus, at $t = 10^{-35}$ sec one needs $\delta\rho/\rho \approx 10^{-48}$. For comparison, one might consider an ordinary hot gas in thermal equilibrium. Such a gas would have $\delta\rho/\rho \approx N^{-1/2} \approx 10^{-39}$, where $N \approx 10^{78}$ is the number of particles (mostly radiation of effectively massless particles) that make up the protogalaxy. Thus, the standard scenario requires the assumption that the matter in the universe began in a peculiar state of extraordinary but not quite perfect uniformity. The necessity for this assumption is called the smoothness problem, and it was pointed out by Dicke and Peebles in the same paper [24] as the flatness problem.

V The New Inflationary Universe

I shall now describe the inflationary universe, beginning with a brief historical overview. The purpose of the overview is mainly to explain the distinction between the original proposal and the new inflationary universe. I shall then discuss only the new inflationary universe.

The key feature of the original inflationary universe scenario [2] was the assumption that the GUT phase transition did not occur quickly, but instead occurred only after a period of extreme supercooling. The peculiar equation of state of the supercooled matter causes the universe to expand exponentially. Eventually, the phase transition must take place. It was assumed that this phase transition would occur by the random nucleation [26] of bubbles of the new phase, and that these bubbles would grow and coalesce to complete the phase transition. If the overall expansion factor exceeded 10^{28}, and if the process of bubble coalescence was effective, then the monopole, horizon, and flatness problems would have been solved. (The smoothness problem had not yet been considered.)

However, the original proposal contained a crucial flaw, which came to be known as the graceful exit problem. The problem was that there was no smooth way to terminate the period of exponential expansion. The randomness of the bubble formation process would produce gross inhomogeneities [2, 5, 6] that are totally unacceptable. The inflationary universe remained tantalizing, but it was obviously in error.

I was of course delighted at Christmas of 1981, when I received Linde's preprint [3] announcing the new inflationary universe. I soon learned that Albrecht and Steinhardt [4] had independently discovered the same scenario. These physicists pointed out that if the parameters of the Higgs field potential were chosen to satisfy the Coleman-Weinberg condition [27], then the phase transition would have a behavior very different from the usual case. In this scenario a single bubble would undergo inflation, and it would grow large enough to encompass easily the entire observed universe. The question of bubble coalescence becomes academic, since our observations probe only a small part of one bubble. The graceful exit problem is solved, and all of the successes of the original model are preserved.

The scenario requires the fine tuning of the mass term in the Higgs field potential to an accuracy of about one part in 10^{11}, which of course is very contrary to the spirit in which the scenario is motivated. The consequences of the scenario are so successful, however, that one is encouraged to go on. One hopes that realistic theories can be found in which a phase transition of this sort occurs without any fine tuning of parameters.

I shall now describe the new inflationary universe scenario in some detail. The scenario is cast in the context of the $SU(5)$ GUT. In this theory the $SU(5)$ gauge symmetry is broken to $SU(3) \times SU(2) \times U(1)$ by means of an adjoint representation Higgs field, which I shall denote Φ. More explicitly, Φ denotes a traceless hermitian 5×5 matrix of scalar fields. Configurations of Φ that break the symmetry to the desired subgroup can be written as

$$\Phi = \left(\frac{2}{15}\right)^{1/2} \phi \, \text{diag}[1, 1, 1, -3/2, -3/2]. \tag{5.1}$$

The factor of $(2/15)^{1/2}$ is chosen so that ϕ is normalized in the standard way for a real scalar field. The parameters of the Higgs field potential are chosen so that the effective potential $V(\phi)$ obeys the Coleman-Weinberg condition:

$$\frac{\partial^2 V}{\partial \phi^2} = \frac{\partial^3 V}{\partial \phi^3} = 0 \qquad \text{at } \phi = 0. \tag{5.2}$$

By including the one-loop gauge field quantum correlations (calculated in the flat-space approximation), the potential takes the form

$$V(\phi) = \tfrac{25}{16}\alpha^2[\phi^4 \ln(\phi^2/\sigma^2) + \tfrac{1}{2}(\sigma^4 - \phi^4)], \tag{5.3}$$

where $\alpha \equiv g^2/4\pi \approx 1/45$ is the gauge coupling and $\sigma \approx 1.2 \times 10^{15}$ GeV. The form of this potential is shown in figure 1. The minimum lies at $\phi = \sigma$, corresponding to the true vacuum. The point $\phi = 0$ is an equilibrium point that is just barely unstable. I shall refer to the field configuration $\phi = 0$ as the false vacuum, even though this term is traditionally reserved for configurations that are classically stable.

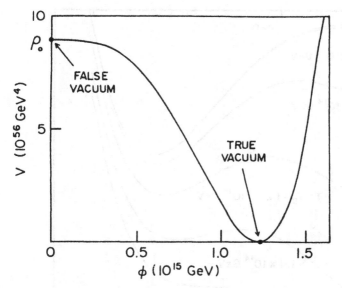

Figure 1
The Coleman-Weinberg potential for the $SU(5)$ Higgs field.

For nonzero values of the temperature, the finite temperature effective potential [28] is given by

$$V(\phi, T) = V(\phi) + \frac{18T}{\pi^2} \int_0^\infty q^2 \, dq \ln\left[1 - \exp\left(-\sqrt{q^2 + \frac{5\pi}{3}\alpha\phi^2/T} \right) \right]. \quad (5.4)$$

The behavior of $V(\phi, T)$ is shown in figure 2. For any temperature $T > 0$, the false vacuum is stabilized by a dip in the finite temperature effective potential. For small T this dip has a depth of order T^4 and a width of order T. There is a phase transition with critical temperature $T_c = 1.25 \times 10^{14}$ GeV, above which the $\phi = 0$ configuration is the global minimum of the potential (and is hence the thermal equilibrium state).

The starting point of a cosmological scenario is somewhat a matter of taste and philosophical prejudice. Some physicists find it plausible that the universe began in some highly symmetrical state, such as a de Sitter space. I prefer to believe that the universe began in a highly chaotic state; one advantage of the inflationary scenario, from my point of view, is that it appears to allow a wide variety of starting configurations. I require only that the initial universe is hot ($T > 10^{14}$ GeV) in at least some places, and

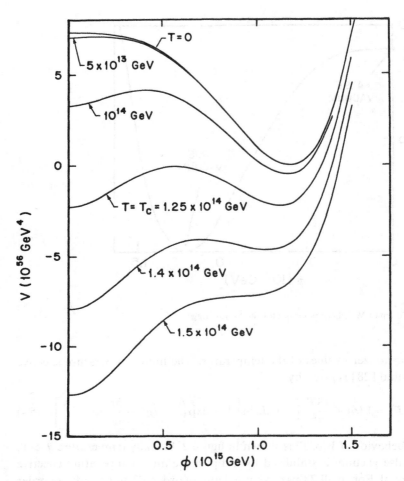

Figure 2
The Coleman-Weinberg finite temperature effective potential for the $SU(5)$ Higgs field.

that at least some of these regions are expanding rapidly enough so that they will cool to T_c before gravitational effects reverse the expansion.

In these hot regions, thermal equilibrium would imply $\langle \Phi \rangle = 0$, where $\langle \Phi \rangle$ denotes the expectation value of Φ. (Actually, though, the universe has not had time at this point to thermalize [29]. Thus, I need to assume that there are some regions of high energy density with $\langle \Phi \rangle \approx 0$, and that some of these regions lose energy with Φ being trapped in the false vacuum.) These regions will cool to T_c, and nucleation rate calculations [30] indicate that they will continue to supercool well below T_c. The energy density ρ will approach $\rho_0 \equiv V(\phi = 0)$. Since this false vacuum state is Lorentz invariant, the energy-momentum tensor must then have the form

$$T_{\mu\nu} = \rho_0 g_{\mu\nu}. \tag{5.5}$$

To the extent that (3.1) and (3.2) are applicable, the universe will approach a space that can be described by the metric (3.1) with

$$R(t) = e^{\chi t}, \tag{5.6}$$

where

$$\chi = \left(\frac{8\pi}{3} G\rho_0 \right)^{1/2}. \tag{5.7}$$

(For our parameters, $\chi \approx 10^{10}$ GeV.) Such a space is called a de Sitter space [31]. Recall, however, that the metric (3.1) embodies the assumption that the universe began in a state that is perfectly homogeneous and isotropic—precisely the assumption that I am trying to avoid. Thus, I must consider perturbations about the metric (3.1). The behavior of these perturbations seems to be governed by a "cosmological no-hair theorem," which states that whenever the energy-momentum tensor is given by (5.5), then any locally measurable perturbation about the de Sitter metric is damped exponentially on the time scale of χ^{-1}. The theorem has been demonstrated [32] in the context of linearized perturbation theory, and it is conjectured to hold even for large perturbations [33, 34]. Thus, a smooth de Sitter metric arises naturally, without any need to fine tune the initial conditions. (Advocates of an initial de Sitter space are welcome to join the scenario at this point.)

As the space continues to supercool and exponentially expand, the energy density is fixed at ρ_0. Thus, the total energy (i.e., all energy other than gravitational) is increasing! If the inflationary model is right, this false vacuum energy is the source of essentially all the matter, energy, and entropy in the observed universe.

This seems to violate our naive notions of energy conservation, but we must remember that the gravitational field can exchange energy with the matter fields. The energy-momentum tensor for matter obeys a covariant conservation equation, which in the Robertson-Walker metric reduces to

$$\frac{d}{dt}(R^3 \rho) = -p \frac{d}{dt}(R^3). \tag{5.8}$$

From (5.5) it follows that the false vacuum has a large and negative pressure, $p = -\rho_0$; (5.8) is then satisfied identically, with the energy of the expanding gas increasing due to the negative pressure. If the space were asymptotically Minkowskian, it would be possible to define a conserved total energy (matter plus gravitational). However, in the Robertson-Walker metric, a global conservation law of this type does not exist, except perhaps in a trivial version in which the total energy vanishes identically.

Let me digress a moment to discuss the quantum numbers of the observed universe. Some approximate values are given as follows: baryon number $\approx 10^{78-79}$, matter energy $\approx 10^{78-80}$ GeV, electric charge ≈ 0, and angular momentum ≈ 0. I have just explained that matter energy is not conserved, but can be increased dramatically by inflation. If baryon number is also nonconserved (as in GUTs), then the universe is, so far as we know, devoid of any conserved quantities [35]. In that case, it is very tempting to believe that the universe began from nothing [36] or from almost nothing. The inflationary model illustrates the latter possibility. Using the inflationary mechanism, it is possible to create the entire observed universe starting with a total matter energy of about 10 kilograms [37].

Now let me return to my region of space that is supercooling into a de Sitter phase. It is a striking fact about de Sitter space that quantum fluctuations mimic thermal effects at a Gibbons-Hawking temperature [33]

$$T_{GH} = \chi/2\pi. \tag{5.9}$$

In our case, $T_{GH} \approx 10^9$ GeV. Calculations on the decay of the supercooling phase were first carried out in Minkowski space, omitting the effects of gravitation (and the Gibbons-Hawking temperature). These calculations [3, 4, 30] indicate that the supercooling continues until $T \approx 10^{7-8}$ GeV, at which point the calculations break down. [They break down when the energy of a critical size bubble becomes comparable to T, resulting in a failure of the steepest descent approximation. The situation becomes even more difficult to analyze when T falls to $\sim 10^6$ GeV, at which point the $SU(5)$ gauge coupling becomes strong.] Since gravitational effects prevent the temperature from ever falling below T_{GH} [34], these Minkowski space calculations are illustrative but inconclusive.

The gravitational effects have now been studied by a number of authors [34, 38, 39]. Within the supercooling region, the Higgs field undergoes fluctuations due to thermal and/or quantum effects. Some fluctuations begin to grow, and at some point these fluctuations become large enough so that their subsequent evolution can be described by the classical equations of motion. I shall use the term "coherence region" to denote a region containing an approximately uniform Higgs field. The coherence regions are irregular in shape [4, 39], and their initial size is typically order χ^{-1}. Note that χ^{-1} is only about 10^{-11} proton diameters; the entire observed universe will evolve from a region of this size or smaller.

The Higgs field ϕ then "rolls" down the potential of figure 1, obeying the classical equations of motion:

$$\ddot{\phi} + 3\frac{\dot{R}}{R}\dot{\phi} = -\frac{\partial V}{\partial \phi}. \tag{5.10}$$

If the initial fluctuation is small, then the flatness of the potential for $\phi \approx 0$ will imply that the rolling begins very slowly. Note that the second term on the left-hand side of (5.10) is a damping term, helping further to slow down the speed of rolling. As long as $\phi \approx 0$, the energy density ρ remains about equal to ρ_0, and the exponential expansion continues. The exponential expansion occurs on a time scale χ^{-1} that is short compared with the time scale of the rolling.

For the scenario to work, it is necessary for the length scale of homogeneity to be stretched from χ^{-1} to at least about 10 cm before the Higgs field ϕ rolls off the plateau in figure 1. This corresponds to an expansion factor of about 10^{25}, which requires about 58 time constants (χ^{-1}) of expansion. It can be shown [40] that such an expansion would result if the initial fluctuation in ϕ were less than about $1.4 T_{GH}$(with no initial fluctuation $\dot{\phi}$).

When the ϕ field reaches the steep part of the potential, it falls quickly to the bottom and oscillates about the minimum. The time scale of this motion is a typical GUT time of $(10^{14} \text{ GeV})^{-1}$, which is very fast compared with the expansion rate. The Higgs field oscillations are then quickly damped by the couplings to the other fields, and the energy is rapidly thermalized [41].

(The Higgs field oscillations correspond to a coherent state of Higgs particles; the damping is simply the decay into other species.) The release of this energy (which is just the latent heat of the phase transition) raises the temperature to the order of 10^{14} GeV. (More precisely, the reheating temperature is about one-sixth of the superheavy vector boson mass, or about 7×10^{13} GeV.)

From here on the standard scenario ensues, including the production of a net baryon number. The length scale of homogeneity increases to $\gtrsim 10^{10}$ light-years by the time T falls to 2.7K.

VI Solutions to the Cosmological Problems

Let me now explain how the four problems of the standard cosmological scenario discussed in section IV are avoided in the inflationary scenario.

First, let us consider the monopole problem. It is easy to see that the Kibble production mechanism, which produced the tremendous excess of monopoles in the standard scenario, is totally ineffective here. The Higgs field is correlated throughout the initial coherence region, and the exponential expansion increases the Higgs correlation length ξ so that it is greater than 10^{10} light-years today. Thus, this mechanism would produce $\lesssim 1$ monopole in the observable universe. Some monopoles would still be produced by thermal fluctuations after reheating, but this number would be negligible in the $SU(5)$ model [18, 42] and presumably in most other models as well.

The horizon problem is clearly avoided in this scenario, since the entire observed universe evolves from a single coherence region. This region had a size of order χ^{-1} at the time when the fluctuation began to grow classically. The region was causally connected, and the Higgs field is expected to be homogeneous on this length scale. The exponential expansion causes this very small region of homogeneity to grow to be large enough to encompass the observed universe.

The flatness problem is avoided by the dynamics of the exponential expansion of the coherence region. As ϕ begins to roll very slowly down the potential, the evolution of the metric is governed by the energy density ρ_0. On the assumption that the coherence region (or a small piece of it) can be approximated locally by a Robertson-Walker metric, the scale factor evolves according to the standard equation

$$\left(\frac{\dot{R}}{R}\right)^2 = \frac{8\pi}{3} G\rho_0 - \frac{k}{R^2},$$

(6.1)

where $k = +1$, -1, or 0 depending on whether the region approximates a closed, open, or flat universe, respectively. [There could also be perturbations, but these would die out quickly, as discussed following (5.7).] In this language, the flatness problem is the problem of understanding why the second term on the right-hand side is no extraordinarily small. But as the coherence region expands exponentially, the energy density ρ remains very nearly constant at ρ_0, while the k/R^2 term falls off as the square of the exponential factor. Thus the k/R^2 term is suppressed by at least a factor of 10^{50}, which provides a "natural" explanation of why its value immediately after the phase transition is less than 10^{-49} times the value of the other terms.

Except for a very narrow range of parameters, this suppression of the curvature term will vastly exceed that required by present observations. This leads to the prediction that the value of Ω today is expected to be equal to one with a high degree of accuracy.

Finally, I come to the smoothness problem. Although this problem is only partially solved, I still consider it a major success of the new inflationary scenario. The problem was not even considered when the scenario was formulated, so it is impressive that the scenario offers a possible solution.

The evolution of density fluctuations in the new inflationary universe was discussed actively by a number of physicists [43] at the Nuffield Workshop on the Very Early Universe, held in the summer of 1982 at Cambridge University. As already discussed, the inflationary process smoothes out any inhomogeneities that may have been present in the initial conditions of the universe, leaving only zero-point quantum fluctuations. During the phase transition, however, quantum fluctuations cause the Higgs field to roll off the plateau of figure 1 at slightly different times at different points in space. This lack of synchronization gives rise eventually to density fluctuations. The calculation of the density fluctuations from the lack of synchronization is an exercise in linearized general relativity. The lack of synchronization can be characterized by a time offset $\delta t(\mathbf{x})$, which is then Fourier transformed with respect to comoving coordinates to give $\delta \tilde{t}(\mathbf{k})$. The density fluctuations are characterized by $\delta \rho(\mathbf{x}, t)$, which is Fourier transformed to give $\delta \tilde{\rho}(\mathbf{k}, t)$. The ratio $\delta \tilde{\rho}(\mathbf{k}, t)/\rho(t)$ grows linearly with t until the Hubble length $H^{-1} = 2t$ grows to be as large as the wavelength corresponding to $\mathbf{k}$.

The ratio then begins to oscillate, with an amplitude that I shall call $(\delta\tilde{\rho}/\rho)_f$. It can then be shown that

$$\left(\frac{\delta\tilde{\rho}(\mathbf{k})}{\rho}\right)_f = 4\chi\,\delta\tilde{t}(\mathbf{k}). \tag{6.2}$$

When quantum physics is used to estimate $\delta\tilde{t}(\mathbf{k})$, it is found that $(\delta\tilde{\rho}/\rho)_f$ is approximately independent of $\mathbf{k}$. This scale-invariant spectrum is precisely the form proposed by Harrison [44] and Zeldovich [45] to give a phenomenological account of galaxy formation. However, the magnitude of $(\delta\tilde{\rho}/\rho)_f$ is much more problematic than the shape of the spectrum. In the minimal $SU(5)$ GUT, it was found [43] to be of order 10^2, while the desired value is about 10^{-4}. (In a revised calculation, Hawking and Moss [46] have obtained a value of order one. I disagree with their methods [39], but in any case the result is still unacceptable.) Thus, the inflationary model and the minimal $SU(5)$ particle theory appear to be incompatible. [It has also been found [38] that the minimal $SU(5)$ theory gives initial fluctuations in ϕ that are too large, giving insufficient inflation.]

Fortunately, it appears that the magnitude of $(\delta\tilde{\rho}/\rho)_f$ is very sensitive to the details of the underlying particle theory, while the shape of the spectrum is not. Over the past year there has been much effort [47] in constructing particle theories that give rise to the desired density fluctuations. It seems that a number of successful models have been constructed, but none of them appear particularly elegant. The issue of density fluctuations and particle physics model building remains an active area of research.

VII Conclusion

In conclusion, I want to say that the basic idea of inflation—the idea that the universe went through a period during which it expanded exponentially while trapped in a false vacuum—appears to me to be probably correct. It is a very simple and natural idea in the context of spontaneously broken gauge theories, and it seems to solve some very fundamental cosmological problems. On the other hand, we clearly do not yet have the details straight. In order to understand the density fluctuations, we must at the same time understand the details of the particle physics at GUT energy scales. We are presumably some distance from that goal. Thus, I expect that the interface between particle theory and cosmology will remain an exciting area of research for some time to come.

Finally, let me say again that probably the most dramatic recent discovery in cosmology is the realization that the universe may be completely devoid of conserved quantum numbers. If so, then even if we do not understand the precise scenario, it become very plausible that our observed universe emerged from nothing or from almost nothing. I have often heard it said that there is no such thing as a free lunch. It now appears possible that the universe itself is the ultimate free lunch.

Acknowledgment

I want to acknowledge the support of the U.S. Department of Energy (DOE) under contract DE-ACO2-76ER03069, and also an Alfred P. Sloan Fellowship.

References

[1]   For a review, see P. Langacker, *Phys. Rep.* C72, 185 (1981).

[2]   A. H. Guth, *Phys. Rev.* D23, 347 (1981).

[3]   A. D. Linde, *Phys. Lett.* B108, 389 (1982).

[4]   A. Albrecht and P. J. Steinhardt, *Phys. Rev. Lett.* 48, 1220 (1982).

[5]   A. H. Guth and E. J. Weinberg, *Nucl. Phys.* B212, 321 (1983).

[6]   S. W. Hawking, I. G. Moss, and J. M. Stewart, *Phys. Rev.* D26, 2681 (1982).

[7]   For the reader who would like a more detailed review of this subject, let me point out that reference [8] contains reviews by both A. D. Linde and by me.

[8]   *The Very Early Universe: Proceedings of the Nuffield Workshop*, G. W. Gibbons, S. W. Hawking, and S. T. C. Siklos, editors (Cambridge University Press, Cambridge, 1983).

[9]   D. A. Kirzhnits and A. D. Linde, *Phys. Lett.* B42, 471 (1972); S. Weinberg, *Phys. Rev.* D9, 3357 (1974); L. Dolan and R. Jackiw, *Phys. Rev.* D9, 3320 (1974); for a review, see A. D. Linde, *Rep. Prog. Phys.* 42, 389 (1979).

[10]   See, for example, the following reviews: P. Goddard and D. I. Olive, *Rep. Prog. Phys.* 41, 1357 (1978); S. Coleman, in *The Unity of the Fundamental Interactions* (Procedings of the International School of Subnuclear Physics, Erice, 1981), A. Zichichi, editor (Plenum, New York, 1983).

[11]   G. 't Hooft, *Nucl. Phys.* B79, 276 (1974); A. M. Polyakov, *Pis'ma Zh. Eksp. Teor. Fiz.* 20, 430 (1974) [*JETP Lett.* 20, 194 (1974)].

[12]   H. Georgi and S. L. Glashow, *Phys. Rev. Lett.* 32, 438 (1974).

[13]   For a review, see M. Yoshimura, in *Grand Unified Theories and Related Topics: Proceedings of the 4th Kyoto Summer Institute*, M. Konuma and T. Maskawa editors (World Scientific, Singapore, 1981). See also E. W. Kolb and M. S. Turner, *Ann. Rev. Nucl. Part. Sci.* 33, 645 (1983).

[14]   For a general background in cosmology, see S. Weinberg, *Gravitation and Cosmology* (Wiley, New York, 1972). At a less technical level, see J. Silk, *The Big Bang* (W. H. Freeman,

San Francisco, 1980), or S. Weinberg, *The First Three Minutes* (Bantam Books, New York, 1977).

[15]   J. P. Preskill, *Phys. Rev. Lett.* 43, 1365 (1979).

[16]   Ya. B. Zeldovich and M. Y. Khlopov, *Phys. Lett.* B79, 239 (1978).

[17]   For a more detailed review of this topic, see [18] and the article by G. Lazarides in the same volume.

[18]   A. H. Guth, in *Magnetic Monopoles* (Proceedings of a NATO Advanced Study Institute on Magnetic Monopoles, Wingspread, Wisconsin, 1982), R. A. Carrigan and W. P. Trower, editors (Plenum, New York, 1983).

[19]   T. W. B. Kibble, *J. Phys.* A9, 1387 (1976).

[20]   J. P. Preskill (private communication); A. H. Guth and S.-H. Tye, *Phys. Rev. Lett.* 44, 631, 963(E) (1980); M. B. Einhorn, D. L. Stein, and D. Toussaint, *Phys. Rev.* D21, 3295 (1980).

[21]   T. Goldman, E. W. Kolb, and D. Toussaint, *Phys. Rev.* D23, 867 (1981).

[22]   W. Rindler, *Mon. Not. R. Astron. Soc.* 116, 663 (1956). See also S. Weinberg, [14, pp. 489–490, 525–526].

[23]   A. H. Guth, in *Asymptotic Realms of Physics: Essays in Honor of Francis E. Low*, A. H. Guth, K. Huang, and R. L. Jaffe, editors (MIT Press, Cambridge, MA, 1983).

[24]   R. H. Dicke and P. J. E. Peebles, in *General Relativity: An Einstein Centenary Survey*, S. W. Hawking and W. Israel, editors (Cambridge University Press, Cambridge, 1979).

[25]   See, for example, D. W. Olson, *Phys. Rev.* D14, 327 (1976). Other relevant references include J. M. Bardeen, *Phys. Rev.* D22, 1882 (1980); W. H. Press and E. T. Vishniac, *Ap. J.* 239, 1 (1980).

[26]   S. Coleman, *Phys. Rev.* D15, 2929 (1977); C. G. Callan and S. Coleman, *Phys. Rev.* D16, 1762 (1977). See also S. Coleman in *The Whys of Subnuclear Physics* (Proceedings of the International School of Subnuclear Physics, Ettore Majorana, Erice, 1977), A. Zichichi, editor (Plenum, New York, 1979).

[27]   S. Coleman and E. J. Weinberg, *Phys. Rev.* D7, 1888 (1973).

[28]   L. Dolan and R. Jackiw, [9].

[29]   G. Steigman, in *Unification of the Fundamental Particle Interactions II* (Proceedings of the Europhysics Study Conference, Erice, Italy, Oct. 6–14, 1981), J. Ellis and S. Ferrara, editors (Plenum, New York, 1983).

[30]   M. Sher, *Phys. Rev.* D24, 1699 (1981).

[31]   The properties of de Sitter space are well described in S. W. Hawking and G. F. R. Ellis, *The Large Scale Structure of Space-Time* (Cambridge University Press, Cambridge, 1973).

[32]   J. A. Frieman and C. M. Will, *Ap. J.* 259, 437 (1982); J. D. Barrow, in [8]; W. Boucher and G. W. Gibbons, in [8]; P. Ginsparg and M. J. Perry, *Nucl. Phys.* B222, 245 (1983).

[33]   G. W. Gibbons and S. W. Hawking, *Phys. Rev.* D15, 2738 (1977).

[34]   S. W. Hawking and I. G. Moss, *Phys. Lett.* B110, 35 (1982).

[35]   D. Atkatz and H. Pagels, *Phys. Rev.* D25, 2065 (1982), and references therein.

[36]   E. P. Tryon, *Nature* 246, 396 (1973); A. Vilenkin, *Phys. Lett.* B117, 25 (1982); A. Vilenkin, *Phys. Rev.* D27, 2848 (1983).

[37]   To begin an inflationary universe with the minimal possible matter energy, one could start with a volume of about χ^{-3} of false vacuum; the total matter energy would be $\rho_0 \chi^{-3} \approx 10$ kg.

[38]   A. Vilenkin and L. H. Ford, *Phys. Rev.* D26, 1231 (1982); A. Vilenkin, *Nucl. Phys.* B226, 504 (1983); A. D. Linde, *Phys. Lett.* B116, 335 (1982).

[39]   A. H. Guth and S.-Y. Pi (in preparation).

[40]   A. H. Guth, in [8].

[41]   A. Albrecht, P. J. Steinhardt, M. S. Turner, and F. Wilczek, *Phys. Rev. Lett.* 48, 1437 (1982); L. F. Abbott, E. Farhi, and M. B. Wise, *Phys. Lett.* B117, 29 (1982); A. D. Dolgov and A. D. Linde, *Phys. Lett.* B116, 329 (1982).

[42]   G. Lazarides, Q. Shafi, and W. P. Trower, *Phys. Rev. Lett.* 49, 1756 (1982).

[43]   A. A. Starobinsky, *Phys. Lett.* B117, 175 (1982); A. H. Guth and S.-Y. Pi, *Phys. Rev. Lett.* 49, 1110 (1982); S. W. Hawking, *Phys. Lett.* B115, 295 (1982); J. M. Bardeen, P. J. Steinhardt, and M. S. Turner, *Phys. Rev.* D28, 679 (1983). See also the contributions of these authors in [8].

[44]   E. R. Harrison, *Phys. Rev.* D1, 2726 (1970).

[45]   Ya. B. Zeldovich, *Mon. Not. R. Astr. Soc.* 160, 1P (1972).

[46]   S. W. Hawking and I. G. Moss, *Nucl. Phys.* B224, 180 (1983).

[47]   A. Albrecht, S. Dimopoulos, W. Fischler, E. W. Kolb, S. Raby, and P. J. Steinhardt, *Nucl. Phys.* B229, 528 (1983); D. V. Nanapoulos, K. A. Olive, and M. Srednicki, *Phys. Lett.* B127, 30 (1983); G. B. Gelmini, D. V. Nanopoulos, and K. A. Olive, *Phys. Lett.* B131, 53 (1983); D. V. Nanopoulos and M. Srednicki, *Phys. Lett.* B133, 287 (1983); B. A. Ovrut and P. J. Steinhardt, *Phys. Lett.* B133, 161 (1983); B. A. Ovrut and S. Raby, *Phys. Lett.* B134, 51 (1984); Q. Shafi and A. Vilenkin, *Phys. Rev. Lett.* 52, 691 (1984); P. Q. Hung, *Phys. Rev.* D30, 1637 (1984); S. Gupta and H. R. Quinn, *Phys. Rev.* D29, 2791 (1984); R. Holman, P. Ramond, and G. G. Ross, *Phys. Lett.* B137, 343 (1984); S.-Y. Pi, *Phys. Rev. Lett.* 52, 1725 (1984).

The New Inflationary Universe Scenario: Problems and Perspectives

A. D. Linde

The present status of the new inflationary scenario is reviewed. A possible realization of this scenario is suggested in the context of $N = 1$ supergravity coupled to matter. It is shown that in this scenario one can obtain a desirable magnitude of density perturbations $\delta\rho/\rho \sim 10^{-4}$, and the primordial monopole problem can be easily solved. It is shown also that inflation can help us to solve the problem of symmetry breaking in SUSY GUTs. A new version of the inflationary universe scenario is suggested. According to this scenario inflation is a natural, and may even be an inevitable, consequence of chaotic initial conditions in the very early universe.

1 Introduction

It is known that the new inflationary universe scenario [1–8] makes it possible to solve simultaneously not only the horizon, flatness, and primordial monopole problems [9, 1], but also the homogeneity, isotropy, and domain wall problems [1], the primoridal gravitino problem [10], and the galaxy formation problem [8, 11–15]. The main idea of this scenario was explained in the talk by Guth at this conference.* After a thorough investigation of the new inflationary universe scenario during the last year (for a review of the present status of this scenario see [16]), two conditions necessary for its realization have been understood. First of all, the effective potential $V(\phi)$ of some scalar field ϕ should be very flat at $\phi \ll \phi_0$, where ϕ_0 corresponds to the minimum of $V(\phi)$. This condition is necessary in order to have a sufficiently large inflation of the universe [1, 6], and to obtain sufficiently small density perturbations appearing after the inflation: $\delta\rho/\rho \sim 10^{-4}$ [8, 12–15]. The second condition is that the effective potential $V(\phi)$ somewhere in the region $\phi \sim \phi_0$ should be sufficiently curved [$d^2V/d\phi^2 > (10^{10} \text{ GeV})^2$], since otherwise the field ϕ oscillating near $\phi \sim \phi_0$ after the inflation does not produce superheavy Higgs bosons and fermions with $M > 10^{10}$ GeV, which are necessary for a subsequent baryon number generation [7].

The second condition can be fulfilled in GUTs. For example, in the $SU(5)$ Coleman-Weinberg theory $d^2V/d\phi^2|_{\phi \sim \phi_0} \sim (10^{14} \text{ GeV})^2$, and the baryon asymmetry, which appears after the inflationary phase transition in this theory, can be one or two orders of magnitude greater than in the standard baryosynthesis scenario [7]. However, it seems very difficult to satisfy the first condition in ordinary GUTs [6, 8, 12–15]. It is especially difficult to

---

* Editors' note: The paper read to the conference by Guth appears in this volume.

satisfy the condition $\lambda(\phi) = d^4 V/d\phi^4 < 10^{-10}$, which is necessary in order to obtain density perturbations $\delta\rho/\rho \sim 10^{-4}$ after the phase transition [8]. The results obtained in [6, 8, 12–15] have stimulated a search for the theories in which $V(\phi)$ is very flat at small ϕ.

A simplest possibility, discussed in [6, 17], was connected with supersymmetric versions of the Coleman-Weinberg model. Indeed, in these theories the effective coupling constant $\lambda(\phi)$ proves to be very small due to (partial) cancellation of the contributions to $\lambda(\phi)$ from bosons and fermions. A first realization of this possibility was considered in [18] in the context of the Witten-Dimopoulos-Raby inverted hierarchy model [19]. It was shown in [18] that under some conditions it is possible to obtain a sufficiently large inflation in this model. However, the effective potential in this model is too flat $[d^2 V/d\phi^2|_{\phi \sim \phi_0} < (10^3 \text{ GeV})^2]$, and therefore no baryon asymmetry is produced after the phase transition [10]. Moreover, density perturbations, which appear after the phase transition, typically are too small [20], and there is also a serious problem with reheating, which is connected with the decoupling of the field ϕ in this theory from all light fields [18]. Therefore at present it seems that the new inflationary universe scenario cannot be implemented in the context of the theories of the type of [19].

Another possibility was suggested in [21], where it was noted that the new inflationary universe scenario can be realized in a more natural way if the inflationary phase transition occurs at a mass scale of the order $M_P \sim 10^{19}$ GeV (primordial inflation). In [22] it was shown that with a proper choice of parameters one may obtain a desirable value $\delta\rho/\rho \sim 10^{-4}$ after the phase transition. Later it was shown [23] that the idea of primordial inflation can be implemented in the context of $N = 1$ supergravity coupled to matter [24].

This version of the new inflationary universe scenario seems very interesting and attractive. It seems plausible that just supergravity and not GUTs may be responsible for the most fundamental features of the structure of the universe. However, the price one could pay for such a philosophy might be very high. Indeed, one of the main motivations for the inflationary universe scenario was the possibility of obtaining a simple solution of the primordial monopole problem. Meanwhile, in [21, 23] it was claimed that primordial inflation occurs *before* the phase transitions in GUTs, and therefore this scenario does not help us to solve the primordial monopole problem. Moreover, as will be shown, in the simplest models considered in [21–23] the inflationary phase transition may not occur at all. Nevertheless, in our opinion the general idea of priomordial inflation is very good. In section 2 of this paper it will be shown that the new inflationary universe scenario actually can be realized in the context of $N = 1$ supergravity coupled to matter [25]. We shall suggest also a simple solution of the primordial monopole problem in this scenario [25, 26]. Section 3 contains a discussion of the symmetry breaking phase transition in SUSY GUTs and suggests a possible solution of the problem of symmetry breaking in these theories. The proposed solution is connected with the generation of the long-wave fluctuations of the scalar fields ϕ in an inflationary universe [26]. In section 4 a novel version of the inflationary universe scenario is suggested [27] that is *not* based on the theory of phase transitions in spontaneously broken gauge theories. According to this scenario inflation is a natural conse-quence of chaotic initial conditions in the very early unvierse. We believe that the chaotic inflation scenario [27] is much more natural and can be much more easily realized than all other versions of the inflationary uni-verse scenario suggested so far. In the conclusion, a brief summary of our results is contained. Finally, in the appendix we suggest a possible solution of the cosmological constant problem, which is based (in part) on the chaotic inflation scenario.

2 Primordial Inflation without Primordial Monopoles

Let us start with the investigation of the first version of the primordial inflation scenario [21], which is based on the theory with a superpotential

$$W(\phi, X, Y) = aX\phi(\phi - \mu) + bY(\phi^2 - \mu^2). \tag{1}$$

Finite temperature effective potential in this theory is given by [21]

$$\begin{aligned}
V(\phi, T) = &(a^2 + b^2)\phi^4 - 2a^2\mu\phi^3 + (a^2 - 2b^2)\mu^2\phi^2 \\
&+ b^2\mu^4 + (a^2 + b^2)T^2\phi^2 - a^2\mu T^2\phi + a^2T^2\mu^2.
\end{aligned} \tag{2}$$

To have a sufficiently large inflation in this theory one should require that $a^2 - 2b^2 \ll a^2, b^2$ [21, 22]. However, in this case one can easily verify that the absolute minimum of $V(\phi, T)$ (2) at *all* temperatures occurs at $\phi(T) \sim \phi_0 \sim \mu$. Therefore the reasons for the inflationary phase transition here are not the usual ones (see, however, section 4). In [21] this question was not analyzed; it was just assumed that with a decrease of temperature the phase transition from small ϕ to $\phi \sim \phi_0$ occurs. The reason for such unusual behavior of $\phi(T)$ in the theory (1) is connected with the presence of a large cubic term ϕ^3 in (2) at $T = 0$, which leads to the appearance of a large term ϕ at $T \neq 0$ (2) and shifts the absolute minimum of $V(\phi, T)$ to large values of ϕ [25].

A similar problem arises in the second version of the primordial inflation scenario based on the theory of a chiral scalar superfield Φ coupled to $N = 1$ supergravity [23]. The effective potential of the first (scalar) component of this field z looks as follows [24]:

$$V(z, z^*) = e^{zz^*/2}\left[2\left|\frac{dg}{dz} + \frac{z^*}{2}g\right|^2 - 3|g^2|\right], \tag{3}$$

where $g(z)$ is some superpotential,

$$g(z) = \mu^3 f(z). \tag{4}$$

Here $f(Z)$ is some arbitrary dimensionless function and μ is some mass parameter. All dimensional quantities in (3) and (4) are written in units $M_p/\sqrt{8\pi}$.

Let us restrict ourselves to the investigation of the phase transition with generation of the real part ϕ of the field z. Due to the arbitrariness of the function $f(z)$, there is great freedom in the choice of the effective potential $V(z, z^*)$. The main restriction is the condition $V(\phi_0) = 0$, which is necessary in order to have zero cosmological constant at the minimum of $V(\phi)$. In [23] it was suggested investigating the theory in which $V(\phi)$ at small ϕ ($\phi \ll \phi_0$) looks as follows:

$$V(\phi) = \mu^6(1 + \alpha^2\phi^2 - \beta^2\phi^3 + \gamma^2\phi^4 + \cdots), \tag{5}$$

where $\alpha^2\gamma \ll \beta^2$. However, in this case, as in the case just considered, the minimum of the effective potential $V(\phi, T)$ at all temperatures occurs at large values of ϕ, $\phi(T) \sim \phi_0$ [25, 28].

Fortunately this disease of the model (5) can be easily cured [25]. Indeed, due to the arbitrariness of the function $f(z)$ one can modify $V(\phi)$ (5) so as to get rid of the dangerous cubic term. A simplest version of $V(\phi)$ without the cubic term that satisfies the condition $V(\phi_0) = 0$ is given by

$$V(\phi) = 3\mu^6\left(1 - \alpha^2\phi^2 + \frac{\alpha^4}{4}\phi^4\right). \tag{6}$$

From this expression it follows that $\phi_0 = \sqrt{2}/\alpha$, $H = \mu^3$, $m_\phi^2(\phi = 0)$

$= d^2 V/d\phi^2|_{\phi=0} = -6\mu^6\alpha^2$, $m_\phi^2(\phi_0) = 12\mu^6\alpha^2$. At large temperatures $V''(0, T) > 0$, and one may assume that the symmetry in this theory was restored, $\phi = 0$ (at least in some domains of the universe; see section 4). With a decrease in temperature, $V''(0)$ becomes negative and the phase transition with generation of a large classical field $\phi = \phi_0$ takes place. To be more precise, one should note that in the case under consideration not the field $\phi = \langle\phi\rangle$ is generated, but the field $\langle\phi^2\rangle$ [6, 8]. However, in this case the leading contribution to $\langle\phi^2\rangle$ is given by fluctuations with exponentially large wavelength, so that these fluctuations practically do not differ from the constant classical field $\phi = \sqrt{\langle\phi^2\rangle}$ [6, 8]. According to [6, 8, 29], the amplitude of the field $\phi = \sqrt{\langle\phi^2\rangle}$ grows as follows:

$$\phi^2 = \frac{\mu^6}{16\pi^2\alpha^2}[e^{4\mu^3\alpha^2(t-t_0)} - 1], \tag{7}$$

where t_0 is the time of the beginning of the phase transition. Corrections to (7) due to the interaction term $\sim(\alpha^4/4)\phi^4$ at $\phi \ll \phi_0$ can be neglected. Let us study now the problem of density perturbations generated after the phase transition. Proceeding along the lines of [8, 12–15], one can easily show that

$$\frac{\delta\rho(\kappa)}{\rho} \sim \frac{\mu^3}{(2\pi)^{3/2}\phi^*\alpha^2}, \tag{8}$$

where κ is the momentum, corresponding to the density perturbation $\delta\rho(\kappa)$ at the end of inflation, and ϕ^* is the value of the field ϕ at the time when the momentum κ correponding to the perturbation $\delta\rho$ mentioned was equal to $H = \mu^3$. By means of (7) one can easily show that

$$\phi^* = \phi_0 \left(\frac{\kappa}{\mu^3}\right)^{2\alpha^2} \quad \left(\text{for } \phi^* \gtrsim \frac{1}{4\pi\alpha}\right)$$

and

$$\frac{\delta\rho(\kappa)}{\rho} \sim \frac{\mu^3}{4\pi^{3/2}\alpha}\left(\frac{\kappa}{\mu^3}\right)^{-2\alpha^2}. \tag{9}$$

For the fluctuations at the galactic scale, $\kappa \sim e^{-50}H$, which yields

$$\frac{\delta\rho}{\rho} \sim \frac{\mu^3\exp(10^2\alpha^2)}{20\alpha}. \tag{10}$$

It is seen that $\delta\rho/\rho \sim 10^{-4}$, e.g., at $\alpha \sim 10^{-1}$, $\mu \sim 10^{-1}$ (to be more precise, $\mu^3 \sim 10^{-4}$). In this particular case $\phi_0 = \sqrt{2/\alpha} \sim 14$ (or, in the standard notation, $\phi_0 \sim 3M_P$) and $\phi^* \sim 0.3\phi_0$. Note that one can considerably diminish ϕ^* by a small increase in α (and a corresponding decrease in μ). The magnitude of inflation and the value of $\delta\rho/\rho$ are determined mainly by the properties of $V(\phi)$ at $\phi \lesssim \phi^*$. Therefore one can obtain the necessary magnitude of inflation and the value $\delta\rho/\rho \sim 10^{-4}$ in a wide class of theories, including the theories in which at $\phi \lesssim \phi^*$ the effective potential is given by (6) with $\alpha \sim 10^{-1}$, $\mu^3 \sim 10^{-4}$; but at $\phi^* \ll \phi < \phi_0$ the effective potential can be more (or less) curved than the potential (6). This fact will be important for us later.

Now let us estimate the duration of inflation Δt. The value of Δt approximately coincides with the time necessary for the field ϕ to grow up to $\phi_0 = \sqrt{2/\alpha}$. Thus from (7) one obtains

$$\Delta t \sim \frac{6}{\mu^3\alpha^2} \sim 600H^{-1} \sim (3.10^{11} \text{ GeV})^{-1}. \tag{11}$$

From (10) and (11) it follows that in the theory (6) one may obtain both sufficient inflation (since $\Delta t \gg 60H^{-1}$) and the desirable value of $\delta\rho/\rho \sim 10^{-4}$. Now let us return to the primordial monopole problem.

In [21–23] it was assumed that the phase transition with breaking of $SU(5)$ and monopole production occurs *after* the primordial inflation, and therefore one should use some other methods to solve the primordial monople problem; in particular one may try to find the theories in which the phase transition with the monopole production occurs at some very small temperature $T_c \lesssim 10^9$ GeV. There exists also another possibility of solving the primordial monopole problem, connected with the confinement of monopoles in the hot Yang-Mills gas [30]. The theory of this effect is not yet finally elaborated, and there exist two main objections to this possibility.

i. In [31] it was claimed that the magnetic field in the hot Yang-Mills gas is not squeezed into a magnetic tube confining two monopoles with opposite charges, but is screened by the Wu-Yang monopoles. We would like to note, however, that the Wu-Yang monopoles are not sourceless solutions of the Yang-Mills equations [32], and therefore they are absent in GUTs and cannot screen any fields in these theories.

ii. In two papers [33] the behavior of magnetic fields in the Yang-Mills gas was studied by means of the Monte-Carlo lattice simulation, and it was shown that magnetic fields in the Yang-Mills gas are screened. From our point of view, however, the results of [33] are not convincing, since with the parametric values used in [33] there should exist many unphysical lattice monopoles (which disappear in the continuum theory). Just these monopoles could screen the magnetic field in the computations in [33].

Therefore in our opinion the possibility of monopole confinement in the hot Yang-Mills gas is still open and deserves further investigation. However, in the theories under consideration the primordial monopole problem can be solved in a much simpler way. In order to study this question let us note first that the period of inflation Δt (11) is very large. The temperature of the universe exponentially decreases due to inflation, and during most part of this period the temperature is much smaller than the critical temperature of the phase transition in the $SU(5)$ theory. A typical time $\tau \sim (10^{15} \text{ GeV})^{-1}$, which is necessary for $SU(5)$ symmetry breaking to occur (if supercooling in this phase transition is not anomalously large), is much smaller than the time of inflation $\Delta t \sim (10^{11} \text{ GeV})^{-1}$. Therefore the $SU(5)$ phase transition with monopole production occurs not *after* the inflation, but long *before* the end of inflation, which solves the primordial monopole problem in the theory under consideration.

It is worth noting that the time $\Delta t(11)$ can be made even much greater than $(10^{11} \text{ GeV})^{-1}$ by a proper choice of α and μ. On the other hand, one can speed up the $SU(5)$ symmetry breaking phase transition, e.g., by adding a term $W' = \lambda Z'(\text{Tr} \phi^2 - m^2)$ to the superpotential. Here ϕ is the $SU(5)$ adjoint Higgs field and Z' is some singlet superfield [21]. In such a theory the phase transition may proceed to the phase $SU(4) \times U(1)$ as well as to the phase $SU(3) \times SU(2) \times U(1)$. However, this will not lead to any troubles with the domain walls between these phases, since a typical size of domains of each phase after inflation exceeds the size of the observable part of the universe, and there will be infinitely many domains of the phase $SU(3) \times SU(2) \times U(1)$ to live in [16].

To be more precise, one should note that in the minimal supersymmetric $SU(5)$ theory without the term $W' = \lambda Z'(\text{Tr} \phi^2 - m^2)$ supercooling is very large indeed. The theory of the symmetry breaking phase transition and the solution of the primordial monopole problem in the context of the minimal $SU(5)$ theory is more tricky and will be discussed in the next section.

One could argue, however, that if the reheating of the universe occurs during the time $\tau_r < H^{-1}$ after the phase transition, as in the $SU(5)$ Coleman-Weinberg theory [7], then all the vacuum energy $\sim V(0)$ transforms into heat, and the temperature of the universe grows up to the reheating temperature $T_r \sim V^{1/4}(0) \sim 10^{17}$ GeV. This temperature may be somewhat greater than the critical temperature of the symmetry restoration in the supersymmetric $SU(5)$ theory, $T_c \sim 10^{16}$–10^{17} GeV. Therefore one could expect that after the reheating the $SU(5)$ symmetry becomes restored, and then with the cooling of the universe the $SU(5)$ breaking phase transition occurs and the monopoles appear again, as in a nightmare.

Fortunately, this process does not actually occur in the theories under consideration because of the very slow reheating in them [25, 34]. According to [7], at the first stage of the reheating process the oscillating field ϕ produces predominately ϕ particles, and the energy density E of the produced particles increases as $dE/dt \sim 10^{-3}m_\phi^5$. In the $SU(5)$ theory the value of m_ϕ in the minimum of the effective potential is $\sim 10^4$ times greater than H, and this was the reason why in that theory the energy of the oscillating field completely transformed into the energy of ϕ particles during the time $\tau_r < H$ [7]. However, in the theory (6) at the beginning of reheating $m_\phi < H$. An investigation similar to that in [7] shows that in this case $T_r < 10^5 H^{-1}$. During this time almost all the energy of the oscillating field ϕ is red shifted away, and it can be shown that the energy density of the ϕ particles created by the oscillating field ϕ never exceeds $E \sim 10^{-10}V_{(0)} \sim (10^{14}$ GeV$)^4$. Even if this energy could be momentarily transformed into the energy of the $SU(5)$ particles, the reheating temperature T_r would be of the order of 3.10^{13} GeV, which is much smaller than the critical temperature of the symmetry restoration in the supersymmetric $SU(5)$ theory, $T_c \gtrsim 10^{16}$ GeV. In fact, the reheating temperature is even much smaller, since the strength of interaction of ϕ particles with all other particles usually is very small [24], and thermalization of the $SU(5)$ particles may occur very late. This may lead to another trouble, since if $T_r \ll 10^{10}$ GeV, then after the phase transition there are no monopoles, but also no heavy Higgs bosons and fermions with $M < 10^{10}$ GeV, which are necessary for the baryon asymmetry generation.

We shall not perform here a detailed study of the process of reheating, since the theory of this process depends crucially on the properties of $V(\phi)$ near $\phi = \phi_0$ and on the strength of interaction between the field ϕ and the $SU(5)$ fields. For our purposes it will be sufficient to note that the temperature of reheating T_r can be made as large as one wishes (though not greater than 10^{17} GeV) just by the choice of a sufficiently curved effective potential $V(\phi)$ near $\phi = \phi_0$. (As we have mentioned, this can be done without modifying $\delta\rho/\rho$ and the magnitude of inflation.) Indeed, at a sufficiently large m_ϕ^2

$= d^2 V/d\phi^2|_{\phi \sim \phi_0}$ the process of production of the ϕ particles proceeds very rapidly, as in the $SU(5)$ Coleman-Weinberg theory [7]. On the other hand, in the theories under consideration [24] an increase of m_ϕ^2 automatically leads to an increase in the strength of interaction between ϕ particles and all other particles. For example, some simple estimates show that the decay rate of ϕ bosons into other particles, like the decay rate of gravitino [35], is of the order $\Gamma_\phi \sim m_\phi^3/M_P^2$ [25, 34]. This means that with a proper choice of m_ϕ^2 one can obtain the value of reheating temperature T_r in the interval between 10^{16} and 10^{10} GeV, which is exactly what is needed in order to have no monopoles and to obtain the baryon asymmetry of the universe after primordial inflation. In particular, from our estimate of Γ_ϕ it follows that even in the simplest model (6) without any modification of $V(\phi)$ near $\phi = \phi_0$ the reheating temperature is $T_r \sim 10^{11}$–10^{12} GeV, which is quite satisfactory [25, 34].

Our model (6) is very simple; at the moment one may consider it a toy model for primordial inflation without making any attempt to link it to any particular version of $N = 1$ supergravity coupled to matter. If one wishes, however, to relate our model to some recent attempts to solve the hierarchy problem with the help of $N = 1$ supergravity (see, e.g., [36] and references therein), one should have not only the heavy field $z(3)$, but also some light chiral fields z_i with masses m_i of the order of the gravitino mass $m_G \sim 10^2$ GeV. In this case one should have $f(z) = 0$ in the minimum of $V(z, z^*)$, since otherwise m_G would be too large [23]. This condition can be fulfilled after some modification of $V(\phi)$ (6) and the details of our scenario, which does not lead, however, to a qualitative modification of our main results concerning the primordial monopole problem and the process of reheating. Note that the reheating temperature obtained previously is very small, which makes it possible to solve the gravitino problem in this scenario [10, 34]. We conclude, therefore, that the new inflationary universe scenario actually can be realized in the context of $N = 1$ supergravity, and that despite some earlier expectations no monopole problem appears in this scenario.

3 Inflation Can Break Symmetry in SUSY GUTS

Attempts to construct a unified supersymmetric theory of all fundamental interactions are of a great interest now. Supersymmetric theories have many advantages over the standard theories. However, there also exist

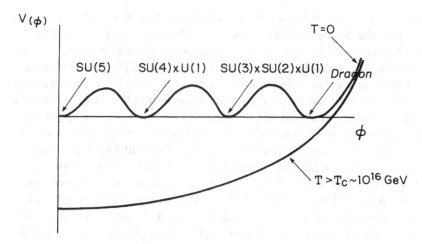

Figure 1
The effective potential $V(\Phi, T)$ in the minimal supersymmetric $SU(5)$ model.

some problems specific to supersymmetric theories. One such problem is that of symmetry breaking in SUSY GUTs.

It is well-known that the effective potentials $V(\phi)$ in supersymmetric theories typically have many different minima of the same depth; see, e.g., a discussion of this question in [37–39]. For example, in the minimal $SU(5)$ supersymmetric theory [40, 41] with the superpotential

$$W = \lambda\left(\frac{m}{2}\operatorname{Tr}\Phi^2 + \frac{1}{3}\operatorname{Tr}\Phi^3\right), \tag{12}$$

where Φ is the adjoint Higgs field and the effective potential $V(\Phi)$ has four degenerate minima: $SU(5)$ invariant minimum, $SU(4) \times U(1)$ minimum, $SU(3) \times SU(2) \times U(1)$ minimum, and the Dragon minimum [38] (figure 1). One can increase the energy of the $SU(5)$ minimum, e.g., by adding the term $W' = \alpha Z'(\operatorname{Tr}\Phi^2 - \mu^2)$ to the superpotential (12) (here Z' is some chiral singlet superfield); however, the degenerate minima $SU(4) \times U(1)$ and $SU(3) \times SU(2) \times U(1)$ are still present in such a nonminimal theory [41]. The number of different degenerate minima of the effective potential increases considerably when one adds to $W(12)$ the superpotential of other Higgs fields ϕ [39].

The degeneracy of energy of different minima of $V(\Phi)$ is removed by gravitational effects, but in a rather unfortunate way: The lowest energy state in the minimal theory (which will be studied in the present paper) becomes the $SU(5)$ invariant state [37]. Moreover, even in the cases in which gravitational effects are small, cosmological considerations peak up not the $SU(3) \times SU(2) \times U(1)$ phase, but just the $SU(5)$ symmetric phase.

Indeed, in the very early universe at the temperature $T \gg m$ the only minimum of $V(\Phi)$ was the $SU(5)$ minimum, corresponding to $\Phi = 0$, and the $SU(5)$ symmetry was restored [42] (figure 1). All other minima appear only at a sufficiently small temperature T. The leading contribution to $V(\phi, T)$ due to high-temperature effects is given by $-(\pi T^4/90)(N_B + \frac{7}{8}N_F)$, where N_B and N_F are the effective numbers of bosons and fermions with masses $M < T$ [42]. The value of $N_B + \frac{7}{8}N_F$ is maximal in the $SU(5)$ symmetric phase, in which $\Phi = 0$ and all vector bosons and fermions are massless [43]. This means that the $SU(5)$ minimum at all temperatures is a global minimum of the effective potential, which makes it absolutely unclear how the phase transition from this phase to the phase $SU(3) \times SU(2) \times U(1)$ could occur.

An interesting possibility for overcoming this difficulty was suggested in [43, 44]. Let us consider, e.g., the phase transition $SU(5) \to SU(3) \times SU(2) \times U(1)$ in the minimal supersymmetric $SU(5)$ theory, which occurs due to the appearance of the classical scalar field

$$\Phi = \phi \sqrt{\frac{2}{15}} \mathrm{diag}(1, 1, 1, -\tfrac{3}{2}, -\tfrac{3}{2}). \tag{13}$$

One-loop effective potential $V(\phi, T)$ in this theory is shown in figure 2. Now let us take into account, following [43, 44], that at $\phi = 0$, $T < 10^9$ GeV all matter in the $SU(5)$ theory should be in the confinement phase. In this phase the one-loop results for $V(\phi, T)$ are unreliable. The number of degrees of freedom in the confinement phase is considerably reduced compared with the normal phase, which leads to an increase of $V(0, T)$ by $O(T^4)$. It was concluded therefore that at $T < 10^9$ GeV the phase transition from $\phi = 0$ to $\phi = \phi_0 \ 10^{16}$ GeV [ϕ_0 corresponds to the minimum of $V(\phi)$ in the $SU(3) \times SU(2) \times U(1)$ phase] becomes possible [50]. Later it was understood that such a phase transition may occur if the effective potential $V(\phi)$ is extremely flat. This is necessary in order to reduce the gravitational corrections to $V(\phi_0, T)$ [37] and to make possible the tunneling through the barrier ΔV of the height Λ^4 between $\phi = 0$ and $\phi = \phi_0$ (figure 2). In [43, 44] it was suggested considering the theories in which the effective coupling constant $\lambda \sim 10^{-12}$ (12) and $\Lambda \sim 10^{10}$ GeV. In such

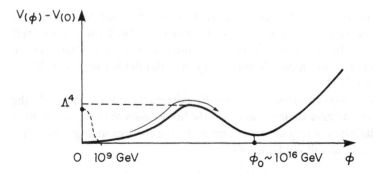

Figure 2
The effective potential $V(\phi, T)$ in the minimal supersymmetric $SU(5)$ theory at
$T < 10^9$ GeV for $\Phi = \phi \, \text{diag}\,(1, 1, 1, -\frac{3}{2}, -\frac{3}{2})$. The broken line corresponds to $V(\phi, T)$
in the $SU(5)$ confinement phase, which may exist at $\phi < 10^9$ GeV. The arrow shows the
behavior of the field $\phi = \sqrt{\langle \phi^2 \rangle}$ in the inflationary universe.

theories gravitational corrections to $V(\phi)$ actually are very small [45].
However, by the use of the methods developed in [46] it can be shown that
the rate of tunneling through the barrier of the height Λ^4 at temperature
$T \lesssim \Lambda$ is suppressed by a factor of the order $\exp(-\lambda^{3/2}) \sim \exp(-\phi_0^3/\Lambda^3)$
$\sim \exp(-10^{18})$. Therefore the phase transition could actually occur only if
there were no barrier between the phases $\phi = 0$ and $\phi = \phi_0$. The situation
becomes even more complicated if one takes into account that the confinement phase at $T \lesssim 10^9$ GeV may exist only at $\phi < 10^9$ GeV, where all
$SU(5)$ vector fields have small masses $M < 10^9$ GeV. At $\phi \gg 10^9$ GeV only
$SU(3) \times SU(2) \times U(1)$ vector particles are massless, whereas all other particles acquire masses $M \gg 10^9$ GeV, and the $SU(5)$ confinement disappears.
Therefore the $SU(5)$ confinement cannot lift up the whole $SU(5)$ minimum;
it can create only a local maximum of $V(\phi, T)$ at $\phi < 10^9$ GeV (figure 2), but
the value of $V(\phi, T)$ at $\phi \sim 10^{10}$ GeV remains smaller than $V(\phi_0, T)$, and the
phase transition to the $SU(3) \times SU(2) \times U(1)$ phase cannot occur.

In our recent paper [26] a possible solution of the problem of symmetry
breaking in SUSY GUTs in the context of the new inflationary universe
scenario was suggested. The main idea of this suggestion is based on the
following observation. As it is shown in [47], fluctuations of a scalar field ϕ
with a small mass $m^2 \ll H^2$ in the de Sitter universe expanding as $a(t) \sim e^{Ht}$
are extremely large:

$$\langle \phi^2 \rangle = \frac{3H^4}{8\pi^2 m^2}. \tag{14}$$

To be more precise, if the field ϕ interacts with the scalar curvature R as $(\xi/2)R\phi^2$, one should write $m^2 + \xi R$ instead of m^2 in (14). However, we shall consider here the theories without such interactions. Note that such an interaction does not appear in the theory of scalar fields coupled to $N = 1$ supergravity [24].

Equation (14) is valid for the externally existing de Sitter universe. In the inflationary universe scenario, in which the hot Friedman universe becomes exponentially expanding at some moment t_0, the corresponding expression for $\langle \phi^2 \rangle$ looks as follows [6, 8, 29]:

$$\langle \phi^2 \rangle = \frac{3H^2}{8\pi^2 m^2}\left(1 - \exp\left[-\frac{2m^2}{3H}(t - t_0)\right]\right). \tag{15}$$

Thus, at $t - t_0 \ll 3H/m^2$ the value of $\langle \phi^2 \rangle$ grows linearly,

$$\langle \phi^2 \rangle = \frac{H^3}{4\pi^2}(t - t_0), \tag{16}$$

and then it approaches the limiting value $3H^4/(8\pi^2 m^2)$ (14). It is very important (and rather unusual) that the leading contribution to $\langle \phi^2 \rangle$ goes from the fluctuations of the field ϕ with the exponentially large wavelength, $\kappa^{-1} \sim H^{-1}e^{H(t - t_0)}$. Therefore at scale $l \ll H^{-1}e^{H(t - t_0)}$ the fluctuations of the field ϕ practically cannot be distinguished from the homogeneous classical field $\phi = \sqrt{\langle \phi^2 \rangle}$. This effect initially served as a basis for the version of the new inflationary universe scenario suggested in [6] and also [8, 15]. However, (4) and (5) actually are valid not only for the field ϕ which is responsible for the inflationary phase transition, but for *any* scalar field with $m^2 \ll H^2$.

The growth of the "classical field" $\phi = \sqrt{\langle \phi^2 \rangle}$ is a rather surprising effect, since the field ϕ grows practically independently of the sign of m^2 (at $|m^2| \ll H^2$). In particular, the field ϕ can grow from the minimum to a maximum (or *over* the maximum) of the effective potential, though it would seem energetically unfavorable. The reason for such a strange effect is analogous to the reason for particle creation in an expanding universe (which at the first glance also would seem energetically unfavorable). Long-range fluctuations in the inflationary universe grow due to the vacuum rearrangement during the transition from the hot Friedman universe, in which long-range fluctuations are suppressed by the high-temperature effects, to the de Sitter universe, in which the density of long-range fluctuations is extremely large [16].

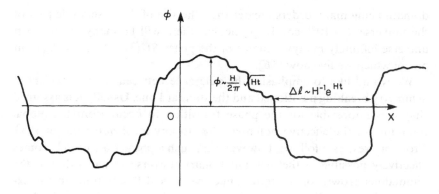

Figure 3
Fluctuations of the scalar field ϕ with mass $m^2 \ll H^2$ in the exponentially expanding universe.

In the inflationary universe scenario, exponential expansion should occur during some time $\Delta t = t - t_0 > 10^2 H^{-1}$ [29, 16]. If the mass of the field ϕ is sufficiently small (one or two orders smaller than H), then during the whole interval Δt the field grows linearly (16). In this case

$$\phi = \sqrt{\langle \phi^2 \rangle} = \frac{H}{2\pi} \sqrt{H \Delta t}, \tag{17}$$

where $H \Delta t \gtrsim 10^2$. From (6) it follows therefore that during the inflation the amplitude of the field ϕ approaches some value $\phi \gtrsim H$. (In the theories in which density perturbations $\delta\rho/\rho$ are of of the order 10^{-4} after inflation, the typical value of $H \Delta t$ is 10^3–10^5, which yields $\phi \gtrsim 10H$). Now let us consider the new inflationary universe scenario with $H \sim 10^{16}$–10^{17} GeV. (Such values of H may appear, e.g., in the primordial inflation scenario [21, 23, 25].) Typical masses of the Higgs fields ϕ in the minimal supersymmetric $SU(5)$ model [40, 41] are 10^{15}–10^{16} GeV, but they can be many orders smaller [44, 45]. Therefore during the inflation the classical fields of all types with the amplitude $\phi \gtrsim H \sim 10^{16}$–$10^{17}$ GeV are generated (figure 3). These fields are practically homogeneous at a scale $l \lesssim H^{-1} e^{H \Delta t}$.

After the end of inflation the field ϕ stops growing and becomes convergently oscillating in the vicinity of a nearest minimum of $V(\phi)$. Therefore after the end of inflation the universe becomes divided into many domains with all possible types of symmetry breaking, the typical size of each domain being many orders greater than the size of the observable part of the universe, $l \sim 10^{28}$ cm. In particular, there will be many (in an open universe infinitely many) domains of the phase $SU(3) \times SU(2) \times U(1)$, in one of which we live now [16].

We would like to emphasize the difference between the mechanism of symmetry breaking mentioned and the standard one. Usually it is assumed that it is impossible for the phase transition to occur from the global minimum of the effective potential $V(\phi)$ to any local minimum of $V(\phi)$. From our results it follows, however, that such a phase transition becomes effectively possible in the new inflationary universe scenario due to the anomalous growth of the long-range scalar field fluctuations in the exponentially expanding universe. Usually the phase transition proceeds to one preferred phase in the whole universe. In our case the phase transition proceeds with a comparable probability to many different phases. Each of these phases fills a domain (miniuniverse) of the size exceeding that of the observable part of our universe. All these phases with a nonnegative vacuum energy density are practically stable. The only phase in which life of our type may exist is the phase $SU(3) \times U(1)$ with vanishing vacuum energy [16]. This phase appears after symmetry breaking in the $SU(3) \times SU(2) \times U(1)$ domains, and proves to be absolutely stable in the theories under consideration [37].

One could argue, however, that after reheating of the universe in the end of inflation, the phase transition with the $SU(5)$ symmetry restoration may occur, and the problem of symmetry breaking in SUSY GUTs may appear again. Fortunately, as it was shown in [25, 32], reheating in the theories with large H is rather ineffective, and the temperature T_r of the universe after reheating typically is many orders smaller than the critical temperature of symmetry restoration in SUSY GUTs, $T_c \ll 10^{16}$ GeV [25]. (For a discussion of this question see section 2).

We would like to note also that the large size of domains, like the large size of bubbles in the first version of the new inflationary universe scenario [1], implies that no monopoles appear in the observable part of the universe after the phase transition. This solves the primordial monopole problem, which was claimed to exist in the primordial inflation scenario [21, 23]. In the nonminimal $SU(5)$ theory containing, e.g., the term $W' = \alpha Z' (\mathrm{Tr}\, \Phi^2 - \mu^2)$ in the superpotential, the primordial monopole problem can be solved even in a more simple way [25]; see section 2.

At present it seems that the new inflationary universe scenario can be completely realized in the context of $N = 1$ supergravity coupled to matter [23, 25]. It is pleasant to note that the inflationary universe is not inhospitable to supersymmetric theories. It provides a possible solution to the gravitino problem [10, 34] and to the problem of symmetry breaking in SUSY GUTs.

4 Chaotic Inflation

As it was argued in section 2, it seems plausible that the new inflationary universe scenario [1 − 8] can be completely realized in the context of $N = 1$ supergravity coupled to matter [23, 25]. This possibility became rather attractive after a simple solution of the primordial monopole problem in this scenario was suggested [25, 26]. Still, it may seem that inflation of the universe is a rather peculiar phenomenon, which may occur only in some restricted class of theories.

In the present paper we would like to show that under some natural assumptions about initial conditions in the very early universe, inflation may occur in a wide class of theories, including the Weinberg-Salam theory and the $SU(5)$ theory. Strong constraints on these theories still may appear if one wishes to obtain small density perturbations after inflation. However, the original problem of obtaining a sufficiently large inflation in a context of some natural theory of elementary particles does not seem to be a problem anymore.

To illustrate the main idea of the new scenario, which we call the chaotic inflation scenario for reasons to be explained shortly, let us first consider an extreme and unrealistic example of a theory of a scalar field ϕ with a degenerate effective potential $V(\phi) = V_0 = \text{constant} > 0$. It is clear that in such a theory there are no reasons to expect that the classical field ϕ is equal to any particular value (say, $\phi = 0$) in the whole universe. On the contrary, one may expect that *all* values of ϕ may appear in different, sufficiently mutually distant regions of space with equal probability. This means that in such a theory the field ϕ in different regions of the early universe may take absolutely arbitrary values varying from $-\infty$ to $+\infty$. In particular, the values $\phi \gg M_P \sim 10^{19}$ GeV are quite legitimate. The only possible constraint on the distribution of the field ϕ in the very early universe is that $(\partial_\mu \phi)^2 \lesssim M_P^4$ for $\mu = 0, 1, 2, 3$, since otherwise the corresponding part of the universe would be in the pre-Planckian era, in which the classical description of space and time is hardly possible. During the expansion of the universe, both the energy density connected with the inhomogeneity of the field ϕ, which is proportional to $(\partial_\mu \phi)^2$, and the thermal energy, $\sim T^4$, rapidly decrease, the energy density of matter reduces to $V(\phi) = V_0 = $ constant 0, and the universe becomes divided into many exponentially large domains containing an almost homogeneous field ϕ. This model, of course, is unrealistic, but it is clear that some very similar effects may appear in any theory with a sufficiently flat effective potential $V(\phi)$.

Indeed, let us consider now a theory $V(\phi) = (\lambda/4)\phi^4$ with $\lambda \ll 1$. A classical description of the universe evolution becomes possible only after the Planck time $t \sim t_P \sim M_P^{-1}$, when the energy density becomes smaller than M_P^4. Before that time the universe is usually assumed to be in some chaotic quantum state; see, e.g., [48–50]. If λ is sufficiently small, there is no reason to expect that at $t \sim t_P$ the field ϕ should be equal to zero everywhere. On the contrary, one may expect that the field ϕ in different regions of space may take any value between $-M_P/\lambda^{1/4}$ and $+M_P/\lambda^{1/4}$, so that $V(\phi) = \lambda\phi^4/4 \lesssim M_P^4$. [Note that the value of $V(\phi)$ at $t \sim t_P \sim M_P^{-1}$ can be measured only with an accuracy $\sim M_P^4$ due to the uncertainty principle.] We shall discuss later which configurations of the field ϕ in the very early universe might be most probable, but now we just note that since the fields ϕ as large as $M_P/\lambda^{1/4}$ are not forbidden by any known laws of nature (quantum gravity effects become important at $\phi > M_P/\lambda^{1/4}$, where $V(\phi) > M_P^4$), in the open (infinite) universe at $t \sim t_P$, there should exist infinitely many locally homogeneous and isotropic domains of the size $l \gg M_P^{-1}$ containing locally homogeneous field ϕ such that $M_P \lesssim \phi \lesssim M_P/\lambda^{1/4}$. One may wonder what is the reason for discussing such nonequilibrium field configurations, which may occur only at some very early stages of the universe evolution and which rapidly disappear due to the rolling of the field ϕ to the minimum of $V(\phi)$. To answer this question let us consider the evolution of the locally homogeneous field ϕ in the early universe.

The part of the universe inside a domain filled with a homogeneous field ϕ expands as de Sitter space with the scale factor $a(t) = a_0 \exp(Ht)$, where

$$H = \sqrt{\frac{8\pi}{3} \frac{V(\phi)}{M_P^2}} = \sqrt{\frac{2\pi\lambda}{3} \frac{\phi^2}{M_P}}. \tag{18}$$

The equation of motion of the field ϕ inside this domain is

$$\ddot{\phi} + 3H\dot{\phi} = -\lambda\phi^3, \tag{19}$$

which implies that at $\phi^2 \gg M_P^2/6\pi$

$$\phi = \phi_0 \exp\left(-\frac{\sqrt{\lambda}M_P}{\sqrt{6\pi}}t\right). \tag{20}$$

This means that at $\lambda \ll 1$ the typical time $\Delta t \sim \sqrt{6\pi}/\sqrt{\lambda}M_P$, during which field ϕ decreases considerably, is much greater than the Planck time $T_P \sim (6M_P)^{-1}$ (see later). During the main part of this period the universe expands exponentially, and the scale factor of the universe after expansion grows as follows:

$$a(\Delta t) \sim a_0 e^{H\Delta t} \sim a_0 \exp\left(2\pi\frac{\phi_0^2}{M_P^2}\right). \tag{21}$$

From (21) it follows that inflation of the universe is sufficiently large $(e^{H\Delta t} > e^{65}$ [9]) if

$$\phi_0 = \phi(t = 0) \gtrsim 3 M_P. \tag{22}$$

Such a value of ϕ_0 is quite possible if $\lambda\phi_0^4/4 < M_P^4$, which implies that

$$\lambda \lesssim 10^{-2}. \tag{23}$$

For a typical initial value $\phi_0 \sim M_P/\lambda^{1/4}$, (21) yields

$$a(\Delta t) \sim a_0 e^{2\pi/\sqrt{\lambda}}. \tag{24}$$

Thus we see that the domains of the field $\phi_0 > 3 M_P$ (22), which inevitably exist in the universe with a chaotic initial distribution of the field ϕ in the theory $(\lambda/4)\phi^4$ with $\lambda \lesssim 10^{-2}$, exponentially grow, like the bubbles in the first version of the new inflationary universe scenario [1], and give rise to miniuniverses whose size exceeds that of the observable part of our universe. The conditions necessary for the existence of intelligent life can hardly appear in the domains with $\phi_0 < M_P$ (if the standard version of the new inflationary universe scenario [1–8] is not realized at $\phi_0 < M_P$). However, for our purposes it is sufficient that in the chaotic inflation scenario there should exist infinitely many domains with $\phi_0 \gtrsim 3 M_P$, in which a very large inflation occurs and life becomes possible [16].

In the previous discussion we have neglected high-temperature corrections to $V(\phi)$, which usually lead to symmetry restoration and to vanishing of the field ϕ in the early universe [42]. However, the high-temperature symmetry restoration does not actually occur in the regions of the universe in which the field ϕ initially was sufficiently large, so that $\phi \gtrsim M_P$ and $V(\phi) \gtrsim 10^{-5}M_P$. Indeed, according to [42], symmetry restoration in the early universe occurs due to the high-temperature contribution to the effective mass squared of the field ϕ [to the curvature of $V(\phi)$]

$$\Delta m^2(T) = CT^2, \tag{25}$$

where C is some constant. In GUTs $C = O(1)$, whereas in the theory of chiral singlet fields coupled to supergravity the value of C may be negligibly small [23, 25]. Let us consider the most dangerous case, $C \sim 1$. In this case the time necessary for the field relaxation near the minimum of $V(\phi)$ should exceed the time $\tau \sim T^{-1}$, which is necessary for one oscillation of the field ϕ to occur. Let us estimate the age of the universe at that moment. Since the value of $V(\phi)$ remains essentially constant during the process, one may expect that initially the energy density of the universe was dominated by relativistic particles. In this case the age of the universe is given by [50]

$$T = \frac{1}{4}\sqrt{\frac{3}{2\pi}}\frac{M_P}{\sqrt{\rho}}, \tag{26}$$

where ρ is the energy density of relativistic particles,

$$\rho = \frac{N\pi^2}{30}T^4. \tag{27}$$

Here N is the effective number of degrees of freedom (of particles) in the theory. Typically in GUTs $N \gtrsim 200$. By comparing $\tau \sim T^{-1}$ and t for $N \gtrsim 200$, one concludes that the field ϕ can be influenced by high-temperature effects only at $T \lesssim M_P/50$. However, at such a temperature the energy density of relativistic particles becomes negligibly small compared with $V(\phi)$ in the domains with $V(\phi) \sim M_P^4$. Therefore much earlier than the temperature T decreases to $M_P/50$, expansion of such domains becomes exponential, temperature inside the domains becomes exponentially small, and all high-temperature effects disappear. This means that the high-temperature effects are irrelevant for the behavior of the field ϕ in the domains with $V(\phi) \sim M_P^4$. (We do not conisder here the effects connected with the Hawking "temperature" $T_H = H/2\pi$, which are important in some theories with comparatively large coupling constants, but can be included in the definition of $V(\phi)$ in curved space [16].)

Now let us try to understand which initial configurations of the field ϕ might be most probable. A classical description of the expansion of the universe becomes possible at $\rho \lesssim \rho_P = M_P^4$, which according to (26) occurs at $t_P \sim (6M_P)^{-1}$. The values of the field ϕ in the domains displaced at a distance l from each other presumably are almost uncorrelated when l is much greater than the size of the horizon, $\sim t_P$. Some small correlation may still exist if one assumes that $(\partial_\mu\phi)^2 \lesssim M_P^4$, since we would like to consider the post-Planckian era, in which energy density is smaller than M_P^4. Thus one may assume that the field ϕ in different regions of space of size t_P takes (almost) random values from the interval $-M_P/\lambda^{1/4} \lesssim \phi \lesssim M_P/\lambda^{1/4}$ in such a way that $|\partial_\mu\phi| \lesssim M_P^2$. In this case a simple statistical analysis indicates that the universe at $t \sim t_P$ is divided into domains filled with the field ϕ of a different sign. The typical amplitude of the field ϕ inside each domain is $O(M_P/\lambda^{1/4}) \gg M_P$, and the typical size of a domain is $O[6/(M_P\sqrt{\lambda})] \gg M_P^{-1} \sim H^{-1}$. Each domain becomes a miniuniverse after the exponential expansion. The only part of the universe that does not expand exponentially is the space between the domains, in which $\phi < M_P$. However, with a decrease of λ most of the universe becomes contained inside the domains with $\phi \gtrsim 3M_P$, and the exponential expansion of most of the universe occurs. Moreover, even if the probability of existence of domains with $\phi > 3M_P$ of the size $l \gg H^{-1}$ is very small [suppose, e.g., that our assumption $(\partial_\mu\phi)^2 \lesssim M_P^4$ is not valid], one may argue that just these domains will cover most of the physical volume of the universe, since other parts of the universe do not expand exponentially. In any case, in the infinite (open) universe at $t \sim t_P$ there should exist infinitely many domains of the desirable type, which give rise to an infinite number of miniuniverses in which life may exist [16].

The main difficulty of our scenario is similar to that of all other versions of the inflationary universe scenario: It is not very easy to obtain density perturbatios $\delta\rho/\rho \sim 10^{-4}$ after the inflation [11, 12–15]. We have estimated the value of $\delta\rho/\rho$ at the galactic scale by the use of the methods of [12–15]. The result numerically is very similar to that obtained in [12–15]: To get $\delta\rho/\rho$ as small as 10^{-4}, one should have $\lambda \sim 10^{-10}$ at $\phi \sim 3M_P$. This is still a rather strong constraint. However, in our scenario this constraint is not as dangerous as before. Many theories of some field ϕ very weakly interacting with itself and with all other fields can be easily suggested if there is no need for the effective potential $V(\phi)$ to satisfy very restrictive and not very natural constraints both on $m^2(\phi) = V''(\phi)$ and on $\lambda(\phi)$ near $\phi = 0$ [1–8, 16], to have a zero-temperature minimum at a scale $\phi \ 10^{16}$– 10^{19} GeV, and to have a high-temperature minimum at $\phi = 0$. In particular, the chaotic inflation scenario can be implemented in the context of

$N = 1$ supergravity [23, 25], and even in the models considered in [21, 23], in which the standard scenario [1–8] cannot be realized [25, 28] (see section 2).

One should note also that the calculation of $\delta\rho/\rho$ in [12–15] originally was based on theories in which reheating of the universe occurs during the time $\Delta t \lesssim H^{-1}$, i.e., immediately after the end of inflation [79]. However, in the models based on the $N = 1$ supergravity the scalar field ϕ remains oscillating near the minimum of $V(\phi)$ during the time Δt, which is many orders greater than H^{-1} [25, 34]. Investigating generation of density perturbations in such theories is more complicated and may lead to a smaller value of $\delta\rho/\rho$ [51], which may relax the constraint $\lambda \sim 10^{-10}$.

Let us summarize the results. We have suggested a novel version of the inflationary universe scenario based on the assumption of chaotic initial conditions in the very early universe. This assumption seems very natural since presumably there was no (or almost no) correlation between the physical processes at the different regions of space displaced at a distance $l \gg t$ from each other. Under this assumption, at $t \sim t_P \sim M_P^{-1}$ there should exist some sufficiently isotropic and homogeneous domains of space of the size greater than M_P^{-1} filled with the classical homogeneous field $\phi \gtrsim M_P$. The amplitude of this field in the early universe decreases very slowly, and during this time the domains filled with the field $\phi \gtrsim M_P$ expand exponentially. In the theories of the type $\lambda\phi^4/4$, the scale factor of space inside these domains increases $\sim e^{2\pi(\phi^2/M_P^2)}$ times during the period of exponential expansion. This suggests that the main part of the physical volume of our universe appears not due to expansion of domains of $\phi < M_P$, but due to the exponential expansion of domains filled with the maximally nonequilibrium field $\phi \gg M_P$. The only possible constraint on the amplitude of the field ϕ at $t \sim t_P$ is connected with the condition $V(\phi) \lesssim M_P^4$, which should be valid in the post-Planckian epoch [though in principle the exponential expansion could start earlier, at $V(\phi) \gg M_P^4$].

Our scenario may resemble many previous attempts to obtain a isotropic and homogeneous universe after expansion of the universe with chaotic initial conditions; see, e.g., [49, 50]. However, the main aim of the previous works was to obtain a *globally* homogeneous and isotropic universe, which is hardly possible. In our approach, as in other versions of the new inflationary universe scenario [1–8], it is sufficient to have small homogeneous and isotropic domains, which after exponential expansion become greater than the observable part of the universe.

On the other hand, the chaotic inflation scenario differs considerably from all other versions of the inflationary universe scenario suggested so far [9, 1–8, 23–25], since it is *not* based on the theory of high-temperature phase transitions in the early universe. According to this scenario, a very large inflation occurs in a wide class of theories in which the effective potential is not too curved, $\phi > 3M_P(\lambda < 10^{-2})$ *at least for one of the scalar fields ϕ*. Unlike the previous versions of the new inflationary universe scenario, in this scenario there is no need for $V(\phi)$ to have a minimum at $\phi \sim 10^{15}-10^{19}$ GeV: The chaotic inflation scenario can be implemented as well in the theories in which there is no symmetry breaking in the equilibrium state at all.

Now let us note that even in the theories with $\lambda \sim 10^{-1}$, a rather large inflation should occur in our scenario (24). Moreover, it is usually believed that in the pre-Planckian epoch there was a singularity, and the energy density was much greater than M_P^4. Therefore it seems natural to assume that large fields ϕ with $V(\phi) \gg M_P^4$ are also possible at the very early stage of the universe evolution. In this case our constraints on λ (23) disappear completely. Note also that our conclusion about the chaotic inflation in the domains with $\phi > M_P$ is almost model independent. Indeed, the main condition for the large inflation is $H^2 \gg M^2(\phi)$ [4, 16], where $H^2 = 8\pi V(\phi)/(3M_P^2)$ and $m^2(\phi) = d^2 V/d\phi^2$.

At large ϕ the value of $m^2(\phi)$ can be roughly estimated as V/ϕ^2, from which it follows that $H^2 \gg m^2(\phi)$ at $\phi^2 \gg M_P^2$, and inflation occurs for all reasonable potentials $V(\phi)$. This suggests that inflation is not a peculiar phenomenom, desirable for a number of well-known reasons [1–9, 16], but is a natural, and may even be inevitable, consequence of the chaotic initial conditions in the very early universe.

5 Conclusions

Let us summarize our results. They show that large inflation in the very early universe is actually possible, and that the necessary magnitude of density perturbations $\delta\rho/\rho \sim 10^{-4}$ can be obtained. Moreover, in the chaotic universe inflation seems not only desirable, but almost inevitable. It appears that after inflation the universe becomes divided into many large domains, in which all possible symmetry breaking patterns can be realized. In particular, in the supersymmetric $SU(5)$ theory the universe becomes divided into many domains whose size exceeds that of the observable part of the universe, some of these domains being filled by the field ϕ in the phase $SU(3) \times SU(2) \times U(1)$, whereas other domains contain the field ϕ in the phase $SU(4) \times U(1)$, etc. In the Kaluza-Klein theories the universe after inflation may become divided into very large domains of different dimensionality d [16]. This fact may serve as a basis for a Weak Anthropic Principle: All possible phases that may appear after inflation and that are sufficiently stable should exist in some of the domains (of the miniuniverses), and then life appears just in those domains that are sufficiently suitable for its existence. For example, life of our type is possible only if we are in the $SU(3) \times SU(2) \times U(1)$ phase with a sufficiently small cosmological constant (!), and if our domain is 4-dimensional (atoms and planetary systems are possible only at $d = 4$ [52]). Of course, it is quite possible that some better type of life may exist in some domains other than ours. However, since the size of each domain is much greater than the size of the observable part of the universe, we probably will be unable to verify this possibility in the near future. Moreover, since we do not know whether life can exist in any other domain, it is worth trying to maintain life at least in our own domain.

In conclusion I would like to say that from my point of view the most importat feature of the new inflationary universe scenario is not the possibility of solving a number of very difficult cosmological problems, but the possibility of obtaining an absolutely unexpected view on the structure of the universe as a whole and at the very early stages of its evolution. It is difficult to predict whether this scenario in its present form will still exist come Shelter Island Conference III. However, it seems very likely that the future theory of the evolution of the universe will incorporate some of the new knowledge that we have gained during the last two years.

Appendix: The Cosmological Constant Problem

As realized after [53], an unbelievably precise fine tuning is necessary in order to have the vacuum energy density (the cosmological constant) smaller than $\sim 10^{29}$ g/cm^3 in the theories with spontaneous symmetry breaking. There exist different possible approaches to this problem; see e.g., [54–56]. Unfortunately, the cosmological constant problem remains unsolved.

A very interesting possibility for solving this problem has been suggested recently by Hawking [57]. This possibility, discussed in his talk at this conference,* is based on the investigation of the space-time foam structure of the universe, which is manifest at very small distances, but which may have observable consequences for the large-scale structure of the universe [57]. However, since the theory of the space-time foam is still not quite complete, it would be interesting also to consider other possibilities for solving the cosmological constant problem. One such possibility, which is also very far from being completely realized, will be suggested now.

Let us assume that the total Lagrangian of matter consists of three parts:

$$\mathscr{L} = -\frac{R}{16\pi G} + L(\phi) - L(\tilde{\phi}),\tag{A.1}$$

where $-R/(16\pi G)$ is the Einstein Lagrangian; $L(\phi)$ is the Lagrangian of matter fields ϕ; and $-L(\tilde{\phi})$ is the Lagrangian of the same functional form as $L(\phi)$, but is a function of other fields $\tilde{\phi}$, and enters into $\mathscr{L}$ (A.1) with a *negative sign*.

At first sight, the theory (A.1) is unstable with respect to generation of infinitely large fields due to the negative sign of $-L(\tilde{\phi})$. However, all Lagrange equations for the field $\tilde{\phi}$ are *the same* as for the field ϕ, and there is *no* instability in the classical theory (A.1). The only difference between these theories is that the field ϕ in the theory $L(\phi)$ will be at the minimum ϕ_0 of the effective potential $V(\phi)$, whereas the field $\tilde{\phi}$ will be at the *maximum* (at $\tilde{\phi} = \tilde{\phi}_0$) of the effective potential $V(\tilde{\phi})$. Since we have $L(\phi)$ and $L(\tilde{\phi})$ of the same functional form, $\phi_0 = \tilde{\phi}_0$ and $V(\phi_0) = -V(\tilde{\phi}_0)$. Therefore the total vacuum energy density of the theory (A.1) will be zero *independently of the value of* $V(\phi_0)$:

$$\varepsilon_{\text{vac}} = V(\phi_0) + V(\tilde{\phi}_0) = 0.\tag{A.2}$$

This is exactly what is needed for a solution of the cosmological constant problem.

One may wonder whether the new particles $\tilde{\phi}$, which we call "down" particles to distinguish them from the usual "up" particles ϕ, are harmless. To answer this question let us note that "up" and "down" particles interact with each other only gravitationally. Therefore only very large, inhomogeneously distributed amounts of "down" matter (like planets) could affect us some way. Moreover, in the inflationary universe scenario one could see

* Editors' note: The paper read to the conference by Hawking appears in this volume.

no "down" particles at all. Indeed, let us consider a domain of the field $\phi \gtrsim 3M_P$ in which the field $\tilde{\phi}$ initially was much smaller than ϕ. Such a domain will exponentially expand (see section 4). If $\tilde{\phi} \ll \phi$, then the process of the of production $\tilde{\phi}$ particles finishes *before* the end of inflation, and then inflation pushes all "down"particles away from the observable part of the universe.

There still exist many problems associated with our suggestion. First of all, there is the vacuum instability due to the gravitational interaction between "up" and "down" particles. Indeed, a pair of "down" and a pair of "up" particles can be produced from vacuum without violating energy conservation. However, this process does not produce any energy density and does not lead to the creation of a preferred reference frame; this is just a vacuum reconstruction process. A preliminary investigation of this question indicates that such a boiling of the vacuum state may not lead to any unacceptable observational consequences.

We realize that the possibility of having "up" and "down" worlds may sound absolutely crazy, but at least at the level at which quantum gravity effects can be neglected we see no reasons why such a scenario could not work. Quantum gravity effects may lead to considerable complications of this scenario, but the general idea of the vacuum energy cancellation between the "up" and "down" vacua seems so simple and natural that it would be a pity to abandon such a possibility without a detailed investigation. We hope to return to a discussion of this question in a separate publication.

Acknowledgments

I am thankful to the Rockefeller University, and especially to N. N. Khuri for the hospitality during Shelter Island Conference II. I am also thankful to G. V. Chibisov, S. Coleman, A. S. Goncharov, A. Guth, S. W. Hawking, R. E. Kallosh, D. A. Kirzhnits, V. F. Mukhanov, P. J. Steinhardt, M. I. Vysotsky, S. Weinberg, and Ya. B. Zeldovich for many valuable discussions.

References

[1]   A. D. Linde, *Phys. Lett.* B108, 389 (1982).
[2]   S. W. Hawking and I. G. Moss, *Phys. Lett.* B110, 35 (1982).

[3] A. Albrecht and P. J. Steinhardt, *Phys. Rev. Lett.* 48, 1220 (1982).

[4] A. D. Linde, *Phys. Lett.* B114, 431 (1982).

[5] A. D. Linde, *Phys Lett.* B116, 340 (1982).

[6] A. D. Linde, *Phys Lett.* B116, 335 (1982).

[7] A. D. Dolgov and A. D. Linde, *Phys. Lett.* B116, 329 (1982).

[8] A. A. Starobinsky, *Phys. Lett.* B117, 175 (1982).

[9] A. H. Guth, *Phys. Rev.* D23, 347 (1981).

[10] J. Ellis, A. D. Linde, and D. V. Nanopoulos, *Phys. Lett.* B118, 59 (1982).

[11] V. F. Mukhanov and G. V. Chibisov, *JETP Lett.* 33, 532 (1981); V. F. Mukhanov and G. V. Chibisov, *ZhETF* 83, 475 (1982).

[12] S. W. Hawking, *Phys. Lett.* B115, 295 (1982).

[13] A. H. Guth and S.-Y. Pi, *Phys. Rev. Lett.* 49, 1110 (1982).

[14] J. M. Bardeen, P. J. Steinhardt, and M. S. Turner, Preprint EFI 83–13 (1983).

[15] S. W. Hawking and I. G. Moss, Cambridge University Preprint (1982).

[16] A. D. Linde, Lebedev Phys. Inst. Preprints 30 and 50 (1983), also in *The Very Early Universe* S. Hawking, G. Gibbons, and S. Siklos, editors (Cambridge University Press, Cambridge, 1983).

[17] J. Ellis, D. V. Nanopoulos, K. A. Olive, and K. Tamvakis, *Phys. Lett.* B118, 335 (1982).

[18] A. Albrecht, S. Dimopoulos, W. Fischler, E. Kolb, S. Raby, and P. J. Steinhardt, Los Alamos Preprint LA-UR-82-2947 (1982).

[19] E. Witten, *Phys. Lett.* B105, 267 (1981); S. Dimopoulos and S. Raby, Los Alamos Preprint LA-UR-82-1282 (1982).

[20] C. E. Vayonakis, Preprint PAR LPTHE 82.17 (1982).

[21] J. Ellis, D. V. Nanopoulos, K.A. Olive, and K. Tamvakis, CERN Preprint TH. 3404.

[22] J. Ellis, D. V. Nanopoulos, K. A. Olive, and K. Tamvakis, *Phys. Lett.* B120, 331 (1983).

[23] D. V. Nanopoulos, K. A. Olive, M. Srednicki, and K. Tamvakis, *Phys. Lett* B123, 41 (1983).

[24] E. Cremmer, B. Julia, J. Scherk, S. Ferrara, L. Girardello, and P. van Nieuwenhuizen, *Nucl. Phys.* B147, 105 (1979).

[25] A. D. Linde, *Pisma ZhETF* 37, 606 (1983); A. D. Linde, Lebedev Phys. Inst. Preprint 151 (1983), submitted to *Phys. Lett.*

[26] A. D. Linde, "Inflation Can Break Symmetry in SUSY," Lebedev Phys. Inst. Preprint (1983), submitted to *Phys. Lett.*

[27] A. D. Linde, "Chaotic inflation," Lebedev Phys. Inst. Preprint (1983), submitted to *Phys. Lett.*

[28] B. Gato, J. Leon, M. Quiros, and M. Ramon-Medrano, CERN Preprint Th. 3538 (1983).

[29] A. Vilenkin and L. H. Ford, *Phys. Rev.* D26, 1231 (1982).

[30] A. D. Linde, *Phys. Lett.* B96, 293 (1980).

[31] D. Gross, R. Pisarski, and L. Yaffe, *Rev. Mod. Phys.* 53, 43 (1981).

[32] R. A. Brandt and F. Neri, *Nucl. Phys.* B145, 221 (1978).

[33]   A. Billoire, G. Lazardies, and Q. Shafi, *Phys. Lett.* B103, 43 (1981); T. A. DeGrand and D. Toussaint, *Phys. Rev.* D25.

[34]   D. V. Nanopoulos, K. A. Olive, and M. Srednicki, CERN Preprint TH. 3555 (1983).

[35]   S. Weinberg, *Phys. Rev. Lett.* 48, 1303 (1982).

[36]   S. Ferrara, D. V. Nanopoulos, and C. A. Savoy, CERN Preprint TH. 3442 (1982).

[37]   S. Weinberg, *Phys. Rev. Lett.* 48, 1776 (1982).

[38]   N. V. Dragon, *Phys. Lett.* B113, 288 (1982).

[39]   P. H. Frampton and T. W. Kephart, *Phys. Rev. Lett.* 48, 1237 (1982); F. Buccella, J. P. Derendinger, S. Ferrara, and C. A. Savoy, *Phys. Lett.* B115, 375 (1982).

[40]   E. S. Fradkin, in Proceedings of the Seminar Quark-80 Sukhumi (April 1980), p. 80.

[41]   N. Sakai, *Z. Phys.* C11, 153 (1982).

[42]   D. A. Kirzhnits, *JETP Lett.* 15, 529 (1982); D. A. Kirzhnits and A. D. Linde, *Phys. Lett.* B42, 471 (1972), *ZhETF.* 1263 (1974), *Ann. Phys. (N.Y.)* 101, 195 (1976); S. Weinberg, *Phys. Rev.* D9, 3357 (1974); L. Dolan and R. Jackiw, *Phys. Rev.* D9, 3320 (1974); A. D. Linde, *Rep. Progr. Phys.* 42, 389 (1979).

[43]   D. V. Nanopoulos and K. Tamvakis, *Phys. Lett.* B110, 449 (1982); M. Srednicki, *Nucl. Phys.* B202, 327 (1982).

[44]   M. Srednicki, *Nucl. Phys.* B206, 132 (1982); D. V. Nanopoulos, K. A. Olive, and K. Tamvakis, *Phys. Lett.* B15, 15 (1982).

[45]   D. V. Nanopoulos, K. A. Olive, M. Srednicki, and K. Tamvakis, CERN Preprint TH. 3503 (1983).

[46]   A. D. Linde, *Nucl. Phys.* B216, 421 (1983).

[47]   T. S. Bunch and P. C. W. Davies, *Proc. R. Soc.* A360, 117 (1978).

[48]   J. A. Wheeler, in *Relativity, Groups and Topology*, B. S. and C. M. DeWitt, editors (Gordon and Breach, New York, 1964); S. W. Hawking, *Nucl. Phys.* B144, 349 (1978).

[49]   M. J. Rees, *Phys. Rev. Lett.* 28, 1669 (1972).

[50]   Ya. B. Zeldovich and I. D. Novikov, *Structure and Evolution of the Universe* (Moscow, Nauka, 1975).

[51]   G. V. Chibisov, private communication.

[52]   P. Ehrenfest, *Proc. Amsterdam Acad.* 20, 200 (1917).

[53]   A. D. Linde, *JETP Lett.* 19, 183 (1974); M. Veltman, University of Utrecht Preprint (1974), *Phys. Rev. Lett.* 34, 777 (1975); J. Dreitlein, *Phys. Rev. Lett.* 33, 1243 (1974).

[54]   A. D. Dolgov, in *The Very Early Universe*, S. Hawking, G. Gibbons, and S. Siklos, editors (Cambridge University Press, Cambridge, 1983).

[55]   S. Hawking, "The Cosmological Constant and the Weak Anthropic Principle," DAMTP Preprint (1982).

[56]   A. Zee, University of Washington Preprint (1983), to appear in the Proceedings of the 1983 Coral Gables Conference.

[57]   S. Hawking, "The Cosmological Constant," DAMTP Preprint (1983).

The Cosmological Constant Is Probably Zero

S. W. Hawking

It is suggested that the apparent cosmological constant is not necessarily zero, but that zero is by far the most probable value. One requires some mechanism like a three-index antisymmetric tensor field or topological fluctuations of the metric that can give rise to an effective cosmological constant of arbitrary magnitude. The action of solutions of the euclidean field equations is most negative, and the probability is therefore highest, when this effective cosmological constant is very small.

The cosmological constant is probably the quantity in physics that is most accurately measured to be zero: observations of departures from the Hubble law for distant galaxies place an upper limit of the order of

$$\frac{|\Lambda|}{m_p^2} < 10^{-120}, \tag{1}$$

where m_p is the Planck mass. On the other hand, one might expect that the zero point energies of quantum fluctuations would produce an effective or induced Λm_p^{-2} of order one if the quantum fluctuations were cut off at the Planck mass. Even if this were renormalized exactly to zero, one would still get a change in the effective Λ of order $\mu^4 m_p^{-2}$ whenever a symmetry in the theory was spontaneously broken, where μ is the energy at which the symmetry was broken. There are a large number of symmetries that seem to be broken in the present epoch of the universe, including chiral symmetry, electroweak symmetry, and, possibly, supersymmetry. Each of these would give a contribution to Λ that would exceed the upper limit (1) by at least forty orders of magnitude.

It is very difficult to believe that the bare value of Λ is fine tuned so that after all the symmetry breakings, the effective Λ satisfies the inequality (1). What one would like to find is some mechanism by which the effective value of Λ could relax to zero. Although there have been a number of attempts to find such a mechanism (see, e.g., [1, 2]), I think it is fair to say that no satisfactory scheme has been suggested. In this paper, I want to propose instead a very simple idea: the cosmological constant can have any value, but it is much more probable for it to have a value very near zero. A preliminary version of this argument was given in (3).

My proposal requires that a variable effective cosmological constant be generated in some manner. One possibility would be to include the value of the cosmological constant in the variables that are integrated over in the path integral. A more attractive way would be to introduce a three-index antisymmetric tensor field $A_{\mu\nu\rho}$. This would have gauge transformations of the form

$$A_{\mu\nu\rho} \to A_{\mu\nu\rho} + \nabla_{[\mu} C_{\nu\rho]}. \tag{2}$$

The action of the field is F^2, where F is the field strength formed from A:

$$F_{\mu\nu\rho\sigma} = \nabla_{[\mu} A_{\nu\rho\sigma]}. \tag{3}$$

Such a field has no dynamics: the field equations imply that F is a constant multiple of the four-index antisymmetric tensor $\varepsilon_{\mu\nu\rho\sigma}$. However, the F^2 term in the action behaves like an effective cosmological constant [4]. Its value is not determined by field equations. Three-index antisymmetric tensor fields arise naturally in the dimensional reduction of $N = 1$ supergravity in 11 dimensions to $N = 8$ supergravity in 4 dimensions. Other mechanisms that would give an effective cosmological constant of arbitrary magniture include topological fluctuations of the metric [3] and a scalar field ϕ with a potential term $V(\phi)$ but no kinetic term. In this last case, the gravitational field equations could be satisfied only if ϕ was constant. The potential $V(\phi)$ then acts as an effective cosmological constant.

I assume that the quantum theory is defined by a path integral over euclidean, i.e., positive definite, metrics. The dominant contributions to the path integral are expected to come from metrics that are near to solutions of the field equations. Of particular interest are solutions in which the dynamical matter fields, i.e., the matter fields apart from $A_{\mu\nu\rho}$ or ϕ, are near their ground state values over a large region. This would be a reasonable approximation to the universe at the present time. The ground state of the matter fields plus the contribution of the $A_{\mu\nu\rho}$ or ϕ fields will generate an effective cosmological constant Λ_e. If the effective value Λ_e is positive, the solutions are necessarily compact and their four-volume is bounded by that of the solution of greatest symmetry, the four-sphere of radius $(3\Lambda_e^{-1})^{1/2}$. The euclidean action $\hat{I}$ will be negative and will be bounded below by

$$-\frac{9\pi m_p^2}{4\Lambda_e} \tag{4}$$

if Λ_e is negative, the solutions can be either compact or noncompact [5]. If they are compact, the action $\hat{I}$ will be finite and positive. If they are noncompact, $\hat{I}$ will be infinite and positive.

The probability of a given field configuration will be proportional to

$$\exp(-\hat{I}) \tag{5}$$

if Λ_e is negative. $\hat{I}$ will be positive and the probability will be exponentially small. If Λ_e is positive, the probability will be of the order of

$$\exp\left[\frac{3\pi m_p^2}{\Lambda_e}\right]. \tag{6}$$

Clearly, the most probable configurations will be those with very small values of Λ_e. This does not imply that the effective cosmological constant will be small everywhere in these configurations. In regions in which the dynamical fields differ from the ground state values there can be an apparent cosmological constant as in the inflationary model of the universe.

References

[1]   F. Wilczek, in *The Very Early Universe*, G. W. Gibbons, S. W. Hawking, and S. T. C. Siklos, editors (Cambridge University Press, Cambridge, 1983).

[2]   A. D. Dolgov, in *The Very Early Universe*, G. W. Gibbons, S. W. Hawking, and S. T. C. Siklos, editors (Cambridge University Press, Cambridge, 1983).

[3]   S. W. Hawking, *Phil. Trans. Roy. Soc. (London)* A310, 303–310 (1983).

[4]   A. Aurilla, H. Nicolai, and P. K. Townsend, *Nucl. Phys.* B176, 509 (1980).

[5]   S. W. Hawking, *Nucl. Phys.* B144, 349 (1978).

A Brief Survey of Superstring Theory

John H. Schwarz

William James once said, "Let no man join together what God has put asunder." Nonetheless, the attempt to uncover the underlying symmetry and unity of the fundamental equations of physics, which are concealed so subtly in the world we observe, is one of the most exciting intellectual challenges of our era. This great adventure, pioneered by Einstein, has made considerable headway in recent years. Our understanding of the electro-magnetic, weak, and strong forces, including their unification, may already be substantially correct. Severe problems arise when the inclusion of gravity is also attempted. All known conventional quantum field theories containing gravity, including $N = 8$ supergravity, are either known to have, or else appear very likely to have, severe nonrenormalizable ultraviolet divergences.

My own efforts in dealing with these issues have centered on the study of supersymmetrical string theories. This work originated in a collaboration with André Neveu in the early 1970s [1] and has been carried much further in a collaboration with Michael Green over the last few years [2]. Super-string theories were initially developed as part of an attempt to describe hadronic physics. However, in 1974 Joël Scherk and I realized that they are better suited to a "grand synthesis" containing gravity [3]. There are two types of superstring theories, both of which demand that the dimension of space-time is 10 (1 time and 9 space dimensions). This is considered an advantage since Yang-Mills gauge symmetries can originate as the symme-try of a compact 6-dimensional manifold (à la Kaluza-Klein). Spontaneous compactification can also give rise to supersymmetry breaking. Both types of superstring theories are free from ghosts and tachyons (problems in earlier "dual string" theories) and inevitably contain supergravity. Type I theories have been shown to be *renormalizable* in one loop, while type II theories have been shown to be *finite* in one loop. There is a fairly good handwaving argument suggesting that these properties are true to all orders in perturbation theory. This, more than anything else, convinces me that the considerable effort required to make the technical leap from fundamental point particles to fundamental 1-dimensional structures is warranted.

It is probably impossible to construct a consistent self-interacting clas-sical field theory containing massless states with spin greater than two.[1] From superalgebra representation theory it follows that the maximum number of conserved supersymmetry charges that can be linearly realized on

1. Massive higher spin states are described consistently in the string theories.

a massless multiplet is 32. For example, $N = 8$ supergravity in 4 dimensions has 8 4-component supercharges, whereas $D = 11$ supergravity has one 32 component supercharge. In 10 dimensions (the required choice for superstring theories) the minimum irreducible spinor satisfies simultaneous Majorana and Weyl conditions [4] and has 16 real components. A theory with 16 conserved supercharges $(N = 1$ in $D = 10)$ is called "type I," whereas one with 32 supercharges $(N = 2$ in $D = 10)$ is called "type II." In the latter case we distinguish cases a and b according to whether the handedness of the two chiral spinors is opposite or the same. In $D = 10$, type I and type IIb theories are "chiral," since they have an intrinsic left-right asymmetry. Type IIa theories are not chiral.

Type I superstring theory contains both open strings (with free ends) and closed strings (loops with no free ends). The mass spectrum corresponding to normal modes, before effects of interaction are taken into account, is given by

$$\alpha'(mass)^2 = 0, 1, 2, \ldots \tag{1a}$$

for the open strings and

$$\alpha'(mass)^2 = 0, 4, 8, \ldots \tag{1b}$$

for the closed strings. The Regge-slope parameter α' is related to the string tension by $T = 1/2\pi\alpha'$. The massless open-string states form a super Yang-Mills multiplet, while the massless closed-string states form a supergravity multiplet. Thus the open strings may be regarded as matter strings and the closed ones as gravity strings. The corresponding coupling constants are the Yang-Mills coupling g [which is dimensionally $(length)^3$ for $D = 10$] and the gravity coupling κ [which is $(length)^4$ for $D = 10$]. Open strings can join their ends to form closed strings, and therefore g and κ are related. Aside from a numerical factor that I have not worked out, the relation is

$$\kappa \sim g^2/\alpha'. \tag{2}$$

It is plausible that κ, α', and radii associated with spontaneous compactification (R) should all be roughly given by the Planck length to the appropriate power. It is not entirely clear, however, that this is absolutely required. If the scale corresponding to α' or R could be reached experimentally, a rich spectrum of new "elementary particles" would be found. In the former case the spins would be unbounded, whereas in the latter case only low spins would occur.

The coordinates of a superstring are its spatial location $x^\mu(\sigma)$ and Grassmann coordinates $\vartheta(\sigma)$, where σ is a parameter labeling points along the string, conventionally taken to range from 0 to π. Such a string is created or destroyed by a functional superfield. For example, in the case of open strings and an $SO(n)$ Yang-Mills group the field is $\Phi_{\alpha\beta}[x(\sigma), \vartheta(\sigma)]$, where α, $\beta = 1, 2, \ldots, n$. The indices are associated with the ends of the string. Thus

$$\Phi_{\alpha\beta}[x(\sigma), \vartheta(\sigma)] = -\Phi_{\beta\alpha}[x(\pi - \sigma), \vartheta(\pi - \sigma)], \tag{3}$$

which implies that the states at even mass levels are described by antisymmetric matrices (adjoint representation) and the ones at odd mass levels by symmetric matrices. A similar construction is possible for symplectic groups also. The closed strings are group-theory singlets created and destroyed by a field Ψ with the constraints

$$\Psi[x(\sigma), \vartheta(\sigma)] = \Psi[x(\pi - \sigma), \vartheta(\pi - \sigma)], \tag{4}$$

$$\Psi[x(\sigma + \sigma_0), \vartheta(\sigma + \sigma_0)] = \Psi[x(\sigma), \vartheta(\sigma)], \tag{5}$$

reflecting the fact that type I superstrings are unoriented and that the origin of σ has no physical meaning for a closed string.

The superstring perturbation expansion is given by a sum of Feynman diagrams that are 2-dimensional surfaces, interpretable as the world sheets of strings. In the case of type I theories the diagrams are classified topologically by

a. orientability of the surface,
b. the number of holes in the surface,
c. the number of handles attached to the surface,
d. the choice of external open-string states and the boundaries to which they are attached, and
e. the choice of external closed-string states attached to the interior of the surface.

Some tree and one-loop diagrams have been evaluated analytically. The result is that the only divergences that occur have precisely the correct form to be absorbed in a renormalization of α'. This result is reasonable since the divergence occurs at the corner of the integration region corresponding to the limit in which the size of a hole in the surface shrinks to a point.

A potentially fatal problem for the *chiral* superstring theories (i.e., I and IIb, but not IIa) is the occurrence of anomalous divergences for physical

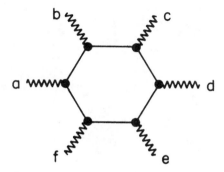

Figure 1
The hexagon diagram.

gauge currents. In 10 dimensions these problems can first occur in the hexagon diagram depicted in figure 1. In fact, if one considers the field theories that describe the massless sectors of the superstring theories, the anomaly calculations are well-defined even though these theories are much too singular in the ultraviolet to be fully consistent. One finds that the super Yang-Mills theory has a gauge-current anomaly given by

$$\partial_\mu J_\mu^a \propto \text{tr}(\Lambda^a \Lambda^b \cdots \Lambda^f)\varepsilon_{\mu_1 \cdots \mu_{10}} F_{\mu_1 \mu_2}^b \cdots F_{\mu_9 \mu_{10}}^f. \tag{6}$$

The group theory trace involves matrices Λ in the adjoint representation (since the fermions are) and is nonvanishing for *any* nonabelian gauge group.[2] The string theory having the super Yang-Mills theory for its massless sector has additional contributions to the anomaly, and a cancellation appears likely. This is possible since anomalies are controlled by short-distance behavior, which is drastically altered in string theories. Any anomalous term has to be real since the loop diagram is constructed to have all the correct discontinuities. However, string-theory loop diagrams invariably exhibit Regge behavior associating nontrivial phases with any term that can be sensibly isolated. For this reason I am optimistic that string theories can be free from bad anomalies.

Type II field theories consist of 11-dimensional supergravity and lower-dimensional theories obtained by dimensional reduction and duality transformations. In general, the physical (nongauge) symmetry group of these

2. This conclusion disagrees with [5]. $D = 10$ anomalies are also discussed in [6], but no results are stated for adjoint Λs.

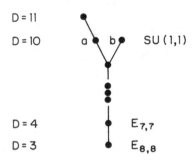

Figure 2
Type II supergravity theories.

theories contains $E_{11-D,11-D} \times SL(D-2,R)$, where the E group is a non-linearly realized internal symmetry and the SL group has to do with the description of a graviton [7]. One exception is the $D = 10$ theory (IIa), for which the $E_{1,1} \equiv SU(1,1)$ factor is missing. However, there is a second $D = 10$ extended supergravity theory (IIb) not obtainable from 11 dimensions—see figure 2—that does have this symmetry [8]. The field content of this theory consists of a complex scalar B, a complex second-rank potential $A_{\mu\nu}$, a complex Weyl spinor λ, a complex Weyl gravitino ψ_μ, the metric tensor $g_{\mu\nu}$ (or the zehnbein e_μ^m), and a real fourth-rank potential $A_{\mu\nu\rho\lambda}$ whose fifth-rank field strength is self-dual. It happens that theories with a self-dual field strength cannot be described by a manifestly covariant action principle [9]. Covariant field equations have been derived, however [10].

Type II supergravity theories, especially the $D = 11$ and $D = 4$ ones, have been seriously proposed as candidates for a complete theory of nature. In addition to all the difficult phenomenological questions, a theoretical objection to these proposals is that the theories have severe ultraviolet divergences. Specifically, I expect that at l loops the ultraviolet divergences set in at

$$D = 2 + 6/l, \tag{7}$$

a result that has been proved for $l = 1$, and follows from plausible power-counting considerations for $l > 1$ [11]. Thus the $D = 11$ theory is cubically divergent at one loop[3] and the $N = 8$, $D = 4$ theory is likely to be singular

---

3. A formal procedure of doubtful validity can be used to associate a finite value to this divergent integral. It is very unlikely to generalize to higher orders.

starting at three loops. Therefore I advocate regarding the $N = 8$ theory as a low-energy approximation to type II superstring theory, which appears to be a finite theory (for either case a or b). The chiral theory (IIb) seems to be the more attractive option if one hopes to understand the chiral structure of the standard model from first principles, although it is difficult to see in detail how this should work.

Type II superstring theory contains oriented closed strings only. In case b an elegant superfield formulation is possible [12]. The field still satisfies the constraint in (5), but no longer is subject to the one in (4), since it is oriented (as in QCD). The action for the cubic coupling of the interacting field theory has been worked out explicitly. Unlike type I theories, no elementary Yang-Mills gauge fields are introduced. Gauge fields can arise from the spontaneous compactification of spatial dimensions. Six dimensions are not quite enough to give $SU(3) \times SU(2) \times U(1)$. However, the physical $SU(1,1)$ or its $U(1)$ subgroup might be dynamically gauged, in which case it could be possible to account for all the required gauge symmetries.

The perturbation expansion of type II superstring amplitudes is remarkably simple. The diagrams consist entirely of oriented closed surfaces characterized by the number of handles and the external particle states. Thus there is *only one diagram* at each order of the perturbation expansion, in striking contrast with the limiting massless supergravity field theory. In fact, it is even conceivable that the perturbation series could be formally summed by a linear integral equation. These features, finiteness, and the richness of $N = 8$ supergravity and Kaluza-Klein compactifications are all incorporated in this intriguing theory. The theoretical and phenomenological hurdles that still remain to be surmounted are substantial, but so is the potential payoff.

Acknowledgment

This work was supported in part by the U.S. Department of Energy under contract DE-AC03-81-ER40050.

References

[1]   A. Neveu and J. H. Schwarz, *Nucl. Phys.* B31, 86 (1971), *Phys. Rev.* D4, 1109 (1971); J. H. Schwarz, *Phys. Rep.* 8C, 269 (1973).

[2]   M. B. Green and J. H. Schwarz, *Nucl. Phys.* B181, 502 (1981), *Phys. Lett.* B109, 444 (1982); J. H. Schwarz, *Phys. Rep.* 89, 223 (1982).

[3]   J. Scherk and J. H. Schwarz, *Nucl. Phys.* B81, 118 (1974), *Phys. Lett.* B57, 463 (1975).

[4]   F. Gliozzi, J. Scherk, and D. I. Olive, *Nucl. Phys.* B122, 253 (1977).

[5]   P. K. Townsend and G. Sierra, *Nucl. Phys.* B222, 493 (1983).

[6]   P. H. Frampton and T. W. Kephart, *Phys. Rev. Lett.* 50, 1343, 1347 (1983).

[7]   M. Goroff and J. H. Schwarz, *Phys. Lett.* B127, 61 (1983).

[8]   J. H. Schwarz and P. C. West, *Phys. Lett.* B126, 301 (1983).

[9]   N. Marcus and J. H. Schwarz, *Phys. Lett.* B115, 111 (1982).

[10]   J. H. Schwarz, *Nucl. Phys.* B226, 269 (1983); P. S. Howe and P. C. West, *Nucl. Phys.* B238, 181 (1984).

[11]   M. B. Green, J. H. Schwarz, and L. Brink, *Nucl. Phys.* B198, 474 (1982).

[12]   M. B. Green and J. H. Schwarz, *Phys. Lett.* B122, 143 (1983); M. B. Green, J. H. Schwarz, and L. Brink, *Nucl. Phys.* B219, 437 (1983).

Fermion Quantum Numbers in Kaluza-Klein Theory

Edward Witten

The problem of obtaining left-right asymmetry of fermion quantum numbers in Kaluza-Klein theory is discussed. In the absence of elementary gauge fields, a theorem by Atiyah and Hirzebruch states that the Dirac equation in $4 + n$ dimensions always leads to vectorlike fermion quantum numbers in 4 dimensions. The proof of this theorem is sketched. It is shown that the same holds for the Rarita-Schwinger operator *on homogeneous spaces*, but a *general* impossibility theorem for the Rarita-Schwinger field is not proved. (However, in view of the apparent restriction of supergravity to $d \leqslant 11$, this line of approach is severely constrained.) Also discussed are some Kaluza-Klein theories with elementary gauge fields, some difficulties in obtaining massless charged scalars, and some speculations about the cosmological constant.

I Introduction

Kaluza-Klein theory [1] has recently attracted increasing interest as a program for unifying gauge interactions with gravity. This theory can be viewed [2] in terms of spontaneous symmetry breaking, the compact and noncompact dimensions being on an equal footing as far as the laws of nature are concerned, just as the photon and the massive vector mesons are treated symmetrically in the standard weak interaction models. This simple observation is one of the chief reasons for the revived interest [3] in Kaluza-Klein theory.

One may start with a general relativistic theory in $4 + n$ dimensions and assume the grand state to be $M^4 \times B$, where M^4 is 4-dimensional Minkowski space and B is a compact space. Continuous symmetries of B^1 will always be observed [4] as gauge symmetries in the effective 4-dimensional world. The gauge fields (which in general can be more numerous than the extra dimensions) originate in the normal mode expansion of the fluctuations in the $(4 + n)$-dimensional metric tensor. For instance, starting in 11 or more dimensions, one can [4, 5] obtain gauge fields of $SU(3) \times SU(2) \times U(1)$. Much of this paper will be devoted to the consequences of assuming that all observed gauge forces originate in this way, as part of the metric tensor, from a theory that originally had no elementary gauge fields. However, this is not necessarily the only attractive possibility. It might be equally attractive to start with a unified theory (perhaps a supergravity [6]

---

1. That is transformations that leave fixed both the geometry of B and the expectation values of any matter fields that may be present.

or superstring [7] theory) that determines the original gauge group. [For instance, one of the $n = 2$ supergravity theories in 10 dimensions requires the existence of a $U(1)$ gauge field.] Our remarks in section VI will be relevant to such theories.

As soon as one begins to think about Kaluza-Klein theory, one faces a bewildering variety of choices. There are many assumptions one might make, and many facts about elementary particle physics one might try to explain. It appears unlikely that at the present time we can guess correctly the whole detailed form of the $(4 + n)$-dimensional laws and all the key points of the dynamics. For these reasons, it seems important to isolate problems that can be addressed without claiming to understand all the details of a theory. As will become apparent, the problem of trying to predict the quark and lepton quantum number is such a problem, and it will be our main interest in this paper. However, we shall also make some remarks on certain other qualitative problems: the gauge hierarchy problem and the problem of the cosmological constant.

Since we shall deal mainly with the problem of the fermion quantum numbers, it is worthwhile to recall briefly some aspects of that problem as it presently appears.

One of the most striking aspects of particle physics is that left-handed fermions transform under $S(3) \times SU(2) \times U(1)$ differently from the way the right-handed fermions transform. (The quantum numbers are not "vectorlike.")

For instance, left-handed quarks are $SU(2)$ doublets, but right-handed quarks are $SU(2)$ singlets. Equivalently, one may say that the fermions of given helicity form a complex representation of $SU(3) \times SU(2) \times U(1)$. The fermions of one generation transform under $SU(3) \times SU(2) \times U(1)$ as $(3, 2)^{1/3} \oplus (\bar{3}, 1)^{-4/3} \oplus (\bar{3}, 1)^{2/3} \oplus (1, 1)^{2} \oplus (1, 2)^{-1}$, which is a so-called complex representation. [In other words, it is not equivalent to its complex conjugate, which is $(3, 2)^{-1/3} \oplus (3, 1)^{4/3} \oplus (3, 1)^{-2/3} \oplus (1, 1)^{-2} \oplus (1, 2)^{1}$; by CPT, this is the representation furnished by the *right-handed* fermions.] This fact is of utmost importance, because it means that bare masses of the quarks and leptons are forbidden by gauge invariance. The quarks and leptons can acquire mass only when $SU(2) \times U(1)$ is spontaneously broken. This, in turn, means that the quarks and leptons cannot be much heavier than the mass scale at which $SU(2) \times U(1)$ is broken; they cannot have masses of order, say, the Planck mass. The relative "lightness" of the fermions would therefore be explained if the "smallness" of the $SU(2) \times$

$U(1)$ breaking scale were understood; it is not an independent problem. In this paper we shall assume, in accord with the "survival hypothesis" [8], that the only light fermions are fermions that are required to be light by gauge invariance; this assumption will not always be explicitly stated.

The fact that the quantum numbers are not vectorlike means that the spectrum of light fermions[2] depends only on the "universality class" of an $SU(3) \times SU(2) \times U(1)$ invariant theory. The lightness of the light fermions and their quantum numbers cannot be modified by any $SU(3) \times SU(2) \times U(1)$ invariant perturbations. We do not know at what length scale the spectrum of light fermions is determined, but it may be that this reflects physics at the smallest length scales.

Of course, there is no experimental *proof* that mirror fermions with $V + A$ couplings to the usual W mesons will not be found, restoring the vectorlike nature of the fermion spectrum. But there are many reasons to doubt that this will occur. If mirror fermions are discovered, we shall lose our theoretical understanding of why the quarks and leptons are $\lesssim M_W$ in mass. (Of course, in any case we don't understand why some of the fermions are so much *lighter* than M_W.) If mirror fermions do exist, they fail to a remarkable extent to mix with the usual fermions; the first-generation fermions are very light and have almost pure $V - A$ weak interactions. If mirrors do exist, it is odd that none of the 14 $SU(3) \times SU(2) \times U(1)$ multiplets observed so far appear to be mirrors. This is all the more remarkable in that the mirrors cannot weigh more than at most a few hundred GeV. [Since they do not mix with usual fermions, their bare masses are forbidden by $SU(2) \times U(1)$.] Finally, and perhaps most convincingly, the triangle anomalies cancel among the observed quarks and leptons. This cancellation appears to be a rather striking confirmation of current ideas, but if mirrors exist, it is just an elaborate and unnecessary charade, since the mirrors would automatically cancel the anomalies of the known fermions, whatever those anomalies might be.

What is more, the fermion representation is complicated [each family consits of five irreducible representations of $SU(3) \times SU(2) \times U(1)$] and redundant (there are three families). On the first point, no doubt the $SU(5)$ and $O(10)$ grand unified theories are the most successful efforts to date. [A

---

2. By the "spectrum of light fermions" we mean the quantum numbers of the light fermions. We shall usually speak of the quarks and leptons as if they were massless, ignoring the $SU(2) \times U(1)$ breaking.

family is $\bar{5}_L + 10_L$ in $SU(5)$, or 16_L in $O(10)$.] On the second point—which is an updated version of Rabi's question, "Who ordered the muon?"—there is no equally convincing answer. It is natural to try to embed the three families as one irreducible representation of a bigger group. Perhaps the most attractive such idea is to use the spinor representation [9] of $O(N)$ for $N \geq 18$. The spinor representation of $O(N)$ is the representation space of N gamma matrices, which automatically furnishes a representation of any subset of the gamma matrices; so the $O(N)$ spinor transforms as a sum of spinor representations of any minimally embedded $O(k)$ subgroup. For instance, the irreducible spinor of $O(18)$ transforms under $O(10)$ as four families plus (unfortunately) four antifamilies. This beautifully achieves the desired multiplicity, but it is not easy to eliminate the antifamilies. One may invoke a "hypercolor" force that becomes strong at ~ 1 Tev, breaking $SU(2) \times U(1)$ and confining the antifamilies. This elegant idea [10] has innumerable difficulties in detail. In this paper, difficulties will be much more conspicuous than phenomenological successes, but we shall note, in section VI, that Kaluza-Klein theory gives an alternative way of avoiding antifamilies in the $O(N)$ approach to the family problem.

A discussion of zero modes of nontrivial Dirac operators in Kaluza-Klein theory was apparently first given in a special situation by Palla [11]. In reference [4], in connection with a discussion of some pseudorealistic models, the detailed proposal was made that the quark and lepton quantum numbers are determined by the topology of a manifold with $SU(3) \times SU(2) \times U(1)$ symmetry. The importance and difficulty of obtaining a complex representation were pointed out. Chapline and Slansky and Manton [12] discussed the problem of obtaining a complex spectrum in Kaluza-Klein theory; they anticipated the kinematical analysis of section II and some of the ideas of section VI. The kinematical analysis has been recently developed in much more detail by Wetterich [13], who worked out all of the kinematical consequences of the mod 8 periodicity of the spinor representation of $O(N)$. He also introduced in a different language the mathematical concept of the character-valued index, which, as we shall see, plays a very important role. Models exhibiting many of the ideas of section VI have been analyzed by Randjbar-Daemi, Salam, and Strathdee [14].[3] As regards

---

3. I understand, in addition, that these authors have considered (unpublished) some of the detailed models in section VI.

fermion quantum numbers, the novelty in the present paper is primarily the restrictions disussed in sections IV and V and the more realistic models in section VI.

Our conclusions will be as follows. If all gauge fields are part of the metric tensor, then a theorem of Atiyah and Hirzebruch [15] states that the Dirac operator in 4 + n dimensions always leads to vectorlike quantum numbers in 4 dimensions. (The relevance of this theorem to Kaluza-Klein theory was first noted in reference [16].) For the Rarita-Schwinger operator the situation is more complicated. We shall show that if the hidden dimensions form a homogeneous space, the Rarita-Schwinger operator likewise always leads to vectorlike quantum numbers. What happens in general for the Rarita-Schwinger operator on spaces that are not homogeneous I do not know. However, the fact that supergravity is apparently restricted to $d \leqslant 11$ [17] and certain other facts discussed in section V indicate that this avenue is not promising. If one is less ambitious and introduces elementary gauge fields in $4 + n$ dimensions, it is possible, but still subtle, to get complex representations. Indeed, as we shall see in section VI, one can naturally get very big, complicated, duplicated representations. In section VII, we discuss some other ways that the assumptions might be modified.

We shall encounter considerable difficulties in our attempts to interpret the fermion quantum numbers as the solution of an index problem. Nevertheless, this seems to be a quite attractive idea.

II Preliminaries

Let us first recall how—in a Kaluza-Klein theory with ground state $M^4 \times B$—massless particles originate as zero modes of appropriate wave operators on B. A massless Dirac particle in $4 + n$ dimensions obeys

$$0 = \not{D}\psi = \sum_{i=1}^{4+n} \Gamma^\mu D_\mu \psi, \tag{1}$$

where Γ^μ, $i = 1, \ldots, 4 + n$, are the gamma matrices. Notice that we may as well use the minimal Dirac equation. Even if nonminimal terms (couplings to matter fields or nonminimal couplings to gravity) are present, they cannot change the quantum numbers of massless fermions in a complex representation of the symmetry group. This is an illustration of the fact that the problem of fermion quantum numbers depends only on the "universality class" of a theory. We do not have to believe we know which Dirac operator is physically relevant. If we define the 4-dimensional Dirac operator $\not{D}^{(4)} = \sum_{\mu=1}^{4} \Gamma^\mu D_\mu$ and the internal Dirac operator $\not{D}^{(n)} = \sum_{j=5}^{4+n} \Gamma^j D_j$, then (1) becomes

$$0 = \not{D}^{(4)}\psi + \not{D}^{(n)}\psi. \tag{2}$$

We see that $\not{D}^{(n)}$ is the mass operator, in effect. Its eigenvalues are observed in 4 dimensions as the particle masses. Its zero eigenvalues are the massless fermions.

The Rarita-Schwinger operator may be discussed similarly. There are many, equivalent ways [6] to write the $(4 + n)$-dimensional Rarita-Schwinger equation. One way is

$$0 = \Gamma^\mu(D_\mu\psi_\nu - D_\nu\psi_\mu), \tag{3}$$

where $\psi_{\mu\alpha}$ is the Rarita-Schwinger field ($\mu = 1, \ldots, 4 + n$ is a vector index; α is a spinor index). In the gauge $\Gamma^\mu\psi_\mu = 0$, (3) reduces to

$$0 = \Gamma^\mu D_\mu\psi_\nu = \not{D}^{(4)}\psi_\nu + \not{D}^{(n)}\psi_\nu. \tag{4}$$

Again zero modes of $\not{D}^{(n)}$ are observed as massless particles in 4 dimensions. The general zero mode is a sum of modes of two special kinds. For $\nu = 1, \ldots, 4$, $\not{D}^{(n)}$ is the ordinary spin $\frac{1}{2}$ Dirac operator, and the zero modes are spin $\frac{3}{2}$ fermions in 4 dimensions. For $\nu = 5, \ldots, 4 + n$, the zero modes have spin $\frac{1}{2}$ as seen in 4 dimensions, while their dependence on the compact dimensions is determined by the gauge condition *and* the Dirac-like equation

$$0 = \sum_{\nu=5}^{4+n} \Gamma^\nu\psi_\nu, \qquad 0 = \sum_{\mu=5}^{4+n} \Gamma^\mu D_\mu\psi_\nu. \tag{5}$$

These conditions, taken together, are equivalent to the gauge invariant internal Rarita-Schwinger equation

$$\sum_{\mu=5}^{4+n} \Gamma^\mu(D_\mu\psi_\nu - D_\nu\psi_\mu) = 0, \qquad \nu = 5, \ldots, 4 + n, \tag{6}$$

in a particular gauge. (We temporarily introduced a gauge fixing condition only to decouple the Minkowski dimensions from the compact ones.) We see that zero modes of the internal Rarita-Schwinger operator become massless spin $\frac{1}{2}$ fermions in 4 dimensions.

Now, can either of these operators have zero eigenvalues? And can the zero eigenvalues form complex representations of a symmetry group?

The crudest problem, which was pointed out in (4), arises in an *odd* number of dimensions. For odd n, the group $O(n)$ only has one spinor representation. Likewise, the group $O(1, 3 + n)$ has only one spinor representation, which transforms under $O(1, 3) \times O(n)$ as the product of the four component spinor of $O(1, 3)$ with the unique spinor of $O(n)$. This being so, fermions that are left- or right-handed in 4 dimensions transform the same way under transformations of the internal space. They obey the same Dirac equation in the internal space (modulo nonminimal terms that cannot affect the quantum numbers), so they have the same quantum numbers and furnish a real representation of any relevant symmetry group.

In an even number of dimensions the situation is more subtle. For even n the operator $\hat{\Gamma} = \Gamma^1 \cdots \Gamma^n$ anticommutes with all Γ^i, so it is a c-number in any representation of the Clifford algebra. Since $\hat{\Gamma}^2 = \pm 1$ (depending on n), the representation space of the Clifford algebra decomposes into two eigenspaces of $\hat{\Gamma}$, the eigenvalues being ± 1 or $\pm i$. Since $\hat{\Gamma}$ commutes with the $O(n)$ group generators $\frac{1}{4}[\Gamma^i, \Gamma^j]$, the group has two inequivalent spinor representations, labeled by the eigenvalue of $\hat{\Gamma}$.[4]

In a world of $4 + n$ dimensions we define

$$\bar{\Gamma} = \Gamma^1 \Gamma^2 \cdots \Gamma^{4+n},$$

$$\Gamma^{(4)} = \Gamma^1 \Gamma^2 \cdots \Gamma^4, \tag{7}$$

$$\Gamma^{\text{Int}} = \Gamma^5 \Gamma^6 \cdots \Gamma^{4+n}.$$

These operators have simple interpretations. $\bar{\Gamma}$ labels the spinor representations of $O(1, 3 + n)$, $\Gamma^{(4)}$ measures the helicity of 4-dimensional fermions, and Γ^{Int} labels the spinor representations of $O(n)$; it measures what might be called the internal helicity. These operators obey the simple relation

$$\bar{\Gamma} = \Gamma^{(4)} \cdot \Gamma^{\text{Int}}. \tag{8}$$

This equation has an important consequence. For fixed $\bar{\Gamma}$, the 4-dimensional

4. For odd n, $\hat{\Gamma}$ commutes with the Γ^i and is a c-number in an irreducible representation of the Clifford algebra. The Clifford algebra thus has two inequivalent representations, labeled by the sign of $\hat{\Gamma}$. They are related to each other by $\Gamma^i \to -\Gamma^i$ (which for odd n yields $\hat{\Gamma} \to -\hat{\Gamma}$), and they are equivalent as representations of $O(n)$ since under $\Gamma^i \to -\Gamma^i$ the group generators $\frac{1}{4}[\Gamma^i, \Gamma^j]$ are unchanged.

and internal chiralities are correlated. If we start with a fermion field restricted to (say) $\bar{\Gamma} = 1$ in $4 + n$ dimensions, it breaks down under $O(1,3) \times O(n)$ to components with[5]

$$\eta\Gamma^{(4)} = \eta^{-1}\Gamma^{\text{Int}} = +1 \qquad \text{or} \qquad \eta\Gamma^{(4)} = \eta^{-1}\Gamma^{\text{Int}} = -1. \tag{9}$$

Fermions of left- or right-handed physical helicity are left- or right-handed in the internal space. They obey different Dirac equations, whose zero modes might have different quantum numbers.

This idea quickly runs into trouble if the number of dimensions is divisible by four. In $4k$ dimensions, $\bar{\Gamma}$ is odd under CPT. This may be seen readily in a Majorana basis, with real gamma matrices. In such a basis CPT acts on spinors just by complex conjugation. $\bar{\Gamma}$ [as defined in (7)] is real in a Majorana basis, but in $4k$ dimensions it may be readily seen to obey $(\bar{\Gamma})^2 = -1$. The eigenvalues of $\bar{\Gamma}$ are $\pm i$. Being complex conjugates, the eigenvalues of $\bar{\Gamma}$ are related to each other by CPT, and CPT requires that there be equal numbers of fields with $\bar{\Gamma} = +i$ and $\bar{\Gamma} = -i$. Hence there is no net correlation between 4-dimensional chirality and internal chirality. Fields of $\bar{\Gamma} = +i$ give one correlation and fields of $\bar{\Gamma} = -i$ give the opposite correlation. Thus, in $4k$ dimensions, CPT requires that the gravitational interactions be vectorlike. Naturally, therefore, if the weak interactions are part of the gravitational force in $4 + n$ dimensions, the weak interactions are also vectorlike. Alternatively, one may say [13, 18] that in $4k$ dimensions, bare masses are possible for any fermions coupled to gravity only. Such a theory will, of course, always reduce to an appropriate 4-dimensional theory in which bare masses are still possible.

In $4k + 2$ dimensions, the situation is very different. In this case $\bar{\Gamma}^2 = +1$, so $\bar{\Gamma}$ has eigenvalues ± 1. CPT leaves $\bar{\Gamma}$ unchanged, and we can consider a theory with fermions of (say) $\bar{\Gamma} = +1$ only. (This option is forced on us in certain situations—for instance, in certain 10-dimensional supersymmetric field theories and string theories.) This corresponds roughly to a theory with $V - A$ gravitational interactions that forbid fermion bare masses; the question is whether $V - A$ gravity can reduce to $V - A$ weak interactions in 4 dimensions.

The special role of $4k + 2$ dimensions in multidimensional field theory was first raised in constructing supersymmetric Yang-Mills theories [19].

---

5. Here η is a phase factor that will be determined momentarily. It can be ignored for the time being.

In the context of analyzing fermion quantum numbers this point was made and developed in references [13] and [14]. Similar observations are important in grand unified model building [9]. In the mathematical literature the periodicity of the spinor representation is an old observation [20].

Let us now work out the phase factor η of equation (9). We note that as defined $\Gamma^{(4)}$ is (in a Majorana basis) a real matrix whose square is -1, so the eigenvalues of $\Gamma^{(4)}$ are $\pm i$. In $4k + 2$ dimensions Γ^{Int} likewise has square -1 and eigenvalues $\pm i$. [In $4k$ dimensions $(\Gamma^{Int})^2 = +1$.] A fermi field that obeys $\bar{\Gamma} = \Gamma^{(4)} \cdot \Gamma^{Int} = +1$ therefore has

$$\text{(A)} \quad \Gamma^{(4)} = -\Gamma^{Int} = +i \qquad \text{or} \qquad \text{(B)} \quad \Gamma^{(4)} = -\Gamma^{Int} = -i. \tag{10}$$

A CPT transformation will complex conjugate the eigenvalues, so eigenvalues of type A and B are exchanged by CPT. This is as it should be. A zero mode of the internal Dirac or Rarita-Schwinger operator with $\Gamma^{Int} = -i$ corresponds to a left-handed massless fermion in 4 dimensions. Its complex conjugate will have $\Gamma^{Int} = +i$ and corresponds to a right-handed massless fermion in 4 dimensions. Massless fermions in 4 dimensions will transform in a complex representation of some symmetry group G if the zero modes of the internal Dirac operator with $\Gamma^{Int} = -i$ form a complex representation of G, or equivalently, if the zero modes of $\Gamma^{Int} = +i$ transform differently from those of $\Gamma^{Int} = -i$.

Since the remainder of this section and the next one will deal with "chiral theories of gravity" with elementary fermi fields of a definite value of $\bar{\Gamma}$, it should be mentioned at the outset that these theories suffer from a major problem. The fermion one-loop diagrams in an external gravitational field are anomalous [18]. (The anomaly first appears in a diagram with $2k + 2$ external gravitons.) In a few special theories in 6 or 10 dimensions the anomalies cancel between fields of different spin; beyond 10 dimensions this is impossible. We shall not limit ourselves to the anomaly-free theories but will investigate the fermion quantum numbers that emerge (at the tree level) in the whole class of chiral gravity theories. There are several justifications for ignoring the anomalies. First, general methods for treating the whole class of chiral gravity theories are as simple as any special methods for analyzing the particular anomaly-free theories. And the general methods are likely to be important for other attempts to calculate fermion quantum numbers in Kaluza-Klein theory; for instance, we shall apply them from a different standpoint in section VI. Second, it may happen that in the future a massless field of some exotic spin might be successfully coupled to gravity.

This could expand the room for cancellation of anomalies without affecting our tree level considerations for spin $\frac{1}{2}$ and spin $\frac{3}{2}$ fields. Third, though it seems unlikely, perhaps there is some way to make sense of anomalous field theories or of other theories whose low energy limit is an anomalous field theory.

A rather simple argument due essentially to Lichnerowicz [21] severely limits the possibility of obtaining zero modes of the Dirac operator in complex representations of a symmetry group. If one squares the internal Dirac operator (which will simply be denoted $i\not{D}$; we henceforth suppress the 4 Minkowskian dimensions), we find

$$(i\not{D})^2 = -\sum D_i D^i - \tfrac{1}{4}[\gamma^i, \gamma^j][D_i, D_j]$$

$$= -\sum D_i D^i - \tfrac{1}{32}[\gamma^i, \gamma^j][\gamma^k, \gamma^l]R_{ijkl} \qquad (11)$$

$$= -\sum D_i D^i + \tfrac{1}{4}R.$$

Since $-\sum D_i D^i$ is a nonnegative operator, this shows that if $R > 0$ everywhere, the Dirac operator has no zero eigenvalues. Of course, in (11) we have considered a minimally coupled Dirac operator. If nonminimal couplings are present, the Dirac operator may have zero eigenvalues. But the fact that the minimally coupled operator has no zero eigenvalues at all, and leads to no massless fermions in 4 dimensions, means that even in the presence of nonminimal couplings, the zero eigenvalues form real representations of whatever symmetry group may be present.[6] A particularly important case of this is the following. Suppose the compact space B has a symmetry group G. In general there will be many G-invariant metrics on B. If even one of them has $R > 0$, then for *any* G-invariant metric on B, the zero modes of the Dirac operator, if any, form a real representation of G. (Of course, under these circumstances the Dirac operator will generically have no zero eigenvalues.)

Much of the literature on Kaluza-Klein theory has concerned homogeneous spaces $B = G/H$, G and H being compact nonabelian groups. These spaces all admit a canonical G-invariant metric of positive scalar curvature, so (even if nonminimal terms are added or a different G-invariant metric is used) they give real representations for zero eigenvalue. More generally, a theorem by Lawson and Yau [22] shows that on any compact space B (not necessarily a homogeneous space) with a nonabelian symmetry G, there is a

6. This should be obvious "physically" from the connection with fermion quantum numbers in 4 dimensions. The precise mathematical argument will be given shortly.

G-invariant metric of positive scalar curvature R. For nonabelian groups such as $SU(3) \times SU(2) \times U(1)$, this rules out the possibility of getting zero modes of the Dirac operator in complex representations.

This simple line of argument does not address the question (of conceptual but probably not of practical interest) whether zero modes of the Dirac operator can form a complex representation of an abelian symmetry group. (Manifolds with a continuous abelian symmetry group in general do not admit an invariant metric of positive scalar curvature.) Much more important, this line of reasoning does not extend to the Rarita-Schwinger operator whose square is more complicated than (11) and is not manifestly positive even if $R > 0$. For this reason, the Rarita-Schwinger operator can have zero eigenvalues more readily than the Dirac operator. For instance, in 4 dimensions there is one compact manifold that is not flat but obeys $R_{\mu\nu} = 0$. It is the K3 surface, and for topological reasons it has two zero eigenvalues of the Dirac operator and 42 zero eigenvalues of the Rarita-Schwinger operator.

In multidimensional supergravity and superstring theories—which are the only known theories in which fermions are really unified with gravity—we inevitably are dealing with Rarita-Schwinger fields. It therefore is important to learn to analyze the zero modes of these fields.

There is another no-go theorem, due to Atiyah and Hirzebruch [15], which for our particular problem is much more restrictive than the reasoning just sketched. They proved precisely that for *any* continuous symmetry group, abelian or nonabelian, the Dirac zero modes form a real representation. As we shall see, their argument has important implications for the Rarita-Schwinger case; for instance, we shall use it to prove that if the compact space is a homogeneous manifold, the Rarita-Schwinger operator always leads to a real representation. We shall present in section IV an elementary proof of the Atiyah-Hirzebruch theorem that is closely related to the original argument.

Why would a wave operator have zero eigenvalues? And why would these zero eigenvalues form complex representations? We shall illustrate the relevant concepts in terms of the Dirac operator. In the rest of this paper, we suppress the 4 Minkowskian dimensions and concentrate on properties of the n-dimensional Kaluza-Klein space B. To streamline notation, gamma matrices are henceforth gamma matrices $\Gamma^1, \Gamma^2, \ldots, \Gamma^n$ of B, and we define an operator $\hat{\Gamma} = i\Gamma^{\mathrm{Int}} = i\Gamma^1 \cdots \Gamma^n$ with eigenvalues ± 1. Indices i, j, k refer to the internal space; indices μ, ν, α refer to all $4 + n$ dimensions.

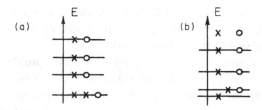

Figure 1
Zero modes of the Dirac operator of positive or negative chirality are indicated by × or ○,
respectively. The number of × s minus ○s at zero energy is invariant under perturbations.

Let us define a "Hamiltonian" $H = (i\not{D})^2$. Since $[\hat{\Gamma}, H] = 0$, H eigenstates
can be chosen to be at the same time $\hat{\Gamma}$ eigenstates. If $H\psi = E\psi$, then
$H \cdot i\not{D}\psi = E \cdot i\not{D}\psi$. So ψ and $i\not{D}\psi$ are degenerate in energy, unless $i\not{D}\psi = 0$.
But since $\not{D}\hat{\Gamma} = -\hat{\Gamma}\not{D}$, ψ and $i\not{D}\psi$ have opposite eigenvalues of $\hat{\Gamma}$. Consequently (figure 1), the H eigenvalues of nonzero energy are paired. For every
state of $\hat{\Gamma} = 1$ there is a state of $\hat{\Gamma} = -1$. The zero eigenvalues need not be
paired in this way. The number of zero eigenvalues of $\not{D}$ with $\hat{\Gamma} = 1$ minus
the number with $\hat{\Gamma} = -1$ is called the *index* of $\not{D}$.

The index is invariant under arbitrary deformations of $\not{D}$ that preserve
the property $\hat{\Gamma}\not{D} = -\not{D}\hat{\Gamma}$, since no smooth distortion of figure 1 that
preserves the pairing at nonzero energy can disturb whatever chirality
imbalance may exist at $E = 0$. In particular the index of $\not{D}$ is a topological
invariant, depending on the topology of B but not on its metric tensor.
Generically, in the absence of some symmetry principles (which can, however, change the situation, as we shall see), zero eigenvalues of $\not{D}$ all have
positive chirality or all have negative chirality. This is so because zero
eigenvalues of equal and opposite chirality would gain nonzero energy
under a generical perturbation.

Although the index is the simplest deformation invariant of the Dirac
operator that is relevant to the occurrence of zero modes, for our purposes
we need a slightly different concept. In $4k + 2$ dimensions the index of the
Dirac operator always vanishes, for the simple reason that the positive and
negative chirality zero modes of the Dirac operator are complex conjugates
of each other (as we have seen earlier) and therefore equal in number. The
concept of interest to us is what mathematicians call the G-index or the
character-valued index of the Dirac operator.

Let the manifold B have a symmetry group G. The eigenstates of H or $i\not{D}$

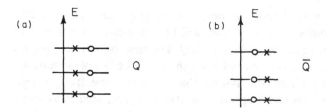

Figure 2
An × or ○ now indicates a positive or negative chirality multiplet in the Q or $\bar{Q}$ representation [(a) and (b), respectively]. In passing from (a) to (b) the × s and ○s are exchanged.

will then form representations of G. Pick a representation Q, and draw the same picture as before (figure 2), but only counting multiplets in the Q representation. We define $\text{index}_Q(\not{D})$ to be the number of zero mode multiplets in the Q representation of positive chirality minus the number of zero mode multiplets in the Q representation of negative chirality. For reasons similar to those given earlier, $\text{index}_Q(\not{D})$ is invariant under arbitrary perturbations that respect G symmetry and preserve the fact that $\not{D}\hat{\Gamma} = -\hat{\Gamma}\not{D}$.

Of course, we can still complex conjugate our eigenstates of H. This still reverses the eigenvalue of $\hat{\Gamma}$, but now it exchanges Q with its complex conjugate representation $\bar{Q}$. By complex conjugation (figure 2b) this implies that $\text{index}_Q(i\not{D}) = -\text{index}_{\bar{Q}}(i\not{D})$. Upon reduction to 4 dimensions, this implies the perfectly valid statement that the number of left-handed massless fermions in the Q representation equals the number of right-handed massless fermions in the $\bar{Q}$ representation.

An equivalent way to define the character-valued index is as follows. The positive and negative chirality zero modes of $i\not{D}$ form representations Λ^+ and Λ^- of G. For $g \in G$ we define

$$\text{index}(g) = \text{tr}_{\Lambda^+}(g) - \text{tr}_{\Lambda^-}(g). \tag{12}$$

Or equally well we define

$$\text{index}(g) = \sum_Q \text{index}_Q(i\not{D}) \chi_Q(g), \tag{13}$$

where $\chi_Q(g)$ is the trace of g in the Q representation.[7]

---

7. This definition of $\text{index}(g)$ makes sense in $4k$ as well as $4k + 2$ dimensions, though it is then not related to our physical problem, and we lose the identity $\text{index}(g) = (\text{index}(g^{-1}))^*$ that follows from complex conjugation in $4k + 2$ dimensions.

If the character-valued index is nonzero, $i\not{D}$ must have zero modes. Generically, the number of zero modes will be the minimum required to yield the right value of $\mathrm{index}_Q(i\not{D})$ for each Q. The spectrum of zero modes required by the character-valued index we shall call the "stable spectrum" of zero modes. We shall usually assume that the actual spectrum of zero modes coincides with the stable spectrum and can be computed by evaluating the character-valued index. From the fact that $SU(3) \times SU(2) \times U(1)$ forbids bare masses for all the known quarks and leptons, it appears that this assumption is valid in nature. A successful model would be one in which the character-valued index consists of three families minus three antifamilies.

Unfortunately, as we shall see in section IV, the Atiyah-Hirzebruch theorem ensures that the character-valued index always vanishes for the Dirac operator in theories without elementary gauge fields. At least on homogeneous spaces, this is also true for the Rarita-Schwinger field. It is not true (even on homogeneous spaces) for fields of spin $\frac{5}{2}$ or larger; we shall discuss a counterexample in section V (but there does not seem to be any way to use massless fields of spin $\geq \frac{5}{2}$ in physics [23]). The character-valued index also need not vanish in the presence of elementary gauge fields, and we shall construct some pseudorealistic models in section VI on the basis of this fact.

As we shall see, there are very powerful methods for calculating the character-valued index of arbitrary operators. It is never necessary to write down and solve an explicit differential equation.

III Operators on Homogeneous Spaces

For our first experience in calculating the character-valued index of various operators, we shall consider the simple case in which the manifold B is a homogeneous space G/H. In that case, there is a particularly elementary way [24] to compute the stable spectrum of zero modes of the Dirac operator (or any other G-invariant operator). One simply expands the spinor fields on B in harmonics (irreducible representations) of the group G. For any representation Q of G, let $n_+(Q)$ be the number of times the Q representation appears in the harmonic expansion of positive chirality spinors, and let $n_-(Q)$ be the number of times the Q representation appears in the expansion of negative chirality spinors. By standard theorems about homogeneous spaces and elliptic operators, $n_+(Q)$ and $n_-(Q)$ are always finite, and $n_+(Q) = n_-(Q)$ for all but finitely many Q. Moreover, $\text{index}_Q(i\slashed{D})$ $= n_+(Q) - n_-(Q)$. After all, the index is the difference between the number of positive and negative chirality multiplets. This difference normally must be regularized since there are infinitely many states in Hilbert space, and the regularization usually involves pairing off the states of equal, nonzero energy. In a homogeneous space, working in a subspace of Hilbert space defined by a definite representation, Q reduces the problem to a finite-dimensional problem. No regularization is needed; $\text{index}_Q(i\slashed{D})$ is just the difference $n_+(Q) - n_-(Q)$ between the number of positive chirality and negative chirality Q multiplets.

Let us illustrate these ideas with some simple examples. Consider first a particle of spin $\frac{1}{2}$ moving on the ordinary 2-dimensional sphere S^2. Let $\mathbf{J}$ be the angular momentum operator, and define the "helicity" of a particle at $\mathbf{x}$ to be the component of angular momentum about the $\mathbf{x}$ axis. It equals $\pm\frac{1}{2}$; it is $+\frac{1}{2}$ for states of positive chirality, $-\frac{1}{2}$ for states of negative chirality.

It is well known that a particle of helicity h can be in a state of total angular momentum $J = |h|, |h| + 1, |h| + 2, \ldots$, with each allowed value of J appearing exactly once in the harmonic expansion. This is why $J = 1$ is the lowest possible value for photons, and $J = 2$ is the lowest possible value for gravitons.

Whether the chirality is positive or negative, the absolute value of the helicity of a spin $\frac{1}{2}$ particle is $\frac{1}{2}$. So the allowed values of angular momentum are the same for each chirality:

$$\hat{\Gamma} = +1: \quad J = \tfrac{1}{2}, \tfrac{3}{2}, \tfrac{5}{2}, \ldots,$$
$$\hat{\Gamma} = -1: \quad J = \tfrac{1}{2}, \tfrac{3}{2}, \tfrac{5}{2}, \ldots. \tag{14}$$

Now we can see that the Dirac operator on the sphere has no stable spectrum of zero modes. Since the Dirac operator commutes with J but reverses chirality, acting on (say) the multiplet of given J and chirality ± 1, the Dirac operator gives either zero or else the multiplet of the same J and chirality ∓ 1 (see figure 3). Since the Dirac operator is hermitian, it either exchanges these two multiplets or annihilates both of them. Therefore the positive chirality zero modes have the same eigenvalues of J as the negative chirality zero modes, and the character-valued index vanishes.

This result could be obtained in various other ways. It follows from the fact that the rotation group $SU(2)$ has no complex representations, or from the fact that a reflection of the two-sphere reverses parity and exchanges the

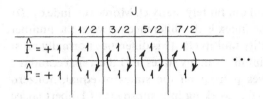

Figure 3
The angular momentum and chirality spectrum of the Dirac operator on a sphere. The Dirac operator permutes the states in the way indicated by the arrows.

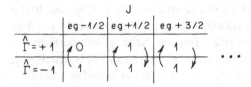

Figure 4
The quantum numbers of a charged Dirac particle on a sphere in the presence of a magnetic monopole field.

two chiralities, or from the fact that the two-sphere with its usual metric has positive scalar curvature so that (by Lichnerowicz's theorem) the Dirac operator has no zero modes at all. Now, however, we shall consider a slightly modified problem in which the character-valued index is nonzero.

Place at the center of the sphere a magnetic monopole of strength $eg = n/2$, for some integer n. The angular momentum operator now acquires an extra piece [25] $eg\hat{x}$ related to the quantization of magnetic charge. This adds eg to the fermion helicity, so that a fermion of chirality $+1$ has effective helicity $eg + \frac{1}{2}$ and a fermion of chirality -1 has effective helicity $eg - \frac{1}{2}$. If, say, $eg > 0$, the allowed values of angular momentum are now

$$\hat{\Gamma} = +1: \quad J = eg + \tfrac{1}{2}, eg + \tfrac{3}{2}, eg + \tfrac{5}{2}, \ldots,$$

$$\hat{\Gamma} = -1: \quad J = eg - \tfrac{1}{2}, eg + \tfrac{1}{2}, eg + \tfrac{3}{2}, eg + \tfrac{5}{2}, \ldots. \tag{15}$$

The crucial point is now that states of $J = eg - \frac{1}{2}$ exist for chirality -1 but not for chirality $+1$. The Dirac operator must annihilate these states, because acting on states of $\hat{\Gamma} = -1, J = eg - \frac{1}{2}$, the Dirac operator would give states of $\hat{\Gamma} = +1$ and $J = eg - \frac{1}{2}$, and such states do not exist. Other multiplets cancel out as before (figure 4), so the stable spectrum of zero modes is a single multiplet of $\hat{\Gamma} = -1$ and $J = eg - \frac{1}{2}$.

We shall obtain this answer in a different way in section IV as an illustration of a much more general and powerful method of calculating the character-valued index of an operator. The technique for harmonic expansions for G-invariant operators on a homogeneous space G/H does not seem to be well known among physicists. It is explained, for instance, in appendix IV of Salam and Strathdee in [3]. In later sections we shall occasionally state without detailed derivation results obtainable from harmonic expansions.

IV The Atiyah-Hirzebruch Theorem

We now turn to the proof of the Atiyah-Hirzebruch theorem, which states that the character valued index of the Dirac operator vanishes on any manifold with a continuous symmetry group (in any even number of dimensions, though our main interest is $d = 4k + 2$). The presentation will parallel a recent treatment of Morse theory [16] and is essentially a more concrete version of the original proof.

Let B be a compact Riemannian manifold of even dimension n. Suppose B admits the action of a symmetry group G. We wish to prove that index $(g) = 0$ for every $g \in G$. Since every element of G can be approximated arbitrarily well by elements of suitably chosen $U(1)$ subgroups of G, it suffices to prove that index$(h) = 0$ whenever h is an element of any $U(1)$ subgroup R of G. We therefore specialize to the case of a $U(1)$ symmetry group R. Since the representations of $R \cong U(1)$ are labeled by an integer or half-integer n,[8] it suffices to show that index$_n(i\slashed{D}) = 0$ for all n.

Let ϕ^i be a local coordinate system of B. Let $K^i(\phi^j)$ be the Killing vector field that generates R. [This means that, infinitessimally, the R transformation is $\phi^i \to \phi^i + \varepsilon K^i(\phi^j)$.] Acting on spinors, the generator of R is the "Lie derivative" operator:

$$\mathscr{L}_K = i\left(K^i D_i + \tfrac{1}{4}\Gamma^{ij}(D_i K_j)\right), \tag{16}$$

where $\Gamma^{ij} = \tfrac{1}{2}[\Gamma^i, \Gamma^j]$. Using the Killing vector equation $D_i K_j + D_j K_i = 0$ and standard identities, it is not difficult to verify that $\mathscr{L}_K$ and $i\slashed{D}$ commute—as should be the case since $\mathscr{L}_K$ generates a symmetry. Therefore we may study simultaneous solutions of the equations

$$\mathscr{L}_K \psi = n\psi, \qquad i\slashed{D}\psi = \lambda\psi. \tag{17}$$

8. The eigenvalues on spinor states are half-integers in certain cases.

Our problem is to show that the Dirac index $\text{index}_n(i\slashed{D})$ vanishes for each sector of Hilbert space labeled by the integer (or half-integer) n.

The basic property of this index is that it is invariant under arbitrary deformations of the operator $i\slashed{D}$ that are $U(1)$ invariant and preserve the property $i\slashed{D}\hat{\Gamma} = -\hat{\Gamma}i\slashed{D}$. Let us therefore perturb the Dirac operator in a way that preserves these properties and simplifies the analysis of it spectrum. Instead of $i\slashed{D}$ we shall study

$$i\slashed{D}_t = i\slashed{D} + t\Gamma^i K_i, \tag{18}$$

where t is a conveniently chosen real number. The character-valued index of $i\slashed{D}_t$ must be independent of t. We shall prove that $\text{index}_n(i\slashed{D})$ vanishes for $n \geqslant 0$ by studying the behavior as $t \to +\infty$, and we shall prove that $\text{index}_n(i\slashed{D})$ vanishes for $n \leqslant 0$ by studying the behavior as $t \to -\infty$.

We define a "Hamiltonian"

$$H_t = (i\slashed{D}_t)^2 = (i\slashed{D})^2 + t^2 K^2 + 2it K^j D_j + it\Gamma^{ij}D_i K_j. \tag{19}$$

If we could show that for sufficiently large t, H_t has no zero eigenvalues, this would establish the vanishing of the character-valued index. The general reason this might be true is that the $t^2 K^2$ term is positive definite and becomes very large for large t. However, the analysis is made subtle by the term $2it K^j D_j$, which is not positive definite and can have large matrix elements.

A crucial observation is that in the sector $\mathcal{L}_K \psi = n\psi$, H_t reduces to

$$H_t^{(n)} = (i\slashed{D})^2 + t^2 K^2 + 2tn + \tfrac{1}{2}it\Gamma^{ij}D_i K_j. \tag{20}$$

Since t is freely at our disposal, we choose $t > 0$ if $n \geqslant 0$ and $t < 0$ if $n < 0$. In this way the term $2tn$, as well as the $t^2 K^2$ term, is positive.

Now, if the Killing vector field K^i has no zeros, then all eigenvalues are of order t^2 as $t \to \infty$. In this case, the character-valued index certainly vanishes. In general, however, K^i vanishes at certain points, and our analysis is more difficult.

For large $|t|$, the spectrum of $H_t^{(n)}$ can be calculated in an asymptotic expansion in powers of $1/|t|$ by expanding near the minima of the potential. The relevant minima (which might give states that do not diverge in energy as $|t| \to \infty$) are zeros of K. For simplicity, we shall treat the case of an isolated zero of K, but the general case is not much different.

We may take our isolated zero of K to be at $\phi^i = 0$. Near $\phi^i = 0$ we can choose the locally euclidean coordinates ϕ^i to be such that $K_i = \omega_{ij}\phi^j +$

$O(\phi^2)$, with ω_{ij} a constant matrix,

$$\omega_{ij} = \begin{bmatrix} 0 & r_1 & & & & & \\ -r_1 & 0 & & & & & \\ & & 0 & r_2 & & & \\ & & -r_2 & 0 & & & \\ & & & & \ddots & & \\ & & & & & 0 & r_{d/2} \\ & & & & & -r_{d/2} & 0 \end{bmatrix} . \tag{21}$$

[Here d is the dimension of our manifold; the r_i are integers since K generates a $U(1)$ group.] For large $|t|$, by keeping only terms that contribute an amount of order t to the energy, $H_t^{(n)}$ simplifies to

$$H_t^{(n)} = -\sum_{i=1}^{d} \frac{\partial^2}{\partial \phi_i^2} + \sum_{l=1}^{d/2} t^2 r_l^2 (\phi_{2l-1}^2 + \phi_{2l}^2) + 2tn$$

$$- t \sum_{l=1}^{d/2} \frac{ir_l}{2} [\Gamma_{2l-1}, \Gamma_{2l}]. \tag{22}$$

The ground state energy of (22) is easily calculated. The first two terms are a sum of commuting harmonic oscillator Hamiltonians. The matrices $(i/2)[\Gamma_{2l-1}, \Gamma_{2l}]$ commute with each other and with the rest of the Hamiltonian, and have eigenvalues ± 1. The ground state energy of (22) is $|t| \sum_{i=1}^{d/2} |r_i| + 2tn$.[9] This is a good approximation for large $|t|$. Considering $t \to +\infty$ if $n \geqslant 0$, and $t \to -\infty$ if $n < 0$, we see that $H_t^{(n)}$ has no zero eigenvalues and consequently that the character-valued index of $i\slashed{D}$ vanishes.

If the zeros of K are not isolated points, the discussion must be changed only slightly. Let F be any connected component of the submanifold on which K vanishes. The potential $t^2 K^2$ vanishes on F, so our spin $\frac{1}{2}$ particle moves freely along F, but the motion orthogonal to F is restricted to a distance of order $1/\sqrt{t}$. The orthogonal motion is governed by an operator similar to (20), and the zero point energy of the orthogonal motion is strictly

---

9. If we restrict ourselves to states of $\mathscr{L}_K \psi = n\psi$ (as we should), the ground state energy of (22) is even larger. This of course does not change the conclusion.

positive if $|t| \to \infty$ with $tn \geqslant 0$, again showing that the character-valued index vanishes. (The discussion of degenerate Morse theory in reference [16] is similar.) This completes the proof of the Atiyah-Hirzebruch theorem.

Let us, however, now look at the preceding formulas from a different viewpoint, the goal being to obtain the fixed point formula associated with the Atiyah-Singer index theorem [26]. As we shall see, this formula is a powerful tool for computing the character valued index when it is not zero.

Let us now study H_t for $t \to +\infty$. Low-lying eigenvalues of H_t are concentrated near zeros of K, which for simplicity we take to be isolated points. Most of these states have energy of order t, but some have energy that vanishes as $t \to \infty$. (This has been obscured in the presentation until now.) Let $a_{n,\pm}^{(i)}$ be the number of state ψ concentrated near the ith zero whose energy does not diverge as $t \to +\infty$ and that obey $\hat{\Gamma}\psi = \pm\psi$ and $\mathscr{L}_K\psi = n\psi$. (We have proved $a_{n,\pm}^{(i)} = 0$ for $tn \geqslant 0$.) Define the "local index" of the Dirac operator at the ith zero as

$$f_i(\theta) = \sum_n e^{in\theta}(a_{n,+}^{(i)} - a_{n,-}^{(i)}). \tag{23}$$

The character-valued index $I(\theta)$ is obtained as the sum of the local indexes,

$$I(\theta) = \sum_i f_i(\theta), \tag{24}$$

since this sum includes the contribution of all states whose energy is not of order $|t|$, and only such states can contribute to $I(\theta)$. Of course, we have proved that the Dirac case has $I(\theta) = 0$, so (24) is a set of restrictions on the $a_{n,\pm}^{(i)}$. However, we shall obtain a formula like (24) for other problems in which $I(\theta) \neq 0$.

To compute the $a_{n,\pm}^{(i)}$, we could simply diagonalize the harmonic oscillator Hamiltonian (22). A more efficient method is as follows. Zero energy states must obey $i\not{D}_t\psi = 0$. Here $i\not{D}_t = i\Gamma^j(D_j - itK_j)$ has a very simple interpretation; it corresponds to a particle interacting with the abelian vector potential $A_j = tK_j$ as well as the metric of the curved manifold under study. For large t, we know the low-lying states are concentrated near zeros of K. If near a zero at (say) $\phi = 0$, $K^i = \omega^{ij}\phi_j$, with ω^{ij} a constant matrix, then the "magnetic field" $F_{ij} = \partial_j A_j - \partial_j A_i$ is just the constant matrix $t\omega_{ij}$. So as $t \to \infty$, our problem reduces to the study of the Dirac equation in a constant magnetic field. As in the usual 3-dimensional case, the ground state energy is zero (but states of zero energy have $n < 0$ if $t \to +\infty$ or $n > 0$ if $t \to -\infty$).

With ω_{ij} in the canonical form (21), the analysis is very simple. Define

$$i\!\!\not{D}_t^{(1)} = i \sum_{i=1,2} \Gamma^i(D_i + itK_i),$$

$$iD_t^{(2)} = i \sum_{i=3,4} \Gamma^i(D_i + itK_i),$$

$$\vdots$$

$$iD_t^{(n/2)} = i \sum_{i=n-1,n} \Gamma^i(D_i + itK_i).$$

(25)

Then $(i\!\!\not{D}_t)^2 = \sum_{j=1}^{n/2} (i\!\!\not{D}_t^{(j)})^2$, so a solution of $i\!\!\not{D}_t\psi = 0$ obeys simultaneously $i\!\!\not{D}_t^{(1)}\psi = \cdots = i\!\!\not{D}_t^{(n/2)}\psi = 0$. Thus, we need only the well known solution of the problem of a constant magnetic field in *two* dimensions.

Choose a basis of gamma matrices

$$\Gamma^1 = \begin{pmatrix} 0 & 1 \\ 1 & 0 \end{pmatrix}, \qquad \Gamma^2 = \begin{pmatrix} 0 & -i \\ i & 0 \end{pmatrix}.$$

If the "rotation angle" is r, so $K_1 = r\phi_2$, $K_2 = -r\phi_1$, then the 2-dimensional Dirac operator is

$$\not{D}_t = \begin{pmatrix} 0 & (\partial_1 - \operatorname{tr}\phi_1) - i(\partial_2 - \operatorname{tr}\phi_2) \\ (\partial_1 + \operatorname{tr}\phi_1) + i(\partial_2 + \operatorname{tr}\phi_2) & 0 \end{pmatrix}.$$

(26)

With $\psi = \begin{pmatrix} u \\ v \end{pmatrix}$, $\not{D}_t\psi = 0$ if

$$u(\phi_1,\phi_2) = (\phi_1 + i\phi_2)^k \exp -\operatorname{tr}(\phi_1^2 + \phi_2^2),$$

$$v(\phi_1,\phi_2) = (\phi_1 - i\phi_2)^k \exp +\operatorname{tr}(\phi_1^2 + \phi_2^2),$$

(27)

where $k = 0, 1, 2, \ldots$. The chirality operator is

$$\hat{\Gamma} = \begin{pmatrix} 1 & 0 \\ 0 & -1 \end{pmatrix}.$$

For $t \to +\infty$, if $\operatorname{tr} > 0$, only positive chirality zero modes are normalizable. If $\operatorname{tr} < 0$, only negative chirality zero modes are normalizable.

The symmetry generator is

$$\mathscr{L}_K = i(\Gamma^i D_i + \tfrac{1}{4}\Gamma^{ij}D_i K_j)$$

$$= -ir(\phi_1\,\partial_2 - \phi_2\,\partial_1) + \frac{r}{2}\begin{pmatrix} 1 & 0 \\ 0 & -1 \end{pmatrix}.$$

(28)

We see that on the states (27), $\mathscr{L}_K = r\hat{\Gamma}(k + \frac{1}{2})$.

The local index in the 2-dimensional problem would be $h(\theta) = \sum_n e^{in\theta}$ $(a_{n,+} - a_{n,-})$, where $a_{n,\pm}$ are the number of zero energy states ψ with $\mathscr{L}_K\psi = n\psi$, $\hat{\Gamma}\psi = \pm\psi$. Using (27) and (28), we see that if $r > 0$,

$$h(\theta) = e^{i(r\theta/2)} + e^{i(3r\theta/2)} + e^{i(5r\theta/2)} + \cdots$$

$$= \frac{e^{ir\theta/2}}{1 - e^{ir\theta}}$$

$$= \frac{i}{2}\frac{1}{\sin\frac{1}{2}r\theta}. \tag{29}$$

If $r < 0$, we get

$$h(\theta) = -e^{-ir\theta/2} - e^{-i(3r\theta/2)} - e^{-i(5r\theta/2)} - \cdots$$

$$= -\frac{e^{-ir\theta/2}}{1 - e^{-ir\theta}} = \frac{i}{2}\frac{1}{\sin\frac{1}{2}r\theta}. \tag{30}$$

It is one of the wonders of analytic functions that these expressions are equal, so we need not worry about the sign of r.

In view of the separation of variables (zero eigenvalues of $\mathcal{D}_t$ are zero eigenvalues of each of the $\mathcal{D}_t^{(n)}$), we can now easily compute the local index $f_i(\theta)$ at the ith zero. It is just

$$f_i(\theta) = \prod_{a=1}^{n/2} \left(\frac{i}{2}\frac{1}{\sin\frac{1}{2}r_a^{(i)}\theta} \right), \tag{31}$$

where $r_a^{(i)}$ is the ath rotation angle at the ith zero of k. The character-valued index of the Dirac operator is therefore

$$I(\theta) = \left(\frac{i}{2} \right)^{n/2} \sum_i \prod_a \frac{1}{\sin\frac{1}{2}r_a^{(i)}\theta}, \tag{32}$$

which is the fixed point formula. Although we obtained it by taking $t \to +\infty$, an analysis for $t \to -\infty$ leads to the same formula.

Since we know that $I(\theta) = 0$, (32) is a set of very restrictive conditions on the $r_a^{(i)}$. Actually the Atiyah-Hirzebruch theorem is an easy consequence of (32). The right-hand side of (32) defines a rational function of $w = e^{i\theta/2}$ that vanishes at $w = 0$ and $w = \infty$. This function has no poles. [Individual terms in (32) have poles at $|w| = 1$. These poles must cancel after summing over fixed points, for the following reason. An elliptic operator always has only finitely many zero modes, so $I(\theta)$ has an expansion $I(\theta) = \sum_n a_n e^{in\theta}$ with only finitely many nonzero a_n; therefore $I(\theta)$ is always nonsingular for real θ]. A rational function without poles that vanishes at infinity is zero, so $I(\theta) = 0$.

For our purposes the virtue of (32) is that it generalizes easily to other problems. Suppose we wish to study a field $\psi_{\alpha A}$ with a spinor index α and some other index A. For instance, A may be a vector index if we wish to study the spin $\frac{3}{2}$ field; or A may be a Yang-Mills index. We wish to calculate the character-valued index of an operator acting on $\psi_{\alpha A}$. The physically relevant operator may not actually be the Dirac operator $i\Gamma^j D_j$, but we shall assume it differs from the Dirac operator only by irrelevant nonminimal terms.

As before, the character-valued index may be computed from the large $|t|$ limit of $i\not{D}_t = i\Gamma^j(D_j + itK_j)$. The index A only enters in the connection used to define D_j, but the connection is irrelevant as $|t| \to \infty$, as shown by the reduction to a flat space harmonic oscillator problem. So our previous determination of the spectrum is still valid.

What is different is the determination of the quantum numbers of the low-lying states. The symmetry generator is now

$$\mathcal{L}_K = \mathscr{L}_K + Q(\phi^k) \tag{33}$$

with an extra term Q (without derivatives) that acts on the A index. (It is a generalization of the extra angular momentum term for a charge interacting with a magnetic monopole.) Let $Q^{(i)}$ be the value of Q at the ith zero of K. The states near the ith zero that have approximately zero energy are still given by (27), but near the ith zero, $\mathcal{L}_k$ is $r\hat{\Gamma}(k + \frac{1}{2}) + Q^{(i)}$. The effect is very simple. In the sums (29) and (30), one has an extra factor $\mathrm{Tr}\, e^{i\theta Q^{(i)}}$ (the trace being over the A index), so now

$$f_i(\theta) = \left(\frac{i}{2}\right)^{n/2} \mathrm{Tr}\, e^{i\theta Q^{(i)}} \prod_{a=1}^{n/2} \frac{1}{\sin \frac{1}{2} r_a^{(i)} \theta} \tag{34}$$

and the fixed point formula is

$$I(\theta) = \left(\frac{i}{2}\right)^{n/2} \sum_i \mathrm{Tr}\, e^{i\theta Q^{(i)}} \prod_{a=1}^{n/2} \frac{1}{\sin \frac{1}{2} r_a^{(i)} \theta}. \tag{35}$$

We shall not write down here the more general formula that holds if the zeros of K are not isolated. This formula involves weighting the factors of

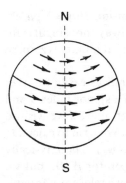

Figure 5
K is taken to generate the rotation of a sphere.

$1/(\sin \frac{1}{2} r\theta)$ by the number of zero modes of a certain Dirac operator on the fixed point set.

One may wonder "why" a fixed point formula exists. The character-valued index is formally $\text{Tr}\,\hat{\Gamma}\, e^{i\theta \mathscr{L}_\kappa}$ (since states of nonzero energy are paired and cancel out of the trace). The trace of a matrix is the sum of the diagonal matrix elements. In the coordinate basis, the diagonal matrix elements of $e^{i\theta \mathscr{L}_\kappa}$ vanish except near the zeros of K. The fixed point formula is similar to the method of Landau and Lifshitz[27] for computing the character of a molecular symmetry group furnished by the molecular vibrations in terms of fixed points of the symmetry group action.

To gain some practice with (35), let us use these methods to retrieve the results of section III. We consider a spin $\frac{1}{2}$ particle moving on the two-sphere. We take K to be (figure 5) the generator of a rotation about the z axis. There are two fixed points, the north pole N and the south pole S. A rotation that is counterclockwise as seen by an observer looking down at N is clockwise to an observer looking down at S. So the rotation angles are $r = 1$ at N and $r = -1$ at S. The fixed point formula gives

$$I(\theta) = \frac{i}{2} \frac{1}{\sin \frac{1}{2}\theta} + \frac{i}{2} \frac{1}{\sin(\frac{1}{2}\theta)} = 0, \tag{36}$$

as expected. Now, as in section III, we assume our spin $\frac{1}{2}$ particle to be charged, and we place at the center of the sphere a magnetic monopole with $eg = n/2$, $n \in Z$. It is well known[25] that the angular momentum operator is shifted: $\mathbf{J} \to \mathbf{J} + eg\mathbf{x}$. In our case $\mathscr{L}_k$ is J_z; the operator Q is the extra piece in

J_z or egz. At N, $z = 1$; at S, $z = -1$. Note that the $Q^{(i)}$ are numbers, $\pm eg$, not matrices, since in the $U(1)$ case the charge index A has only one value. The fixed point formula is now

$$I(\theta) = \frac{i}{2}e^{ieg\theta}\frac{1}{\sin(\frac{1}{2}\theta)} + \frac{i}{2}e^{-ieg\theta}\frac{1}{\sin(\frac{1}{2}\theta)}$$

$$= -\left(e^{ieg\theta} - e^{-ieg\theta}\right)\frac{e^{-i\theta/2}}{1 - e^{-i\theta}} \qquad (37)$$

$$= -\sum_{n=-[eg-(1/2)]}^{eg-(1/2)} e^{in\theta}.$$

This agrees with our result from section III, since the sum in (37) is the trace of $e^{i\theta J_z}$ in the representation of $J = eg - \frac{1}{2}$.

The proof of the Atiyah-Hirzebruch theorem that we have given is closely related to the original argument. It has the virtue of yielding the fixed point formula, which has many other applications, as we shall see. If one is only interested in the vanishing of the character-valued index of the Dirac operator, the following alternative argument may be sketched. For a space B that admits action of a nonabelian group G, the Lawson-Yau theorem[22] states that B admits a G-invariant metric of positive scalar curvature. (The basic idea of the proof is as follows. *Any* G-invariant metric g can be decomposed as $g = g_1 + g_2$, where g_1 is the metric transverse to the directions of the group action and g_2 is the metric along the group action. Lawson and Yau show that the metric $g^\varepsilon = g_1 + \varepsilon g_2$ has positive scalar curvature if ε is suitably small and positive and g_1 obeys some mild conditions at the fixed points of G.) This, combined with the Lichnerowicz theorem (no zero modes of the Dirac operator if the scalar curvature is positive), implies the vanishing of the character-valued Dirac index for manifolds with nonabelian symmetry groups. For manifolds with only an abelian symmetry group, one may reason as follows. Let B be a manifold with $U(1)$ symmetry that violates the Atiyah-Hirzebruch theorem in n dimensions. Let $\tilde{B}$ be a manifold of dimension $n + 2$ defined to be a nontrivial fiber bundle over S^2 with fiber B. [To construct this bundle, consider the space S^3 of pairs of complex numbers $\binom{z_1}{z_2}$ with $|z_1|^2 + |z_2|^2 = 1$. In the product $S^3 \times B$, make the identification

$$\left(\binom{z_1}{z_2}, \phi^i\right) \simeq \left(\binom{z_1 e^{i\alpha}}{z_2 e^{i\alpha}}, e^{in\alpha\mathscr{S}_\kappa}\phi^i\right)$$

for suitable n.] The space $\tilde{B}$ has $SU(2) \times U(1)$ symmetry, and has a nontrivial character-valued index of the Dirac operator if B does. (To see this, choose on $\tilde{B}$ a Kaluza-Klein metric with large radius in the S^2 directions and tiny radius in the B directions and solve the Dirac equation on $\tilde{B}$ in a Born-Oppenheimer–Kaluza-Klein approximation.) Since the Lawson-Yau and Lichnerowicz theorems imply that the character-valued index must vanish on $\tilde{B}$, it must also vanish on B.

V Rarita-Schwinger Fields

In this section we shall study the character-valued index of the Rarita-Schwinger field. We shall not be able to reach a comprehensive result similar to the Atiyah-Hirzebruch theorem for the Dirac case. We shall prove that the zero modes of the Rarita-Schwinger operator on any homogeneous space G/H form a real representation of G. In other words, we shall show that on any homogeneous space of dimension $4k + 2$, the character-valued index vanishes. For homogeneous spaces of dimension $4k$, the same argument shows that the character-valued index vanishes except for the trivial character (the topological index). Unfortunately, I do not know a general result for the Rarita-Schwinger field on spaces that are not homogeneous spaces. (I also have been unable, despite many attempts, to find a case in which the character-valued index is nontrivial, and I believe that if such manifolds exist, they are rather complicated manifolds with rather small symmetry groups in relation to the number of dimensions.) Because our results will not be entirely conclusive, we shall return after discussing the theorem on homogeneous spaces to a discussion of various problems in the use of the Rarita-Schwinger operator.

Our basic tool will be the fixed point formula discussed in the previous section. In particular, we shall not use the local supersymmetry of the Rarita-Schwinger field; it may be possible to find a stronger result by using this property.

In an appropriate gauge, the Rarita-Schwinger field is simply a vector spinor field $\psi_{\mu\alpha}$ (μ is a vector index, α a spinor index) that obeys, up to irrelevant nonminimal terms, a Dirac equation $\not{D}\psi_{\mu} = 0$. Of course, we wish to discard zero modes of ψ_{μ} that can be gauged away or that violate gauge conditions. Physically, in quantizing a theory, zero modes that are gauge artifacts are canceled by zero modes of the spin $\frac{1}{2}$ ghost fields. Therefore, we must subtract from the character-valued index of the Rarita-Schwinger field the corresponding index of the spin $\frac{1}{2}$ ghosts. (This is the general logic, but actually the ghost index vanishes, by the Atiyah-Hirzebruch theorem.)

Let us work out the fixed point formula for the spin $\frac{3}{2}$ field. Consider an isolated zero of the Killing vector field K. Suppose that near the zero (which we assume to be at $\phi^i = 0$), $K_i = \omega_{ij}\phi^j$, where ω is a constant antisymmetric matrix,

$$
\omega = \begin{pmatrix}
0 & r_1 & & & & & \\
-r_1 & 0 & & & & & \\
& & 0 & r_2 & & & \\
& & -r_2 & 0 & & & \\
& & & & \ddots & & \\
& & & & & 0 & r_{n/2} \\
& & & & & -r_{n/2} & 0
\end{pmatrix}. \tag{38}
$$

The symmetry generator $\mathcal{L}_K$ for a spin $\frac{3}{2}$ field is

$$
\begin{aligned}
\mathcal{L}_L\psi_j &= i(K^i D_i\psi_j + \tfrac{1}{4}\Gamma^{lm}(D_l K_m)\psi_j) - i(D_j K^i)\psi^i \\
&= \mathcal{L}_K\psi_j - i(D_j K^i)\psi^i,
\end{aligned} \tag{39}
$$

where $\mathcal{L}_K$ is the generator for a spin $\frac{1}{2}$ field and the extra piece acting on the vector index is $-i(D_j K^i)$. At $\phi^i = 0$, $D_j K_i = \omega_{ij}$, so $\mathcal{L}_K$ acts on the vector index of ψ_i by multiplication by the matrix $i\omega$. The factor of $\operatorname{Tr} e^{i\theta Q^{(i)}}$ in formula (34) for the local index is here to be replaced by

$$
\operatorname{Tr} e^{i\theta(i\omega)} = 2 \sum_{a=1}^{n/2} \cos\theta r_a. \tag{40}
$$

The fixed point formula for the character-valued index of the Rarita-Schwinger field is then

$$
I(\theta) = \left(\frac{i}{2}\right)^{n/2} \sum_i \left[\left(\sum_{a=1}^{n/2} \frac{1}{\sin\frac{1}{2}r_a^{(i)}\theta} \right) \left(\sum_{b=1}^{n/2} 2\cos r_b^{(i)}\theta - 1 \right) \right]. \tag{41}
$$

Here i runs over the fixed points or zeros of K; $r_a^{(i)}$, $a = 1, \ldots, n/2$, are the rotation angles at the ith zero; and the minus one in the last factor in (41) is chosen to subtract the index of the ghost fields, as discussed earlier. (Minus one equals minus two plus one; for the quantization of ψ_i, there are[6] two ghosts with the same chirality as ψ_i and one of opposite chirality.) Equation (41) has a generalization when the fixed points are not isolated.

Equation (41) places very severe restrictions on the possibility of obtaining a complex representation of Rarita-Schwinger zero modes; it shows that $I(\theta)$ is a rational function of $w = e^{i\theta/2}$ with poles only for $|w| = 1$ or at $w = 0$ or $w = \infty$. The poles at $|w| = 1$ must cancel upon summing over fixed points, as in the spin $\frac{1}{2}$ case. [Note the discussion following (32).] The rational function $I(w)$ must be a constant unless there really are poles at $w = 0$ or $w = \infty$. If the largest rotation angle at the ith fixed point is $r_b^{(i)}$, then the contribution of this fixed point to $I(w)$ behaves for $w \to \infty$ or $w \to 0$ as

$$f_i(w) \sim w^{(r_b^{(i)} - \sum_{a \neq b} r_a^{(i)})}. \tag{42}$$

There are poles at $w = 0$ and $w = \infty$ if and only if the largest rotation angle $r_b^{(i)}$ is bigger than the sum of the others:

$$r_b^{(i)} > \sum_{a \neq b} r_a^{(i)}. \tag{43}$$

If (43) is not obeyed for any i, $I(w)$ has no singularities. Even if (43) is obeyed, the poles may vanish in summing over i. It is difficult to satisfy (43) in a multidimensional space with many rotation angles at each fixed point.

If for each i and each b

$$r_b^{(i)} \leqslant \sum_{a \neq b} r_a^{(i)}, \tag{44}$$

then $I(w)$ is a rational function without poles and bounded for $w \to \infty$, and therefore is a constant. The constant is an integer—the ordinary or topological index of the Rarita-Schwinger field. If (44) is always a strict inequality, $r_b^{(i)} < \sum_{a \neq b} r_a^{(i)}$ for all i and b, then $I(w)$ vanishes as $w \to \infty$, so $I(w) = 0$ and the topological index vanishes.

We shall use these considerations to prove that the character-valued index of the Rarita-Schwinger field on a homogeneous space always vanishes except possibly for the trivial character (which may appear in the case of $4k$ dimensions).

The homogeneous space G/H is defined as follows. It is the space of all $g \in G$ with g and gh considered equivalent for any $h \in H$. Because of the equivalence relation, the dimension of G/H equals the dimension of G minus the dimension of H. The space G/H is invariant under $g \to ug$ for any $u \in G$. A fixed point of this transformation is an element g of G such that $ug = gh$ for some $h \in H$; in other words $g^{-1}ug \in H$. If u is not equivalent up to similarity to an element of H, then the symmetry transformation $g \to ug$ has no fixed points.

If the rank of H is less than the rank of G, then the generical generator A

of the Cartan subalgebra of G is not equivalent (up to similarity) to any generator of H. Then $e^{i\theta A}$ acts on G/H without fixed points, so the character-valued index vanishes. Hence we need only consider the case rank H = rank G.

Let us now make a brief detour. In general, given two spaces M, N the character-valued Dirac and Rarita-Schwinger (R.S.) indexes of M, N, and the product $M \times N$ are related by

$$\text{index}_{\text{Dirac}}(M \times N) = \text{index}_{\text{Dirac}}(M) \cdot \text{index}_{\text{Dirac}}(N),$$

$$\begin{aligned}\text{index}_{\text{R.S.}}(M \times N) = {} & \text{index}_{\text{R.S.}}(M) \cdot \text{index}_{\text{Dirac}}(N) \\ & + \text{index}_{\text{Dirac}}(M) \cdot \text{index}_{\text{R.S.}}(N) \\ & + \text{index}_{\text{Dirac}}(M) \cdot \text{index}_{\text{Dirac}}(N).\end{aligned} \tag{45}$$

These equations hold because the Dirac and Rarita-Schwinger equations on $M \times N$ can be solved by separation of variables. The second equation in (45) (which is our real interest) arises because the vector index of ψ_i must be tangent to either M or N, so a Rarita-Schwinger solution on $M \times N$ obeys the Dirac equation on M and the Rarita-Schwinger equation on N or vice versa. [The last term in the second equation in (45), which is not intuitively obvious, arises in subtracting the ghost contributions from the Rarita-Schwinger indexes.] In particular, (45) implies that if M and N have vanishing Dirac index, then $M \times N$ has vanishing Rarita-Schwinger index. Since a homogeneous space has vanishing Dirac index (by the Lichnerowicz or Atiyah-Hirzebruch theorems), a product of two homogeneous spaces has vanishing Rarita-Schwinger index.

From these facts it follows that the Rarita-Schwinger index of G/H vanishes unless G is simple. For suppose $G = G_1 \times G_2$ with nontrivial $G_1 \times G_2$. A subgroup of G of maximal rank is then necessarily $H = H_1 \times H_2$, where H_1 and H_2 are maximal rank subgroups of G_1 and G_2, respectively. Then $G/H = (G_1/H_1) \times (G_2/H_2)$ is a product of homogenous spaces, and has vanishing character-valued Rarita-Schwinger index.

We still must study G/H with G simple and H a subgroup of maximal rank. Let us first calculate the rotation angles at the fixed points for a typical infinitessimal transformation $g \to (1 + i\varepsilon A)g$, A being a generator of G. Since H is maximal, A is equivalent (by conjugation) to a generator of H, so we may assume A is actually such a generator. A typical fixed point is then $g = 1$; because of the equivalence under right multiplication by an element of h, the transformation $g \to (1 + i\varepsilon A)g$ is equivalent to

$$g \to (1 + i\varepsilon A)g(1 - i\varepsilon A). \tag{46}$$

This again makes it clear that $g = 1$ is a fixed point. The fixed point is isolated if A is chosen generically. The rotation angles at $g = 1$ may be computed as follows. The Lie algebra $\mathscr{G}$ of G can be decomposed as the Lie algebra $\mathscr{H}$ of H plus an orthogonal complement $\mathscr{K} : \mathscr{G} = \mathscr{H} \oplus \mathscr{K}$. Near $g = 1$, the generic element of G/H is $1 + ik$, with $k \in \mathscr{K}$. The transformation (46) acts on k by $k \to k + i\varepsilon [A, k]$. The rotation angles at $g = 1$ are therefore just the eigenvalues of A acting on $\mathscr{K}$ by conjugation.

The other fixed points are at points g_i such that $g_i^{-1} A g_i \in \mathscr{H}$. The rotation angles are the eigenvalues of $g_i^{-1} A g_i$ acting by conjugation on $\mathscr{K}$.

Now we are ready to prove that the Rarita-Schwinger index vanishes on homogeneous spaces, except for the trivial character. Before stating the argument in a general way let us first consider the case that G is $SU(N)$. Let Y_0 be the $SU(N)$ generator

$$Y_0 = \begin{pmatrix} \frac{1}{2} & & & & & \\ & -\frac{1}{2} & & & & \\ & & 0 & & & \\ & & & 0 & & \\ & & & & \ddots & \\ & & & & & 0 \end{pmatrix}. \tag{47}$$

It generates an $SU(2)$ subgroup of $SU(N)$, so its eigenvalues in any representation are integers or half-integers. Let $Y = Y_0 + \varepsilon Q$, where Q is a generic $SU(N)$ generator that commutes with Y_0 and ε is a suitably small real number.

Consider the Rarita-Schwinger equation on some space homogeneous under an $SU(N)$ action. Let Λ^+ and Λ^- be the $SU(N)$ representations of positive and negative chirality zero modes. Define $I(\theta) = \text{Tr}_{\Lambda_+} e^{i\theta Y} - \text{Tr}_{\Lambda_-} e^{i\theta Y}$. In any nontrivial $SU(N)$ representation the biggest eigenvalue of Y_0 is at least $\frac{1}{2}$. Hence, if Λ_+ differs from Λ_- by a nontrivial $SU(N)$ representation, the most positive power of $e^{i\theta}$ appearing will be at least $e^{i\theta\alpha}$, where $\alpha = \frac{1}{2} + O(\varepsilon)$. (For suitable Q and sufficiently small ε, there is no accidental cancellation of the highest power. This is the only role of Q and ε in the discussion.) This means that with $w = e^{i\theta/2}$, $I(w)$—if not a constant—diverges for $w \to \infty$ at least as $w^{1+O(\varepsilon)}$. Reference to (42) shows the necessary condition to achieve this; we need at one of the fixed points of the fixed points P of the transformation generated by Y

$$r_b - \sum_{a \neq b} r_a \geqslant 1, \tag{48}$$

where r_b is the largest rotation angle at P and r_a are the other rotation angles at P.

What are the rotation angles of the transformation Y that are not of order ε? They are one, corresponding to the $SU(N)$ generator

$$A = \begin{pmatrix} 0 & 1 & 0 & 0 & 0 & 0 \\ 0 & 0 & 0 & . & . & 0 \\ . & . & . & . & . & . \\ . & . & . & . & . & . \\ 0 & . & . & . & . & 0 \end{pmatrix} \tag{49}$$

and its adjoint, and $\frac{1}{2}$ corresponding to the generators

$$X_\alpha = \begin{pmatrix} 0 & 0 & a_1 & \ldots & a_k \\ 0 & 0 & b_1 & \ldots & b_k \\ 0 & 0 & 0 & \ldots & 0 \\ 0 & 0 & 0 & \ldots & 0 \end{pmatrix} \tag{50}$$

and their adjoints. Of course, in general we do not count A and all the X_α; we only count those that are in $\mathcal{K}$ (the complement in G of the H Lie algebra). The only way to obey (48) is to include A but none of the X_α. This means H must be a subgroup of $SU(N)$ that includes all the X_α and their adjoints and all the diagonal generators (since H has maximal rank) but not A. There is no such subgroup of $SU(N)$. Hence Λ_+ and Λ_- differ at most only by the trivial character.

The same argument goes through with $SU(N)$ replaced by any simple Lie group G. One simply replaces Y_0 by the generator of the Cartan subalgebra parallel to a root E of maximum length. E plays the role of A; the roots not orthogonal to E play the role of the X_α. The rest of the argument is unchanged.

Thus, we have shown that on any homogeneous space G/H (of dimension $4k$ or $4k + 2$) the character-valued index of the Rarita-Schwinger field is a constant, a multiple of the trivial character. It is equal simply to the ordinary index, the difference between the total number of right-handed and left-handed zero modes. The ordinary Rarita-Schwinger index certainly vanishes in $4k + 2$ dimensions, but (unlike the ordinary Dirac index) it need *not* vanish on homogeneous spaces of dimension $4k$. It equals one on $SU(4)/(SU(2) \times SU(2) \times U(1))$, HP^2, and $G_2/O(4)$. This may be seen by the methods of sections III and IV, or from facts in reference [28], where properties of these spaces are described.

Interestingly, for spin greater than $\frac{3}{2}$, it *is* possible to obtain complex representations of zero modes on homogeneous spaces. Consider a spin $\frac{5}{2}$ field, which we may represent as a tensor spinor $\psi_{ij\alpha}$, i and j being vector indices and α a spinor index (we suppose $\psi_{ij} = \psi_{ji}$, $\psi i_i = 0$). The wave equation $\not{D}\psi_{ij} = 0$ *can* have zero modes in complex representations on homogeneous spaces. For instance, on the 6-dimensional manifold CP^3 this operator has the following stable spectrum of zero modes: one multiplet of left-handed zero modes in the symmetric tensor representation S^{ij}, and one multiplet of right-handed zero modes in the complex conjugate representation S^*_{ij}. This may be seen with the methods of sections III and IV.

Of course, physically sensible couplings of a massless spin $\frac{5}{2}$ field to gravity do not seem to exist[23]. However, this arises only because the timelike components ψ_{0i} of the tensor-spinor field have the wrong metric; because of the apparent nonexistence of spin $\frac{5}{2}$ gauge invariance, there is no way to cancel or remove them. Purely as a euclidean equation, with i and j tangent to the positive signature Kaluza-Klein space, the equation $\not{D}\psi_{ij} = 0$ makes perfect sense and has the properties just stated.

Our result about the Rarita-Schwinger operator is much less sweeping than the Atiyah-Hirzebruch theorem of the spin $\frac{1}{2}$ case. I do not know whether on some spaces that are not homogeneous the Rarita-Schwinger operator may have zero modes in complex representations. [I am convinced, from many unsuccessful attempts to find them, that if such spaces exist they are rather complicated. It is difficult to satisfy (42).]

There is actually a strategy that might very plausibly lead eventually to a general proof that the character-valued index of the Rarita-Schwinger field always vanishes except for the trivial character. The topological index of an operator is a cobordism invariant; this means that it vanishes for any manifold M of dimension n that is the boundary of a manifold of dimension $n + 1$. The character-valued index is likewise invariant under equivariant cobordism; this means that if M admits the action of a group G and is the boundary of a manifold of dimension $n + 1$ to which the G action on M can be extended, then the character-valued index vanishes for any operator on M. If a set of generators of the $U(1)$ spin bordism ring [the ring of spin manifolds with $U(1)$ symmetry modulo those which are boundaries] were found, our conjecture about the Rarita-Schwinger field could be proved by showing it to hold for all the generators. The mathematical problem of determining a set of generators for the (oriented) $U(1)$ spin bordism ring has not been solved. However, the analogous problem has been solved for the unoriented[44] and unitary[45] $U(1)$ bordism rings. These rings are generated by very simply spaces (essentially, homogeneous spaces and fiber bundles in which the fiber is a homogeneous space). If the $U(1)$ spin bordism ring is found to be generated by equally simple spaces, it will be possible to use the methods described previously to prove (or disprove) the conjecture that the character-valued Rarita-Schwinger index is always a constant.

Because the situation for spin $\frac{3}{2}$ fields is not completely clear, some general remarks on the subject may be useful. It is believed that massless spin $\frac{3}{2}$ fields can be consistently coupled to gravity only in locally supersymmetric theories. This apparently means[17] that we are limited to 11 dimensions or less. In addition, beyond 10 dimensions the chiral Rarita-Schwinger field has one loop anomalies[18] that spoil general covariance and cannot be canceled by the anomalies of any known fields that can be consistently coupled to gravity. For both of these reasons, it appears that 6 and 10 dimensions are the relevant cases for chiral Rarita-Schwinger fields. This corresponds to 2 or 6 compact dimensions, respectively.

With 2 compact dimensions, the only manifolds with continuous symmetry are the sphere, torus, and Klein bottle; the first two are homogeneous spaces, and on the last two the continuous symmetries have no fixed points, so on all of them the character-valued Rarita-Schwinger index vanishes. We turn then to the case of 6 compact dimensions.

Six compact dimensions are unfortunately too few to admit $SU(3) \times SU(2) \times U(1)$ symmetry[4]. One may be willing to postulate an elementary $U(1)$ gauge field and to try to obtain only $SU(3) \times SU(2)$ as the symmetry group of a six-manifold.[10] The unique six-manifold with $SU(3) \times SU(2)$ symmetry is $CP^2 \times S^2$. This is a homogeneous space to which our theorem applies; the character-valued index could not be nonzero. Even worse, this space does not admit spinors, so the Rarita-Schwinger equation on $CP^2 \times S^2$ cannot be defined[29].

However, if the $U(1)$ gauge field has a magnetic monopole expectation value on CP^2, spinors can be introduced (this is the so-called spin$_c$ structure); all fermi fields, including the Rarita-Schwinger field, must have

10. This suggestion was made independently by M. Gell-Mann.

nonzero (half-integral) $U(1)$ charges. The nonzero $U(1)$ charge of the spin $\frac{3}{2}$ field introduces new anomalies (the mixed gauge-gravity anomalies of reference [18]), which cannot cancel among themselves unless there are many more than two Weyl gravitinos (two is the maximum of any known or conjectured 10-dimensional supergravity theory) and which cannot be canceled by anomalies of spin $\frac{1}{2}$ fields (because of a different tensor structure). If we ignore this and proceed, we can calculate the spectrum of the Rarita-Schwinger operator on $CP^2 \times S^2$. The $SU(3) \times SU(2)$ invariant expectation value of the $U(1)$ field strength on $CP^2 \times S^2$ depends on two "monopole numbers"—a half-integer p on CP^2 (half-integer so as to get a spin$_c$ structure) and an arbitrary integer q on S^2. The resulting zero mode spectrum can be computed by the methods of sections III and IV. One obtains nontrivial complex representations, depending on p and q, but these representations have little resemblance to physics and are anomaly ridden [because the 10-dimensional theory with $U(1)$ coupling to the Rarita-Schwinger field is anomalous].

These accumulated difficulties may encourage us to give up on accommodating $SU(3) \times SU(2)$ symmetry in 6 compact dimensions. We may simply try for $SU(3)$ symmetry. The six-manifolds with $SU(3)$ symmetry are quite restricted and can be seen from the fixed point formula to have vanishing Rarita-Schwinger character-valued index. It may be possible to accommodate $SU(2) \times U(1)$ in 10 dimensions and to obtain leptons but not quarks as Rarita-Schwinger zero modes.

Evidently, whatever is the behavior of the Rarita-Schwinger operator on spaces that are not homogeneous, to obtain physics in this way would not be easy.

VI Elementary Gauge Fields

We have so far assumed (except for some brief remarks at the end of the last section) that gauge fields are not elementary but arise as components of the metric tensor in $4 + n$ dimensions. As this assumption has led to difficulties, let us reconsider it. In this section we shall suppose that elementary gauge fields do play a crucial role. Thus we are considering a much less ambitious program. We can no longer hope to unify all forces or calculate all couplings, but we still can hope to explain the fermion quantum numbers; we can try to start with a simple fermion representation in $4 + n$ dimensions and obtain a complicated, repetitive one in 4 dimensions after compactification.

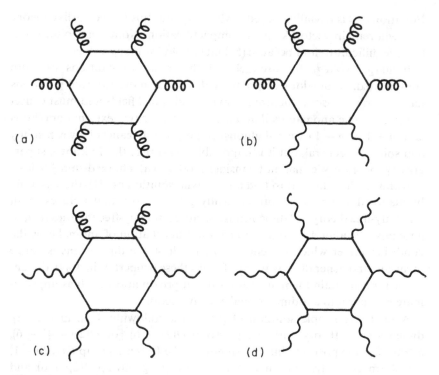

Figure 6
The anomalous diagrams in (say) 10 dimensions. Loopy lines are gravitons; wavy lines are gluons.

We must start in $4 + n$ dimensions with gauge fields that couple differently to left- and right-handed fermions. For if the gauge quantum numbers of the fermions are vectorlike, they will remain vectorlike after any compactification.

Non-vector-like couplings of elementary gauge fields raise an immediate and serious problem of canceling anomalies. In $2p$ dimensions the diagrams with $p + 1$ external gluons, $p - 1$ external gluons and two gravitons, $p - 3$ external gluons and four gravitons, etc., are all anomalous[18] (figure 6). If the gauge field A_μ^a couples to left-handed and right-handed spin $\frac{1}{2}$ fermions via matrices M_L^a and N_R^a, respectively, the condition for anomaly cancellation is

$$\mathrm{Tr}(M_L^a)^r = \mathrm{Tr}(N_R^a)^r,$$

$$r = p + 1, p - 1, p - 3, p - 5, \dots. \tag{51}$$

Equation (51) is trivially obeyed if $M_L^a = N_R^a$, but this is a vectorlike theory that will remain vectorlike after compactification. Nontrivial solutions of (51) are difficult to find because (51) must hold for many values of r.

If charged spin $\frac{3}{2}$ fields are included, the anomaly conditions are more complicated. Anomalous diagrams with four or more external gravitons have different tensor structure for spin $\frac{1}{2}$ and spin $\frac{3}{2}$ fields and must cancel separately. The anomalous diagrams with zero or two external gravitons (and $p + 1$ or $p - 1$ external gluons) could possibly cancel between spin $\frac{1}{2}$ and spin $\frac{3}{2}$, in general, but it is impossible to arrange this in known supergravity theories. We shall not consider models with charged spin $\frac{3}{2}$ fields.

Although it is difficult to find a nontrivial solution of (51), there is a big bonus for doing so. One automatically gets a theory that reduces to an anomaly-free theory in 4 dimensions, because anomaly-free theories remain anomaly-free after any compactification. Cancellation of anomalies is the condition under which the effective action in $4 + n$ dimensions is gauge invariant and generally covariant. If it has these properties in $4 + n$ dimensions, it must retain them after any valid approximation, such as approximate reduction to a 4-dimensional effective action.

A relatively simple solution of (51) is the following. Consider, in $2p$ dimensions, a theory with gauge group $O(2p + 6)$ [or $O(2p + 4k + 6)$, $k \geqslant 0$]. Let the positive chirality spinors of the Lorentz group $O(1, 2p - 1)$ transform as positive chirality spinors of the gauge group $O(2p + 6)$; and let the negative chirality $O(1, 2p - 1)$ spinors transform as negative chirality spinors of $O(2p + 6)$. For $p = 2$ this is the usual $O(10)$ model in 4 dimensions. For any p, this theory is anomaly-free. It obeys (51) because the first $p + 2$ Casimir operators of the left- and right-handed spinor representations of $O(2p + 6)$ are equal.[11] The fermion representation of the $O(2p + 6)$ model in $2p$ dimensions is irreducible in the sense that a combined parity and internal parity operation exchanges the left- and right-handed fermions. This is the simplest theory with nontrivial anomaly cancellation that I can find in $d > 4$.

Despite starting with a non-vector-like theory, we are not assured of keeping this property after dimensional reduction. The familiar relation

$$\Gamma_1 \cdots \Gamma_{4+n} = \Gamma_1 \cdots \Gamma_4 \cdot \Gamma_5 \cdots \Gamma_{4+n} \tag{52}$$

shows that the $(4 + n)$-dimensional chirality operator $\Gamma_1 \cdots \Gamma_{4+n}$ differs from the 4-dimensional operator $\Gamma_1 \cdots \Gamma_4$ by a factor $\Gamma_5 \cdots \Gamma_{4+n}$ that may be

---

11. This is so because $\operatorname{Tr} \sigma_{i_1 j_1} \sigma_{i_2 j_2} \cdots \sigma_{i_k j_k} \cdot \tilde{\Gamma} = 0$ for $k \leqslant p + 2$, if $\tilde{\Gamma}$ is the product of all $2p + 6$ gamma matrices of $O(2p + 6)$ and $\sigma_{ij} = [\gamma_i, \gamma_j]$ are the group generators.

plus or minus one. The non-vector-like nature is washed out after naive dimensional reduction.

Indeed, if the gauge fields have zero expectation value, they are of no help whatsoever in avoiding the problems discussed in previous sections. They simply do not play any role. But vacuum expectation values of gauge fields that can be smoothly turned on or off are likewise irrelevant, for the usual reasons. We therefore must consider gauge field expectation values that cannot be smoothly turned on or off. The prototype of such a thing is the Dirac magnetic monopole, which because it carries topological information and obeys a Dirac quantization condition cannot be smoothly turned on or off. We must place a generalized monopole "inside" the Kaluza-Klein space. Models of this general type have been considered independently by Randjbar-Daemi, Salam, and Strathdee [14].

For illustrative purposes we set $p = 5$ and consider an $O(16)$ theory in 10 dimensions. As explained earlier, the left- (or right-) handed $O(1, 9)$ spinors will be chosen as left- (or right-) handed spinors of $O(16)$. We shall consider three models.

In the first two models we take the 6 compact dimensions to be CP^3—the unique 6-dimensional space with $SU(4)$ symmetry. We shall assume that only a single (abelian) component of the $O(16)$ gauge field has an expectation value.

On CP^3 there is a topologically nontrivial $U(1)$ gauge field. [It is a $U(1)$ connection on the basic nontrivial line bundle over CP^3.] This $U(1)$ gauge field can be chosen to be $SU(4)$ invariant in a unique way. Its strength, like the Dirac monopole charge, must be an integer n in certain units.

Our first two models correspond to two ways of embedding $U(1)$ in $O(16)$.

i. The embedding

$$\begin{pmatrix} 1 & & & & & & \\ -1 & & & & & & \\ & & 1 & & & & \\ & & & -1 & & & \\ & & & & \cdot & & \\ & & & & & 1 & \\ & & & & & & -1 \end{pmatrix} \tag{53}$$

breaks $O(16)$ to $SU(8) \times U(1)$. The 4-dimensional gauge group is therefore $SU(4) \times SU(8) \times U(1)$, where $SU(4)$ originates from gravity and $SU(8) \times U(1)$ from $O(16)$.

The quantum numbers of fermion zero modes can be computed using the methods of section III or IV. One finds rather complicated anomaly-free representations. We shall here consider only the minimal case of monopole number $n = 1$. The representation of $SU(4) \times SU(8) \times U(1)$ that emerges is as follows. Let V^i and W^j be the fundamental 4- and 8-dimensional representations of $SU(4)$ and $SU(8)$, respectively. Let S^{ij} be the symmetric product of two V^i, and $\bar{S}_{ij}$ its complex conjugate; and let A^{jk} be the antisymmetric product of two W^j and $\bar{A}_{jk}$ its complex conjugate. Then the left-handed zero modes transform as $(V^i, W^j)^3 \oplus (1, A_{ij})^{-2} \oplus (\bar{S}_{ij}, 1)^{-4}$; the superscript is the $U(1)$ charge. The right-handed massless fermions transform of course in the conjugate representation $(\bar{V}_i, \bar{W}_j)^{-3} \oplus (1, A^{ij})^2 \oplus (S^{ij}, 1)^4$. Despite its complexity, this representation is anomaly-free. For reasons that are not at all clear, this representation is closely related to the supergroup $SU(4|8)$ [30].

ii. We can instead embed $U(1)$ in $O(16)$ as

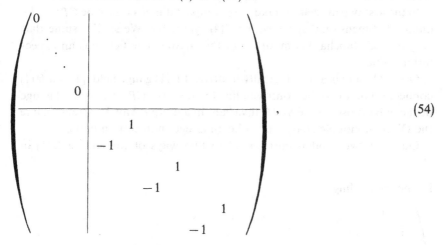

$$\tag{54}$$

breaking $O(16)$ to $O(10) \times SU(3) \times U(1)$. In this case the monopole number must be even, for topological reasons [to ensure proper Dirac quantization for particles in the *spinor* as well as *vector* representations of $O(10)$]. Thus we take the minimal case $n = 2$.

The zero modes can be shown in this case to consist of eight $O(10)$ families (left-handed 16 and right-handed $\overline{16}$) and no antifamilies. Under $SU(4)$

(from gravity) $\times SU(3) \times U(1)$ [from $O(16)$] they transform as follows: they are neutral under $SU(3)$ and transform as $4^3 \oplus \overline{4}^{-3}$ under $SU(4) \times U(1)$.

iii. For our last model, we consider a case in which a nonabelian subgroup of $O(16)$ has a vacuum expectation value. On any 6-dimensional Riemannian manifold there is a natural $O(6)$ gauge field. One simply takes the spin connection $\omega_\mu{}^i{}_j$ and regards it as a gauge field $A_\mu{}^i{}_j$. We embed $O(6)$ in $O(16)$ in the obvious way:

$$\left(\begin{array}{c|c} 0 & 0 \\ \hline 0 & O(6) \end{array} \right). \tag{55}$$

This breaks $O(16)$ down to $O(10)$.

We shall *not* assume the 6 compact dimensions to be CP^3. The Dirac equation for this system can be analyzed in general. The zero modes are always ordinary families, never antifamilies and never other representations of $O(10)$. The families are neutral under any continuous symmetries of the 6-dimensional space B.

The number of families always equals the Euler characteristic of B. Here are some examples:

$$
\begin{array}{ll}
S^6: & \text{2 families,} \\
S^2 \times S^4: & \text{4 families,} \\
S^2 \times S^2 \times S^2: & \text{8 families,} \\
CP^3: & \text{4 families,} \\
S^3 \times S^3: & \text{0 families.}
\end{array} \tag{56}
$$

In 10 dimensions, the number of families is always even. A similar model in 8 or 12 dimensions can give an odd number of families.

Why is the number of families equal to the Euler characteristic of B? With $A_\mu{}^i{}_j \sim \omega_\mu{}^i{}_j$, the $O(16)$ spinor index behaves as an extra Lorentz spinor index of B. The fermi field is therefore a spinor-spinor, a field with two independent 6-dimensional spinor indices. Such a field is equivalent to the de Rham complex of antisymmetric tensor fields. The Dirac operator becomes the d and d^* operators on differential forms, and the number of zero modes (weighted by chirality) is the Euler characteristic.

VII Other Contexts

The preceding discussion has made clear that in Riemannian geometry it will be very difficult to obtain realistic fermion quantum numbers as zero modes of physically acceptable wave operators. One way out, considered in the last section, is to introduce elementary gauge fields. In the context of Kaluza-Klein theory, this is a rather disappointing possibility. What other alternatives might there be?

We must modify Riemannian geometry in some way. One possibility is that the tangent space group in $4 + n$ dimensions is not $O(1, 3 + n)$ but a smaller group G. (This possibility has been considered by Davidson [30] and in much detail by Weinberg [31].) The smaller group will have more representations, in general; some of the new representations may correspond to new options for the spins of massless particles.

For instance, G may be $O(1, 3) \times O(n)$. This group has a representation with spin $\frac{1}{2}$ under $O(1, 3)$ and (say) spin $\frac{5}{2}$ under $O(n)$. Since the "true" spin is $\frac{1}{2}$, there are no timelike modes with wrong metric and no difficulty in writing a sensible wave operator. In fact, in the ground state the spin $(\frac{1}{2}, \frac{5}{2})$ field may be regarded as a tensor spinor $\psi_{ij\alpha}$ with i and j constrained to be tangent to the Kaluza-Klein dimensions; a satisfactory wave equation is $\not{D}\psi_{ij} = 0$. (Once one considers fluctuations away from the ground state these formulas do not have a simple generalization, but *some* generalization would emerge in any theory with restricted tangent space group.) Since the spin $\frac{5}{2}$ and higher spin operators in the internal space can readily have complex zero modes (note the discussion of CP^3 in sections V), this would enable us to obtain non-vector-like theories after compactification.

If n is even, $n = 2k$, one could consider a theory with tangent space group $O(1, 3) \times U(k)$. The $U(k)$ group of the internal space is the tangent space group in Kahler geometry. In Kahler geometry, one may write many variants of the Dirac equation [corresponding to the many representations of $U(k) \cong SU(k) \times U(1)$ that do not extend to representations of $O(2k)$] that lack analogs in Riemannian geometry. This is the subject of $\bar{\partial}$ cohomology, a major subject in Kahler geometry. The modified Dirac equations of Kahler geometry can readily have complex zero modes. For instance, the equations for a charged spin $\frac{1}{2}$ field interacting with a magnetic monopole on S^2 or CP^3 can be regarded as equations in Kahler geometry; therefore, the model of section three and the first two models of section VI can be viewed in this light.

Of course, it is disappointing to consider tangent space groups like $O(1,3) \times O(n)$ or $O(1,3) \times U(k)$ that are product groups. One would much prefer a unified group, even if smaller than $O(1,3+n)$. However, Weinberg [31] has shown that if one desires Lorentz invariance in 4 dimensions, product groups are the only possibilities.

If one is willing to envisage a product group G, one must still find a sensible equation—replacing the Einstein equation—for the time dependence of the G connection. And presumably one must face at some point the unrenormalizability of quantum field theory in $4 + n$ dimensins, which is likely to persist in this context.

A more drastic modification of Riemannian geometry would be to assume that the underlying theory is not a field theory of the usual kind but a theory of some other type. For instance, at present the supersymmetric string theories in 10 dimensions [7] would appear to be very attractive candidates—especially the $n = 2$ theory, which is chirally asymmetric and anomaly-free. This theory naively reduces at low energies to 10-dimensional supergravity, but unlike that theory [7, 32], it is likely to be a finite theory to all orders.

Naive compactification of the string theory proceeds via 10-dimensional field theory and suffers from the problems of 10-dimensional field theory in describing fermion quantum numbers. However, there may be "inherently stringy" ways to compactify the string theory directly to 4 dimensions, without 10-dimensional field theory as an intermediate stage. The rules and problems of Riemannian geometry might not apply in such a case—though I hope some of the concept of this paper would be relevant. I would consider this the most attractive possibility, but unfortunately with the present incomplete understanding of the string theory, it is difficult to pursue this possibility.

I shall, however, discuss one aspect of the problem. The string theory has a single dimensionless coupling constant λ. (It is essentially Newton's constant written in units of the Regge slope.) It is generally believed that λ is an arbitrarily adjustable constant. If so, the fermion quantum numbers cannot depend on λ. I believe that instead, when the string theory is more fully understood, it will be seen that (with proper normalization of λ) the mathematical consistency of the theory will require that $1/\lambda$ be an integer. The action of the string field theory that has been partly constructed [33] is rather analogous to the effective action of the large N expansion in QCD. We now know [34] that the large N effective action is multivalued, defined only modulo $2\pi N$. This means that, internal to the $1/N$ expansion, N must be an integer; this is analogous to the quantization of coupling constants in some $(2 + 1)$-dimensional field theories [35]. The close analogy between λ and $1/N$ strongly suggests that the string field theory action is likewise multivalued, defined only modulo $2\pi/\lambda$. This would mean that $1/\lambda$ would have to be an integer n; the proper quark and lepton quantum numbers might emerge only for a definite value of n.

VIII Massless Scalars

The main focus of this paper is the question of obtaining massless fermions as zero modes in Kaluza-Klein theory. However, massless bosons are also important. Massless spin 1 and spin 2 bosons have a well-known origin in Kaluza-Klein theory, and arise for a simple reason, as reflections of un- broken local symmetries. Massless scalars do not have such a simple rationale. Yet a good explanation of the existence of massless charged scalars would be of utmost importance: it would offer a solution [36] to the problem of the existence of widely disparate mass scales in physics.

Any bose field of a Kaluza-Klein theory might have modes that would be seen as charged scalars in 4 dimensions. We must consider then the gravita- tional field, the antisymmetric tensor fields of certain supergravity and other theories, and gauge fields. [More generally, in suitable backgrounds, different fields may mix; mixed modes may be considered under (i) or (iii), to follow.]

In general terms, we do not want scalars that are massless for reasons of symmetry. The only scalars kept massless by any symmetry argument of the usual sort are Goldstone bosons, which are always neutral under any unbroken gauge symmetries and hence are no help in solving the hierarchy problem. Moreover, for a Goldstone boson any potential at all is forbidden; there is no reason for a Goldstone boson field to acquire tiny but nonzero vacuum expectation values. We wish a more subtle argument for bosonic zero modes, perhaps a topological argument, which will forbid mass terms but allow quartic self-couplings.

Let us consider in turn the cases of gravitational, antisymmetric tensor, and Yang-Mills zero modes.

i. It seems that very little is known about the conditions under which some oscillations in the geometry of a compact space B will correspond to massless scalars. If B is Ricci flat, $R_{\mu\nu} = 0$, a "breathing mode" in which the geometry of B is uniformly dilated corresponds to a massless scalar, because the equation $R_{\mu\nu} = 0$ has a scaling symmetry and does not determine the radius of B. This mode is always neutral under continuous symmetries, so is not helpful in solving the hirarchy problem. The scaling symmetry of the classical equation $R_{\mu\nu} = 0$ is not a symmetry of quantum Kaluza-Klein theories, and therefore [37] (unless there is an unbroken supersymmetry) these modes get nonzero mass at the one-loop level.

In models with an unbroken supersymmetry at the tree level, there are some known cases in which some oscillations in the metric of B correspond to charged massless scalars at the tree level.[12] Little is known about the possibility of eventual supersymmetry breaking in these models.

ii. Many supergravity theories contain antisymmetric tensor fields—for instance, the third-rank antisymmetric tensor field A_{ijk} of 11-dimensional supergravity. The Lagrangian is constructed from the curl of A, $F_{ijkl} = \partial_i A_{jkl} \pm$ cyclic permutations. This curl is invariant under the gauge transformation $A_{ijk} \rightarrow A_{ijk} + (\partial_i \Lambda_{jk} + $ cyclic permutations). The Lagrangian for a kth rank antisymmetric tensor gauge field is

$$\mathscr{L} = \frac{1}{2(k+1)!} \int d^n x \left(F_{i_1 \cdots i_{k+1}} \right)^2. \tag{57}$$

The field equations derived from this Lagrangian may readily have zero modes for topological reasons.

The physical interpretation of these modes depends on how many indices of $A_{i_1 \cdots i_k}$ are tangent to B and how many are tangent to the space-time directions. If all indices are tangent to B, we get a massless scalar in 4 dimensions. If all indices but one are tangent to B, we get a massless spin 1 particle in 4 dimensions. If all indices but two are tangent to B, we get a massless antisymmetric tensor in 4 dimensions that again describes a massless scalar. Other cases do not give rise to propagating modes in 4 dimensions.

In general, the number of zero modes (modulo gauge transformations) of $A_{i_1 \cdots i_k}$ with q indices tangent to B is equal to a topological invariant known as the qth Betti number of B. As an example, one may consider in 11 dimensions the spaces M^{pqr} with $SU(3) \times SU(2) \times U(1)$ symmetry [4]. For most of these spaces, the first and third Betti numbers vanish, so one does

12. I thank M. Duff for a discussion of this point.

not get massless scalars. (The exceptions are $CP^2 \times S^3$ and $CP^2 \times S^3/Z^k$, for which the third Betti number is one and one gets one massless scalar in 4 dimensions; and $CP^2 \times S^2 \times S^1$, for which the first and third Betti numbers are one and one gets two massless scalars.) However, for the M^{pqr} the second Betti number is one (except for $CP^2 \times S^2 \times S^1$, where it is two), so one would get one (or two) massless spin one particles in 4 dimensions [in addition to the gauge fields of $SU(3) \times SU(2) \times U(1)$ coming from the metric tensor]. These massless spin 1 particles do not have minimal couplings to any matter fields. They interact through derivative couplings, such as magnetic moment couplings to fermi fields. They would give rise to long range spin-spin forces of roughly gravitational strength; presumably this is far too weak to be detectable.

There is, however, an old theorem that zero modes of antisymmetric tensor fields are always neutral under any continuous symmetries. Essentially, this is true because, by the de Rham-Hodge theory, zero modes of the qth antisymmetric tensor field on B correspond to topological classes of closed q-dimensional submanifolds of B. A continuous symmetry cannot change the topological class of a submanifold, so it leaves invariant all of the zero modes of antisymmetric tensor fields.

Here is an analytical proof.[13] Let d be the curl operator, so the curl of the antisymmetric tensor field A will be denoted $F = dA$. [Thus, $(dA)_{i_1 \cdots i_{k+1}} = (\partial_{i_1} A_{i_2 \cdots i_{k+1}} \pm$ cyclic permutations).] The change in A under a gauge transformation is $A \to A + d\Lambda$, where Λ is an antisymmetric tensor field with one less index than A. What is a massless mode of the A field? Setting the momentum in the Minkowski directions to zero, we calculate the energy (per unit volume) of an antisymmetric tensor field A by integrating over the compact dimensions. From (57), the integral is $\int_B d\phi (dA)^2$, so a massless mode is an antisymmetric tensor field defined on B such that $dA = 0$. Actually, we want zero modes that cannot be gauged away, so we want solutions of $dA = 0$ modulo gauge transformations $A \to A + d\Lambda$. (Since $d^2 = 0$, any pure gauge $A = d\Lambda$ obeys $dA = 0$.)

Now let $K^i(\phi^j)$ be an arbitrary Killing vector field, generating the infinitesimal symmetry transformation $\phi^i \to \phi^i + \varepsilon K^i(\phi^j)$. To show that the zero modes of A are neutral under arbitrary continuous symmetries, we must show that the transformation generated by K^i leaves A unchanged, or more exactly that it leaves A unchanged up to a gauge transformation.

---

13. See, for instance, reference [38] for further material on antisymmetric tensor fields.

The generator of the symmetry is known as the Lie derivative. Acting on antisymmetric tensor fields, it takes a particularly simple form and can be conveniently defined as follows. Let i_K be the operation of contraction with K; explicitly, for any antisymmetric tensor field $A_{i_1 \cdots i_k}$, $i_K A$ is an antisymmetric tensor field with one index less: $(i_K A)_{i_1 \cdots i_{k-1}} = K^{i_0} A_{i_0 i_1 \cdots i_{k-1}}$. Then $\mathscr{L}_K$, on differential forms, can be defined as follows:

$$\mathscr{L}_K = d i_K + i_K d. \tag{58}$$

Thus, the infinitesimal change of A under the transformation generated by K is $A \to A + \varepsilon \mathscr{L}_K A$. Equation (58), which may appear unfamiliar, can be seen to agree with the formulas in books on general relativity.

What we must show is that acting on any zero mode, $\mathscr{L}_K$ vanishes up to a gauge transformation. This is not difficult. A zero mode A obeys $dA = 0$, so $\mathscr{L}_K A = (d i_K + i_K d) A = d(i_K A)$, but $d(i_K A)$ is the change in A under a gauge transformation (with gauge parameter $\Lambda = i_K A$).

The Atiyah-Hirzebruch theorem, which is more delicate, may be seen as a generalization of this classical result to the spin $\frac{1}{2}$ case.

iii. Now we consider zero modes of gauge fields. Abelian gauge fields are a special case of the antisymmetric tensor fields just considered, but elementary nonabelian gauge fields $A_i{}^a$ that might be present in the $(4 + n)$-dimensional theory raise different issues.

Zero modes of $A_i{}^a$ in which the vector index i is tangent to the compact dimensions will be observed as massless scalars in 4 dimensions. It is certainly possible—classically—to obtain charged massless scalars in this way. The simplest example is the original Kaluza-Klein theory, the sole extra dimension being a circle S^1. If elementary nonabelian gauge fields are posited, then the mode in which $A_5{}^a$ is a constant, independent of position on the circle, gives rise in 4 dimensions to a multiplet of massless scalars in the adjoint representation.

In this case, there are (classically) no quartic interactions, but that is not a general property. In $4 + n$ dimensions, if the compact dimensions are circles $S^1 \times \cdots \times S^1$, then the constant modes of all $A_i{}^a$, $i = 5, \ldots, 4 + n$, are massless scalars, and one obtains at the tree level the quartic potential

$$V(A_i{}^a) = \frac{e^2}{4} \sum_{i, j = 5, \ldots, 4+n} \mathrm{Tr}[A_i, A_j]^2. \tag{59}$$

There are other, more elaborate scenarios that lead at the classical level to charged massless scalars as zero modes of Yang-Mills fields. Here is one such scenario. Consider an 8-dimensional world in which the 4 extra dimensions are $S^2 \times S^2$, giving rise to an $SU(2) \times SU(2)$ symmetry. Let there be an elementary $SU(2)$ gauge field, one of whose components has an expectation value, breaking $SU(2)$ to $U(1)$. The total unbroken gauge group in 4 dimensions is then $SU(2) \times SU(2) \times U(1)$. Suppose that the gauge field expectation value on $S^2 \times S^2$ is of the monopole type, with equal monopole number p on each two-sphere. (According to Dirac, p is an integer or a half-integer.) Then, if $p \geqslant 1$, it can be shown that for topological reasons the Yang-Mills equations have massless zero modes at the tree level; they have $U(1)$ charge ± 1 and transform under $SU(2) \times SU(2)$ as $(p, p - 1) \oplus (p - 1, p)$. Although these modes are massless, they have at the tree level a nontrivial quartic potential.

What these examples have in common is that in all examples I am aware of, there is no reason for the massless modes to remain massless when loop corrections are considered. Indeed, in the first example considered (the massless mode being a constant on the circle), a one-loop calculation has been carried out in the abelian case[37], showing that a nonzero mass does arise unless there is a bose-fermi cancellation.

IX The Cosmological Constant

Until now we have considered exclusively the zero modes of wave operators. However, the spirit of this paper is to study qualitative problems that might be solved without full understanding of the details of a Kaluza-Klein theory. In that spirit, we shall here consider another qualitative problem of outstanding significance: the apparent vanishing in 4 dimensions of the cosmological constant.

There has been much interest in recent years[39] in the possibility of a dynamical explanation of the vanishing of the cosmological constant—the possibility of a theory in which regardless of the value of the bare parameters, the cosmological constant spontaneously relaxes to zero. We want, in other words, a mechanism analogous to the axion mechanism for avoiding strong CP violation. Some ideas in this section have been introduced independently in work cited in reference [39].

We would like to find a theory in which the classical equations do not determine the effective cosmological constant—the actual, macroscopic curvature of 4-dimensional space. The classical equations should admit for any values of the bare parameters a one-parameter family of solutions, depending on an integration constant. The effective cosmological constant should depend on this integration constant. We shall look for a mechanism by which the integration constant spontaneously relaxes to the value at which space-time is macroscopically flat.

Of course, it is easy to find a theory in which there is an undetermined integration constant at the classical level. Consider a theory of scalar fields ϕ_i. If the potential energy $V(\phi_i)$ is independent of one of these fields ϕ, the vacuum expectation value of that field will be undetermined at the classical level. However, precisely because $V(\phi^i)$ is independent of ϕ, the effective cosmological constant will be independent of ϕ.

To find a theory with an undetermined integration constant *upon which the cosmological constant depends* requires a different approach. The only way I know to do this in $3 + 1$ dimensions is to introduce a third-rank antisymmetric tensor gauge field $A_{\mu\nu\alpha}$. As in section VIII, the Lagrangian is

$$\mathscr{L} = -\frac{1}{48} \int d^4 x \, (F_{\mu\nu\alpha\beta})^2, \tag{60}$$

where $F_{\mu\nu\alpha\beta}$ is the gauge invariant curl, $F_{\mu\nu\alpha\beta} = (\partial_\mu A_{\nu\alpha\beta} \pm \text{cyclic}$ permutations). If we define the scalar $F = (1/24)\varepsilon^{\mu\nu\alpha\beta} F_{\mu\nu\alpha\beta}$ ($\varepsilon^{\mu\nu\alpha\beta}$ being the 4-dimensional Levi-Civita symbol), then the equation of motion from (60) is $\partial_\mu F = 0$. Thus, F is a constant—but the constant is a constant of integration, not determined by the classical equations. And the cosmological constant definitely depends on F. It equals its value at $F = 0$, plus $F^2/8$.

However, this example is too trivial. The equation of motion $\partial_\mu F = 0$—which is an exact statement, even quantum mechanically—appears to tell us that the integration constant F cannot possibly relax to the value at which the effective cosmological constant would vanish.

A less trivial example of the same kind arises if the third-rank antisymmetric tensor field is considered in $4 + n$ dimensions. We consider the Lagrangian

$$\mathscr{L} = \int d^{4+n} x \, \sqrt{g} \left(\frac{1}{16\pi G} R - \frac{1}{48}(F_{\mu\nu\alpha\beta})^2 - \Lambda_0 \right). \tag{61}$$

For $n = 7$, this differs from the bosonic part of 11-dimensional supergravity by the omission of an FFA term (its inclusion would not affect our discussion) and the inclusion of a nonzero bare cosmological constant Λ_0 (forbidden by supersymmetry in 11 dimensions but needed in our discussion to get a solution in which 4 dimensions are flat). The solution we will discuss is the Freund-Rubin solution[40], generalized to $\Lambda_0 \neq 0$.

We look for a solution of the classical equations derived from (61) in which space-time takes the form $D^4(\lambda) \times S^7(R)$, where $D^4(\lambda)$ is a 4-dimensional de Sitter space of positive, negative, or zero curvature λ, and $S^7(R)$ is a seven-sphere of radius R.

In looking for such a solution, we encounter—as Freund and Rubin did—the possibility of a nonzero value of $F = F_{0123}$. As in the 4-dimensional case, the equations determine only the *derivatives* of F, not F itself. One may assume an arbitrary value of F, and use the classical equations to solve for λ and R in terms of F. For a whole continuous range of the bare parameters in (61), there is a value of F at which λ—the curvature of ordinary space—vanishes.

The gain in going from 4 to $4 + n$ dimensions is that in $4 + n$ dimensions $F_{\mu\nu\alpha\beta}$ has a nontrivial dynamics with propagating modes as well as an integration constant. One can at least imagine that there may be a quantum mechanical mechanism by which the integration constant F spontaneously relaxes to some special value—one hopes the value at which $\lambda = 0$. But what might this mechanism be?

In condensed matter physics there are some fascinating systems with the following properties. (For recent theoretical discussions and references to previous work see reference [41].) The macroscopic equations have a one-parameter family of solutions depending on an integration constant x. There is a critical value of x, say $x = x_0$, such that the classical solution is stable against small oscillations for $x \geqslant x_0$ and unstable for $x < x_0$.

What value of x would be observed physically? One would hardly expect to observe the unstable solutions of $x < x_0$, but one might expect that, depending on initial conditions, any stable solution with $x \geqslant x_0$ would be accessible.

The surprise is that it is claimed[41] that a whole class of systems spontaneously relaxes—by means that are not well understood—to the threshold of stability, $x = x_0$. The fact that the mechanism is so little understood in the condensed matter context invites the speculation that a similar phenomenon could occur in the case of the cosmological constant. Although $\lambda = 0$ (flat space) is not exactly a threshold of stability in any obvious sense, it is certainly the dividing point between two qualitatively different regimes, de Sitter space and anti-de Sitter space. Anti-de Sitter space has a positive energy theorem, which de Sitter space does not[42]; but de Sitter space has a global initial value hypersurface, which anti-de Sitter space does not[43]. They are certainly very different. Perhaps the little-understood mechanism by which the condensed matter systems relax has a "cosmological" analog—though it is not yet clear whether it is de Sitter or anti-de Sitter space that should correspond in the analogy to $x > x_0$.

Acknowledgments

I would like to thank R. Stong for useful discussions and for drawing my attention to references [44] and [45], and to thank W. Browder and especially W. Hsiang for explaining various aspects of cobordism theory.

This research was supported in part by the National Science Foundation under grant PHY80-19754.

References

[1] Th. Kaluza, *Sitzungsber. Preuss. Akad. Wiss. Berlin, Math. Phys.* K1, 966 (1921); O. Klein, *Z. Phys.* 37, 895 (1926), *Arkiv. Mat. Astron. Fys.* B A34 (1946), *Helv. Phys. Acta Suppl.* IV, 58 (1956); A. Einstein and P. Bergmann, *Ann. Math.* 39, 683 (1938); B. De Witt, in *Relativity, Groups, and Topology*, B. De Witt and C. De Witt, editors (Gordon and Breach, New York, 1964), and *Dynamical Theory of Groups and Fields* (Gordon and Breach, New York, 1965); J. Rayski, *Acta Phys. Pol.* 27, 89 (1965); R. Kerner, *Ann. Inst. H. Poincaré* 9, 143 (1968); A. Trautman, *Rep. Math. Phys.* 1, 29 (1970); Y. M. Cho, *J. Math. Phys.* 16, 2029 (1975); Y. M. Cho and P. G. O. Freund, *Phys. Rev.* D12, 1711 (1975); L. N. Chang, K. I. Macrae, and F. Mansouri, *Phys. Rev.* D13, 235 (1976).

[2] J. Scherk and J. H. Schwarz, *Phys. Lett.* B57, 463 (1975); E. Cremmer and J. Scherk, *Nucl. Phys.* B103 393 (1976), B108, 409 (1976).

[3] A complete list is impossible, but here are some of the recent papers: Z. Horvath, L. Palla, E. Cremmer, and J. Scherk, *Nucl. Phys.* B122, 57 (1977); J.-F. Luciani, *Nucl. Phys.* B135, 111 (1978); M. J. Duff and D. Toms, in *Unification of Fundamental Interactions II*, S. Ferrara and J. Taylor, editors (Cambridge University Press, Cambridge 1982); M. J. Duff and C. N. Pope, Imperial College Preprint ICTP/82–83/7 (1983); M. J. Duff, B. Nilsson, and C. Pope, University of Texas Preprint UTTG-5 (1983); F. Englert, M. Rooman, and P. Spindel, *Phys. Lett.* B127, 47 (1983); C. Orzalesi, University of Parma Preprint IFPR/TH63 (1980); W. Mecklenburg, *Phys. Rev.* D21, 2149 (1980); A. Salam and J. Strathdee, *Ann. Phys.* 131, 316 (1982); C. Destri, C. Orzalesi, and P. Rossi, New York University Preprint NYU/TR8/82 (1982); S. Weinberg, *Phys. Lett.* B125, 65 (1983); P. Candelas and S. Weinberg, University of Texas Preprint UTTG-6-83 (1983); M. F. Awada, *Phys. Lett.* B127, 413 (1983); F. Mansouri and L. Witten, *Phys. Lett.* B127, 341 (1983).

[4] E. Witten, *Nucl. Phys.* B186, 412 (1981).

[5] L. Castellani, R. D'Auria, and P. Fré, University of Torino Preprint (1983).

[6] For reviews, see P. Von Nieuwenhuizen, *Phys. Rep.* 68, 191 (1981); J. Wess and J. Bagger, *Supersymmetry and Supergravity* (Princeton University Press, Princeton, NJ, 1983); J. Gates, M. Grisaru, M. Rocek, and W. Siegel, *Superspace* (Addison-Wesley, Reading, MA, 1983).

[7]   M. B. Green and J. H. Schwarz, *Nucl. Phys.* B181, 502 (1981), B198, 252, 441 (1982); M. B. Green, J. H. Schwarz, and L. Brink, *Nucl. Phys.* B198, 474 (1982); J. H. Schwarz, *Phys. Rep.* 89, 223 (1982).

[8]   This concept and some others discussed in this section are largely due to H. Georgi—see, for instance, *Nucl. Phys.* B156, 126 (1979).

[9]   M. Gell-Mann, P. Ramond, and R. Slansky, in *Supergravity* P. van Nieuwenhuizen and D. Z. Freedman, editors (North-Holland, New York, 1979); F. Wilczek and A. Zee, Princeton Report (1979) (unpublished), *Phys. Rev.* D25, 553 (1982); F. Wilczek, in *Proceedings of the 1979 International Symposium on Lepton and Photon Interactions at High Energies*, T. Kirk and H. Abarbanel, editors (Fermilab, Batavia, Il, 1980); H. Georgi and E. Witten (unpublished); H. Sato, *Phys. Rev. Lett.* 45, 1997 (1980); A. Davidson, K. C. Wali, P. D. Mannheim, *Phys. Rev. Lett.* 45, 35 (1980); J. Maalampi and K. Enqvist, *Phys. Lett.* B97, 217 (1980); R. Cahn and H. Harari, *Nucl. Phys.* B176, 135 (1980); R. Mohapatra and B. Sakita, *Phys. Rev.* D21, 1062 (1980); R. Gatto and J. Casabuoni, *Phys. Lett.* B93, 46 (1980); J. F. Kim, *Phys. Rev. Lett.* 45, 1916 (1980).

[10]   M. Gell-Mann, P. Ramond, R. Slansky, in reference [9].

[11]   L. Palla, Proceedings 1978 Tokyo Conference on High Energy Physics, p. 629.

[12]   N. S. Manton, *Nucl. Phys.* B193, 391 (1981); G. Chapline and N. Manton, *Nucl. Phys.* B184, 391 (1981); G. Chapline and R. Slansky, *Nucl. Phys.* B209, 461 (1982).

[13]   C. Wetterich, *Nucl. Phys.* B187, 343 (1981), *Phys. Lett.* B110, 379 (1982), B113, 377 (1982), CERN Preprints TH. 3310 (1982), 3488 (1982), 3517 (1983).

[14]   S. Randjbar-Daemi, A. Salam, and A. Strathdee, *Nucl. Phys.* B214, 491 (1983), *Phys. Lett.* B124, 349 (1983), University of Trieste Preprint (1983).

[15]   M. Atiyah and F. Hirzebruch, in *Essays on Topology and Related Topics*, A. Haefliger and R. Narasimhan, editors (Springer-Verlag, New York, 1970), p. 18.

[16]   E. Witten, *J. Diff. Geom.* 17, 661 (1982).

[17]   W. Nahm, *Nucl. Phys.* B135, 149 (1978).

[18]   L. Alvarez-Gaumé and E. Witten, University of Harvard Preprint (1983).

[19]   L. Brink, J. H. Schwarz, and J. Schwarz, *Nucl. Phys.* B121, 77 (1977).

[20]   M. F. Atiyah, R. Bott, and A. Shapiro, *Topology* 3 (Suppl. 1), 3 (1964).

[21]   A. Lichnerowicz, *C. R. Acad. Sci. Paris, Ser. A-B* 257, 7 (1963).

[22]   H. B. Lawson, Jr., and S.-T. Yau, *Comm. Math. Helv.* 49, 232 (1974).

[23]   M. T. Grisaru, H. N. Pendleton, and P. van Nieuwenhuysen, *Phys. Rev.* D15, 996 (1977); F. A. Berends, J. W. van Holten, B. de Wit, and P. van Nieuwenhuysen, *Nucl. Phys.* B154, 261 (1979), *Phys. Lett.* B83, 188 (1979), *J. Phys.* A13, 1643 (1980); C. Aragone and S. Deser, *Phys. Lett.* B85, 161 (1979); *Phys. Rev.* D21, 352 (1980); K. Johnson and E. C. G. Sudarshan, *Ann. Phys.* 13, 126 (1961); G. Velo and D. Zwanziger, *Phys. Rev.* 186, 1337 (1967).

[24]   R. Bott, in *Differential and Combinatorial Topology* (Princeton University Press, Princeton, NJ, 1965).

[25]   See, for example, S. Coleman, in *The Unity of the Fundamental Interactions*, A. Zichichi, editor (Plenum Press, New York, 1983).

[26]   M. F. Atiyah and I. M. Singer, *Ann. Math.* 87, 484, 546 (1968); M. F. Atiyah and G. B. Segal, *Ann. Math.* 87, 531 (1968); M. F. Atiyah and R. Bott, *Ann. Math.* 87, 456 (1968).

[27]   L. D. Landau and E. M. Lifshitz, *Quantum Mechanics*, J. B. Sykes and J. S. Bell, translators (Pergamon Press, Oxford, 1965), p. 371.

[28]   A. Borel and F. Hirzebruch, *Amer. J. Math.* 80, 458 (1958), 81, 315 (1959).

[29]   For an exposition of the topological obstruction to defining spinors, see, for example, S. W. Hawking and C. N. Pope, *Phys. Lett.* B73, 42 (1978).

[30]   A. Davidson, Ph.D. Thesis (Technion, Israel), and private communication.

[31]   S. Weinberg, *Phys. Lett.* B138, 47 (1984) and University of Texas Preprint (1984).

[32]   J. H. Schwarz, *Phys. Rep.* 89, 223 (1982), and references therein; C. H. Ragadiakos and J. Taylor, *Phys. Lett.* B124, 201 (1983).

[33]   M. Green and J. H. Schwarz, Caltech Preprints (1982–1983).

[34]   E. Witten, *Nucl. Phys.* B223, 422 (1983).

[35]   S. Deser, R. Jackiw, and S. Templeton, *Phys. Rev. Lett.* 48, 975 (1982), *Ann. Phys. (NY)* 140, 372 (1982).

[36]   S. Weinberg, *Phys. Lett.* B62, 111 (1976); S. Weinberg and E. Gildener, *Phys. Rev.* D13, 333 (1976).

[37]   T. Appelquist and A. Chodos, *Phys. Rev. Lett.* 50, 141 (1983); D. Pollard, ICTP Preprint 82–83/11; M. A. Rubin and B. D. Roth, *Phys. Lett.* B127, 55 (1983); T. Appelquist, A. Chodos, and E. Meyers, *Phys. Lett.* B127, 51 (1983); T. Appelquist and A. Chodos, Yale University Preprint YTP 83-05 (1983); Y. Hosotani, University of Pennsylvania Preprint (1983); M. A. Awada and D. J. Toms, Imperial College Preprints (1983).

[38]   R. Bott and L. W. Tu, *Differential Forms in Algebraic Topology* (Springer-Verlag, New York, 1982).

[39]   S. W. Hawking, *Nucl. Phys.* B144, 349 (1978), in *Unified Theories of Elementary Particles* (Springer-Verlag, New York, 1982), talk at Shelter Island Conference (June 1983);* L. F. Abbott, Brandeis University Preprint (1981); V. A. Rubakov and M. E. Schaposnikov, *Phys. Lett.* B125, 139 (1983).

[40]   P. G. O. Freund and M. A. Rubin, *Phys. Lett.* B97, 223 (1980); A. Aurelia, H. Nikolai, and P. K. Townsand, *Nucl. Phys.* B176, 509 (1980).

[41]   G. Dee and J. S. Langer, *Phys. Rev. Lett.* 50, 383 (1983); R. C. Brower, D. A. Kessler, J. Koplik, and H. Levine, *Phys. Rev. Lett.* 51, 1111 (1983).

[42]   L. Abbott and S. Deser, *Nucl. Phys.* B195, 76 (1982); P. Breitenlohner and D. Z. Freedman, *Ann. Phys. (NY)* 144 (1982); G. W. Gibbons, C. M. Hull, and N. P. Warner, *Nucl. Phys.* B218, 173 (1983).

[43]   S. W. Hawking and G. F. R. Ellis, *The Large Scale Structure of Space-Time* (Cambridge University Press, Cambridge, 1973).

[44]   C. Kosniowski, *Actions of Finite Abelian Groups* (Pitman, London, 1978).

[45]   O. R. Musin, *Math. USSR Sbornik* 44, 325 (1983).

* Editors' note: The paper read to the conference by Hawking appears in this volume.

Theory of Lepton Anomalous Magnetic Moments, 1947–1983

Toichiro Kinoshita

I Introduction

The 1947 Shelter Island Conference marked the beginning of a new era in particle physics [1]. It was there that the first experimental evidences for the hydrogen Lamb shift and the electron anomalous magnetic moment were reported and discussed extensively. They provided a timely stimulus to the renormalization theory of quantum electrodynamics (QED), which was then being developed by Tomonaga, Schwinger, and Feynman. Within a few months after the conference Schwinger demonstrated the power of the new theory by calculating the electron magnetic moment anomaly, $a_e = (g_e - 2)/2$, in the second-order perturbation theory. His result [2],

$$a_e = \alpha/(2\pi), \tag{1}$$

where α is the fine structure constant, was in agreement with experiment [3], to within a few percent. Thus began a very fruitful collaboration of theory and experiment concerning a_e. After 36 years of hard work and remarkable progress it is still producing interesting and exciting physics, in spite of the tremendous transformation that our knowledge and understanding of the physical world has since undergone.

Initially, the anomalous moment of the electron, a_e, and that of the muon, a_μ, served as testing grounds of QED. The electron anomaly is still primarily concerned with QED, although it has now acquired the second role as a testing ground of quantum mechanics of condensed matter physics [4]. Meanwhile, a_μ has gradually evolved into a sensitive laboratory for studying hadronic vacuum polarization effects. Its role as an important probe of the electroweak effect is just around the corner. Once these effects are understood, a_e and particularly a_μ might yield crucial information on the possible internal structure of leptons [5], along with useful constraints on the mass of sleptons in supersymmetric theories [6].

In this talk I should like to review briefly the progress of theoretical work on lepton moments from 1947 to this date and also discuss the outlook for the immediate future. Of course the progress of experimental techniques over this period has been truly spectacular. Without it theory would have had no incentive to push ahead this far. However, I shall refer to specific experiments only to the extent that they are relevant to particular theoretical points. To gain some perspective of how theory and experiment advanced together under mutual influence, however, I list major experimental and theoretical developments chronologically in table 1, which I have

Table 1
Major experimental and theoretical developments concerning the lepton anomalous magnetic moments[a]

Year	Authors	Description
1947	Nafe, Nelson, Rabi [70]; Nagle, Julian, Zacharias [71]	hfs in H, D at variance with the Dirac theory
1947	Breit [72]	Suggests $g \neq 2$
1947	Kusch, Foley [3]	First measurement of a_e using atomic beam technique to measure Zeeman splittings in Ga, In, Na $a_{e^-} = 0.00119(5)$
1948	Schwinger [2]	First QED calculation; $a_e^{(2)} = \alpha/2\pi = 0.00116$
1949	Gardner, Purcell [73]	Measure μ_p/μ_o; this times μ_e/μ_p gives a_e
1950	Karplus, Kroll [8]	Attempt to calculate $a_e^{(4)}$; $a_e^{(4)} = -2.97(\alpha/\pi)^2$, implying $a_e = 0.001147$
1953	Louisell, Pidd, Crane [74]	First direct measurement of g_e using a Mott double-scattering technique; result, $g_e = 2.00 \pm 0.01$
1956	Franken, Liebes [9]	Improved measurement of μ_p/μ_o, which gives $a_{e^-} = 0.001168(5)$, disagreeing with Karplus, Kroll [8]
1957	Sommerfield [10]; Petermann [11]	$a_e^{(4)} = -0.32848(\alpha/\pi)^2$
1957	Petermann [13]; Suura, Wichman [14]	$a_\mu^{(4)} - a_e^{(4)} = 1.08(\alpha/\pi)^2$
1958	Dehmelt [75]	First direct observation of spin transitions and measurement of a_e for free electrons; $a_{e^-} = 0.001116(40)$
1961	Schupp, Pidd, Crane [76]	Modify 1953 technique to measure a_e directly for free electrons; $a_{e^-} = 0.0011609(24)$; agree with Petermann [11], Sommerfield [10]
1961	Charpak et al. [24]	First direct measurement of a_μ; $a_{\mu^+} = 0.001145(22)$; later value = 0.001162(5)
1961	Bouchiat, Michel [16]	Hadronic contribution to a_μ
1963	Wilkinson, Crane [17]	Refine 1961 technique [74]; $a_{e^-} = 0.0011159622(27)$; a_e(1963 theory) = 0.001159615

Table 1 (continued)

Year	Reference	Description
1966	Rich, Crane [77]	$a_{e^+} = 0.001168(11)$. Test TCP theorem; 1969 refinement $= 0.0011602(11)$
	Farley et al. [25]	CERN muon storage ring; $a_\mu = 0.001165(3)$
1967	Kinoshita [29]	Calculation of $a_\mu^{(6)} - a_e^{(6)}$ by renormalization group and mass singularity theorem
1968	Parker, Taylor, Langenberg [78]	Measurement of α by the ac Josephson effect; $\alpha^{-1} = 137.0359(4)$
	Veltman [39]; Hearn [40]; and others	Development of SCHOONSCHIP and other algebraic computation programs which enable large-scale calculation on computers
	Rich [19]	Recalculation of a_e from 1963 data [17], using corrected orbit motion theory and data analysis procedures; $a_{e^-} = 0.001159557(30)$; a_e(1968 theory) $= 0.001159641$ using α (1967 ac Josephson value [78])
1968	Bailey et al. [26]	Refinement of 1966 CERN muon experiment [25]; $a_{\mu^-} = 0.0011616(31)$; a_μ(theory) $= 0.0011656(1)$
	Gräff, Major, Roeder, Werth [79]	Direct observation of spin and cyclotron resonances of free thermal electrons in Penning trap; $a_{e^-} = 0.001159(2)$; later result $= 0.001159660(30)$
1969	Aldins et al. [31]	Large light-by-light contribution to a_μ
1971	Wesley, Rich [21]	Extend the 1961 technique [74] to high fields; $a_{e^-} = 1,159,657,700(3,500) \times 10^{-12}$; disagrees with revised 1968 result [19]
1969–1982	Brodsky, Kinoshita [30]	Complete α^3 calculation; $a_e^{(6)} = 1.1765(13)(\alpha/\pi)^3$; $a_\mu^{(6)} = 24.45(6)(\alpha/\pi)^3$
	Levine, Wright [37], Roskies [38]; Calmet et al. [80]; Mignaco, Remiddi [35]; Cvitanovic, Kinoshita [36]; Lautrup, de Rafael [30]; and others	
1972	Jackiw, Weinberg; Bars, Yoshimura; Bardeen, Gastmans, Lautrup; Fujikawa, Lee, Sanda [44]	Electroweak effect on a_μ; a_μ(weak) $\simeq 2 \times 10^{-9}$

1977	Van Dyck, Jr., Schwinberg, Dehmelt [46]	Measurement of a_e of a free electron in a Penning trap using continuous Stern-Gerlach effect; $a_{e^-} = 1,159,652,410(200) \times 10^{-12}$
	Bailey et al. [27]	CERN muon storage ring experiment using magic γ; $a_{\mu^-} = 1,165,937(12) \times 10^{-9}$, $a_{\mu^+} = 1,165,911(11) \times 10^{-9}$
1980	Von Klitzing, Dorda, Pepper [44]	Quantized Hall effect; α^{-1}(quantized Hall) = 137.0353(4)
1981	Kinoshita, Lindquist [48]	$a_e^{(8)} = -0.8(2.5)(\alpha/\pi)^4$
1983	Kinoshita et al. [49]	$a_\mu^{(8)} = 140(6)(\alpha/\pi)^4$

a. $a_e^{(n)}$ ($a_\mu^{(n)}$) means the nth-order QED contribution to a_e (a_μ).

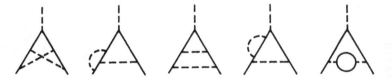

Figure 1
Examples of fourth-order diagrams contributing to C_2. Electron and photon are represented by a continuous and dotted lines, respectively.

borrowed from the excellent review of the early development of lepton moment physics by Rich and Wesley [7] and updated to include more recent developments. It is not meant to be comprehensive, and I should like to apologize in advance if I have made some inadvertent omissions.

II Brief Historical Review

Soon after the initial formulation of QED, Dyson was able to show (except for an important detail that was filled in later) the renormalizability of QED to all orders of perturbation theory. Thus, for instance, a_e can be written as a power series in α/π:

$$a_e = C_1(\alpha/\pi) + C_2(\alpha/\pi)^2 + C_3(\alpha/\pi)^3 + \cdots, \tag{2}$$

where the coefficients C_i are finite and calculable for all i. Schwinger's result corresponds to $C_1 = 0.5$.

The first attempt to calculate the higher order terms in this expansion is that of Karplus and Kroll [8]. They reported in 1950 a calculation of C_2, represented by 7 fourth-order Feynman diagrams, some of which are shown in figure 1. Their result was $C_2 = -2.97$. However, measurement of g_{e^-} by Franken and Liebes [9] in 1956 indicated a significantly different value for C_2. Recalculations of C_2 by Sommerfield [10] and Petermann [11], which began before the experimental information became available, revealed that a portion of the Karplus-Kroll calculation was in error. The revised result is

$$C_2 = \frac{197}{144} + \frac{\pi^2}{12} - \frac{\pi^2}{2}\ln 2 + \frac{3}{4}\zeta(3)$$

$$= -0.328478966\ldots, \tag{3}$$

where ζ is the Riemann ζ function.

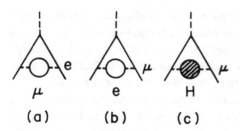

Figure 2
(a) Contribution to a_e from a muon loop. (b) Contribution to a_μ from an electron loop.
(c) Contribution to a_μ from hadronic vacuum polarization, denoted H.

To be precise, one must also include in this order the contribution of the mixed diagram in figure 2a containing a muon vacuum polarization loop, which adds [12]

$$\frac{1}{45}\left(\frac{m_e}{m_\mu}\right)^2 + \cdots = 520 \times 10^{-9} \tag{4}$$

to C_2. The tauon loop contribution is 280 times smaller than (4) and thus totally negligible at present.

The magnetic moment anomaly of the muon a_μ can also be written as a power series in α/π:

$$a_\mu = C_1'(\alpha/\pi) + C_2'(\alpha/\pi)^2 + C_3'(\alpha/\pi)^3 + \cdots. \tag{5}$$

In the second order there is no difference between a_e and a_μ:

$$C_1' = C_1 = 0.5. \tag{6}$$

In the fourth order, however, the diagram in figure 2b, which has an electron loop within a muon vertex, plays a dominant role. This was first noted by Petermann [13] and Suura and Wichmann [14]. They observed that this contribution to a_μ has a term proportional to $\ln(m_\mu/m_e)$:

$$\frac{1}{3}\ln\frac{m_\mu}{m_e} - \frac{25}{36} + \frac{\pi^2}{4}\frac{m_e}{m_\mu} + \left(3 - 4\ln\frac{m_\mu}{m_e}\right)\left(\frac{m_e}{m_\mu}\right)^2 + \cdots \tag{7}$$

Small remaining terms not shown in (7) can be found in [15]. The total α^2 contribution to a_μ, which is the sum of (3) and (7), is given by

$$C_2' = 0.76578222\ldots. \tag{8}$$

Since photons interact indiscriminately with all charged particles, a_e and a_μ will also have contributions from hadronic loops (see figure 2c). As was first noted by Bouchiat and Michel [16], the existence of low energy hadronic resonances such as ρ, ω, and φ makes this contribution far more important than is expected from the phase space consideration alone.

The first experiment with sufficient accuracy for a precision check of C_2 was that of Wilkinson and Crane [17] carried out at the University of Michigan. This experiment measures the $g - 2$ precession in a magnetic mirror trap in which the electrons undergo several hundred $g - 2$ precessions between initial (polarizing) and final (analyzing) Mott scatterings. Their measurement led to an experimental value of C_2 that agreed with the result (3) within 1.5% if the then current value of α [$\alpha^{-1} = 137.0391(5)$, where the numerals enclosed in parentheses represent the uncertainties in the final digits of the measured values] was used. However, subsequent revision of the theoretical value of a_e owing to changes in α [18] (caused by the new technique made possible by the discovery of the ac Josephson effect) coupled with the refined analysis [19] of the original Wilkinson-Crane data led to a disagreement between theory and experiment. This was resolved later by inclusion of further correction terms discovered in a comprehensive reanalysis of the spin motion theory used to interpret the experiment [20].

The best result obtained by a series of Michigan precession experiments is that of Wesley and Rich [21]:

$$a_{e^-} = 11{,}596{,}577(35) \times 10^{-10}. \tag{9}$$

This result is accurate enough to confirm the theoretical value (3) and to provide also an estimate of 1.68(33) for the sixth-order coefficient C_3, in strong disagreement with the dispersion theory estimates [22] ~ 0.1.

The initial experimental information on the muon anomalous magnetic moment was obtained by Coffin et al. [23], who measured the spin precession frequency of stopped muons in a known magnetic field. To derive the g factor g_μ from this one needs the muon magneton $eh/4\pi m_\mu c$, which in turn requires the knowledge of m_e/m_μ. Uncertainty in m_e/m_μ at that time prevented this experiment from giving an accurate value of a_μ.

A direct determination of a_μ avoids the necessity of knowing m_e/m_μ. In an experiment in which positive muons were confined by precisely controlled gradient drift in a long linear magnet, Charpak et al. [24] at CERN were able to determine a_{μ^+} to two significant figures. In further refinements of this

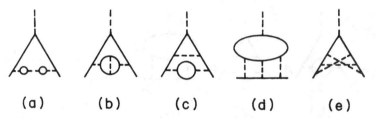

Figure 3
Examples of sixth-order diagrams contributing to C_3.

method by Farley et al. [25] and by Bailey et al. [26], the linear magnet was replaced by a muon storage ring in order to increase the total $g - 2$ precession angle of the trapped muons. In this manner they were able to determine a_μ to a precision of 270 ppm.

The latest of this sequence of experiments takes advantage of the fact that the precession frequency is independent of the electric field for $\gamma = (1 - v^2/c^2)^{-1/2} = 29.3$. This means that, by working at this "magic γ," it is possible to use electric fields for focusing without affecting the precession frequency. The final results obtained by this technique are [27]

$$a_{\mu^-} = 1,165,937(12) \times 10^{-9},$$
$$a_{\mu^+} = 1,165,911(11) \times 10^{-9}. \tag{10}$$

The computation of the coefficient C_3, represented by 72 sixth-order Feynman diagrams, some of which are shown in figure 3, is considerably more complicated than that of C_2. As a matter of fact, it was the muon coefficient C_3' rather than C_3 that was first investigated in the sixth order. This is because C_3' is expected to be dominated by diagrams of the type shown in figure 4, which are obtained by insertion of closed electron loops into second- and fourth-order muon vertex diagrams. As a consequence of electron loop insertion, $\Delta C_3' = C_3' - C_3$, or $\Delta C_n'$ in general, will have the form

$$\Delta C_n' = \sum_{k=1}^{n-1} b_{n,k} \ln^k \left(\frac{m_\mu}{m_e} \right) + O(1), \tag{11}$$

where $\ln(m_\mu/m_e) \simeq 5.33$ is a "large" parameter. Now, the important point is that many (though not all) of the coefficients $b_{n,k}$ can be determined by the renormalization procedure due to the fact that unrenormalized amplitudes

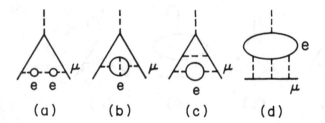

Figure 4
Examples of sixth-order diagrams contributing to $C_3' - C_3$.

for muon vertices are convergent in the limit in which the mass m_e of the vacuum polarization loop vanishes [28]. In this manner I was able to determine all coefficients of $\ln(m_\mu/m_e)$ for diagrams of figures 4a–4c by simple manipulation of known quantities of second and fourth orders [29]. Of course the nonlogarithmic terms, and hence the complete expression, cannot be determined in such a simple manner. They have been evaluated one by one by numerical or analytic means [30].

For those diagrams that cannot be reduced to known lower-order diagrams by shrinking of electron loops to points, however, it is not possible to take advantage of the renormalization procedure. In the sixth order the light-by-light scattering contribution shown in figure 4d gives such an example. Numerical evaluation [31] revealed that this gives a surprisingly large contribution to $\Delta C_3'$. This is due to a $\ln(m_\mu/m_e)$ term whose presence is expected from the mass singularity theorem [32]. It was shown later [33] that the coefficient of $\ln(m_\mu/m_e)$ of this contribution is equal to $2\pi^2/3$.

Collecting all these contributions one finds [34]

$$C_3' = 24.45(6), \tag{12}$$

where the error is mostly due to the remaining uncertainty in the numerical integration of the contribution of figure 4d. [Recent reevaluation of C_3', which is incorporated in (28), deviates considerably from (12).]

The evaluation of contributions of figures 3a–3d to a_e is similar to that of corresponding diagrams for a_μ and has been carried out by several groups [30, 31, 35]. As was noted already, the contribution of 50 diagrams of the type shown in figure 3e to a_e is much harder to evaluate. By the efforts of many theorists over a number of years, however, all these Feynman diagrams have been evaluated numerically and many of them analytically, too

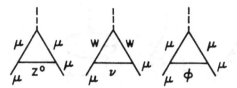

Figure 5
Diagrams representing the weak interaction contribution to a_μ.

[36, 37]. The best present value of the complete C_3 is given by [38]

$$C_3 = 1.1765(13), \tag{13}$$

where the error represents the 90% confidence limit calculated by the integration routine and comes mainly from the numerical uncertainties of 5 diagrams.

I should like to emphasize here that, in view of the size and complexity of sixth-order integrals, such a calculation would have been virtually unthinkable without the help of large scale electronic computers that had just become available.

Equally important is the timely development of programs for algebraic manipulation, such as SCHOONSCHIP [39] and REDUCE [40], and efficient multidimensional integration routines, such as RIWIAD [41] and VEGAS [42].

Another crucial development is the discovery of the ac Josephson effect [43] in 1962 and the quantized Hall effect [44] in 1980, which made it possible to determine α very accurately independently of the more conventional determination by atomic physics.

It is also during this period that the theory of electromagnetic interaction underwent a profound change and evolved into a unified theory of electroweak interaction. One of the consequences is that the previously divergent contribution of the weak interaction to the lepton moments has now become finite and calculable. The result, represented by the Feynman diagrams in figure 5 according to the standard Weinberg-Salam model, is [45]

$$a_\mu(\text{weak}) \simeq 2 \times 10^{-9}, \tag{14}$$

which is only 5 times smaller than the present experimental precision. The weak effect on a_e is too small to be of interest at present:

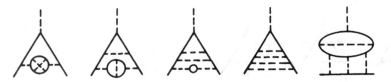

Figure 6
Examples of eighth-order diagrams contributing to C_4.

$$a_e(\text{weak}) \simeq 0.05 \times 10^{-12}. \tag{15}$$

When the Michigan result (9) became available, the theoretical value of a_e was already known several times more accurately than the experimental value. However, the complacency of theory was shattered by the remarkable experiment conceived and carried out by Dehmelt and his coworkers [46], which is based on the rf transitions of electrons caught in a Penning trap. (This experiment may be regarded as the culmination of the series of resonance-type experiments listed in table 1.) Although their initial accuracy was 200×10^{-12} [which was already 17 times more accurate than the Michigan measurement (9)], it was apparent that the experimental error would further be reduced by an order of magnitude before long. In that event, it was clearly necessary to know the eighth-order coefficient C_4, in order to test QED to the accuracy allowed by experiment, since

$$(\alpha/\pi)^4 \simeq 29 \times 10^{-12}. \tag{16}$$

After a preliminary feasibility study [47] and four years of hard work, Lindquist and I managed to set up integration programs for all 891 Feynman diagrams, some of which are shown in figure 6, that contribute to C_4 and carried out rough evaluations of their values. Based on several hundred hours of Monte Carlo integration on CDC-7600, we reported a very preliminary result [48],

$$C_4 = -0.8(2.5). \tag{17}$$

Although this is still very crude, it is already more accurate than the uncertainty arising from the error in the measurement of α by the ac Josephson effect.

The present experimental precision of a_μ does not require a knowledge of C_4' as yet. However, it is important to calculate C_4' and also improve the evaluation of hadronic corrections to a_μ in order to test the prediction (14) of the Weinberg-Salam theory of electroweak interaction. Some progress has been made recently on these refinements [49].

As is evident from the brief account presented, QED has been a very successful theory, meeting all experimental challenges in stride. However, the progress of this theory has been based exclusively on perturbative approach. It is certainly important to ask whether higher order perturbation terms would not upset the good agreement, and whether the perturbation series is summable at all. Some study has been made on these matters, especially about the behavior of large order terms of perturbation expansion [50]. However, further study is needed before we can draw a conclusion. Another problem worth exploring is that of computing a_e or a_μ by nonperturbative means, perhaps along the lines recently developed in the lattice gauge theory calculation of baryon magnetic moments [51].

Finally, I shall comment on the efforts to gain intuitive understanding of the phenomenon of magnetic moment anomaly. It would certainly be aesthetically appealing if we could give it a qualitative and handwaving explanation without going through detailed calculations. Indeed such an attempt was begun soon after Schwinger's result (1) was obtained. Welton [52] tried to picture it as a consequence of vacuum fluctuation acting on the electron spin. He obtained a result that was correct in magnitude but wrong in sign. Koba [53] managed to obtain the correct sign exploiting the *Zitterbewegung* of the electron. Unfortunately, his derivation had a flaw [54]. Since then similar attempts have been made from time to time. As far as I know, no satisfactory picture has yet emerged.

III Present Status of Electron Anomaly

The latest measurements of a_e reported by Dehmelt and coworkers for the electron and positron are [55]

$$a_{e^-} = 1,159,652,200(40) \times 10^{-12},$$
$$a_{e^+} = 1,159,652,222(50) \times 10^{-12}.$$
$$(18)$$

The agreement between these values confirms the validity of the *TCP* theorem, which is one of the fundamental theorems of quantum field theory and predicts the equality of g factors for the electron and positron, to the level of 10^{-10}.

The best QED estimate of a_e at present is

$$a_e = 0.5(\alpha/\pi) - 0.328478966(\alpha/\pi)^2$$
$$+ 1.1765(13)(\alpha/\pi)^3 - 0.8(1.4)(\alpha/\pi)^4, \tag{19}$$

which is obtained from (1), (3), (13), and an update [56] of (17). Contributions from other known sources are the vacuum polarization effects due to the muon, tauon, and hadrons [57],

$$a_e(\text{muon}) = 2.8 \times 10^{-12},$$

$$a_e(\text{tauon}) = 0.01 \times 10^{-12}, \tag{20}$$

$$a_e(\text{hadron}) = 1.6 \times 10^{-12},$$

as well as the weak interaction effect (15).

To compare the theoretical value of a_e, which is the sum of (19), (20), and (15), with experiment we need an accurate value of α. If we use the value [58]

$$\alpha^{-1} = 137.035963(15) \tag{21}$$

determined from the ac Josephson effect, we find

$$a_e^{\text{th}} = 1,159,652,460(44)(127) \times 10^{-12}, \tag{22}$$

where the first error represents the theoretical error in (19) and the second is due to the error of α in (21). We see that theory and experiment are in agreement for the first seven digits. Their difference,

$$a_e^{\text{exp}} - a_e^{\text{th}} = -251(140) \times 10^{-12}, \tag{23}$$

is somewhat less than two standard deviations.

As is seen from (22), the value of α is the largest source of error in a_e^{th}. It prevents us from testing the validity of QED to the extent commensurate with experimental precision. Thus it would be more appropriate to test QED by an alternative method in which one calculates α from measurement (18) and theory (19) of a_e and compares it with other determinations of α. In the following I list the value of α determined from a_e together with the three most accurate values of α determined by the hfs of the muonium ground state [59], the ac Josephson effect [58], and the quantized Hall effect [60]:

$$\alpha^{-1}(a_e) = 137.035993(5)(5), \tag{24}$$

$$\alpha^{-1}(\text{muonium hfs}) = 137.035988(20)(15), \tag{25}$$

$$\alpha^{-1}(2e/h \text{ and } \gamma'_p) = 137.035963(15)(?), \tag{26}$$

$$\alpha^{-1}(\text{quantized Hall}) = 137.035968(23)(?). \tag{27}$$

Here the first errors are experimental and the second theoretical. If we ignore temporarily the theoretical errors in (26) and (27), these values of α agree with each other to within a few parts in 10^7.

On the other hand, we detect slight disagreement between αs determined by QED and condensed matter methods, although it is less than two standard deviations and not yet alarming. Such a discrepancy used to be regarded as a sign of possible breakdown of QED. However, now that theory and measurement of a_e are more accurate than the measurement of α in (26) or (27), comparison of these αs may instead be regarded as a test of quantum mechanics of many-body systems [4]. In this context I should like to call attention to the fact that the numerical precision of the theoretical prediction of the ac Josephson effect and the quantized Hall effect is not yet fully known. Over the years some sources of errors have been examined and found to be zero or extremely small. It may well be that the remaining corrections are also much smaller than the present experimental errors. Nevertheless, we shall never be sure whether any correction is present or not until *all* sources of errors are examined *consciously* and evaluated explicitly. It is to emphasize this that I have used question marks for the theoretical errors in (26) and (27). Careful analysis of this problem is urgently needed.

At present work is in progress to improve the precision of α by the ac Josephson effect approach (which requires an improved measurement of the gyromagnetic ratio of the proton) and by the quantized Hall effect.

We are continuing our effort to reduce the error of the α^4 term further. Our progress is slow because of an enormous amount of computation required. Meanwhile, Dehmelt and his coworkers have succeeded in improving the precision of the measurement of a_e by a substantial factor [61]. They have also started a different experiment in which the auxiliary magnetic bottle needed for the continuous Stern-Gerlach effect is eliminated [62]. This may enable them to improve the precision by up to two orders of magnitude.

IV Present Status of Muon Anomaly

In contrast with the electron anomaly, there has been no recent progress in the measurement of the muon anomaly a_μ. The most accurate measurements of a_μ are still those given in (10) obtained in 1977 at the CERN muon storage ring [27].

The precision of these measurements is at the level of 10^{-5}. This means that a_μ cannot compete with a_e as far as testing QED is concerned. On the other hand, a_μ is much more sensitive than a_e to physics at small distances because of its larger mass scale. Consequently, in order to compare theoretical and experimental values of a_μ it is necessary to evaluate not only the QED effects (including effects due to electron, muon, and tauon vacuum polarization loops) but also the effects of strong and weak interactions.

As a consequence of the recent improvement of theory [49], the present theoretical estimate of a_μ is as follows:

$$a_\mu(\text{QED}) = 0.5(\alpha/\pi) + 0.76578223(\alpha/\pi)^2$$
$$+ 24.09(2)(\alpha/\pi)^3 + 140(6)(\alpha/\pi)^4$$
$$= 11{,}658{,}478(4) \times 10^{-10}, \tag{28}$$

$$a_\mu(\text{tauon}) = 4.2 \times 10^{-10},$$

$$a_\mu(\text{hadron}) = 702(19) \times 10^{-10},$$

$$a_\mu(\text{weak}) \simeq 20 \times 10^{-10},$$

using (21) for α. Summing up these contributions, we find that

$$a_\mu^{\text{th}} = 11{,}659{,}204(20) \times 10^{-10}, \tag{29}$$

which is in excellent agreement with the measurement given in (10) and verifies the presence of the hadronic term to 15%.

The errors in $a_\mu(\text{QED})$ come from the α^3 and α^4 terms. Their magnitudes are 2.5×10^{-10} and 1.8×10^{-10}, respectively. They are still being improved; thus, the published values may differ slightly from (28).

The largest theoretical error comes from the hadronic effects represented by the diagrams in figure 7. This is dominated by the hadronic vacuum polarization effect represented by figure 7a. This contribution can be written as

$$a_\mu(\text{had.a}) = \left(\frac{\alpha m_\mu}{3}\right)^2 \int_{4m_\pi^2}^\infty \frac{ds}{s^2} K(s) R(s), \tag{30}$$

where $K(s)$ is a slowly varying function of s close to 1 except near the threshold, and

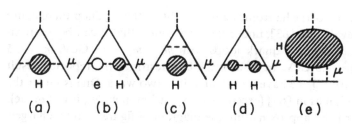

Figure 7
Examples of hadronic contributions to a_μ. The symbol H denotes hadronic vacuum polarization.

$$R(s) = \sigma_{tot}(e^+e^- \to \text{hadrons})/\sigma(e^+e^- \to \mu^+\mu^-). \tag{31}$$

Thus $a_\mu(\text{had.a})$ is calculable from the measured value of $R(s)$. Using the latest data available, we evaluated this integral and found

$$a_\mu(\text{had.a}) = 707(6)(17) \times 10^{-10}, \tag{32}$$

where the first error is statistical and the second systematic. Note that the former is more than 10 times smaller than the previous errors [63, 64]. The latter is mostly due to the systematic error of the $\pi^+\pi^-$ channel.

Clearly the systematic error of $a_\mu(\text{had.a})$ must be improved further. A very encouraging development in this direction is the new experiment at CERN [65] in which 300 GeV e^+ from π^0 decays collide with the e^- in target atoms producing a $\pi^+\pi^-$ pair in the range $2m_\pi \leqslant s^{1/2} \lesssim 600$ MeV. (This range will be pushed to ~ 1 GeV if the 1-TeV proton beam at the Fermilab is used.) In these experiments, in which high energy $\pi^+\pi^-$ and $\mu^+\mu^-$ pairs are counted simultaneously, systematic errors can be controlled more tightly than in conventional colliding beam experiments. Thus it may not be impossible to reduce the overall error of $a_\mu(\text{had.a})$ to less than 3×10^{-10}, removing a major obstacle for the experimental test of the electroweak effect (14).

Higher order hadronic corrections to a_μ arising from diagrams of figures 7b–7d have been estimated to be [63]

$$a_\mu(\text{had.b}) = 110(14) \times 10^{-11},$$

$$a_\mu(\text{had.c}) = -207(29) \times 10^{-11}, \tag{33}$$

$$a_\mu(\text{had.d}) \approx 2 \times 10^{-11}.$$

The contribution of the hadronic light-by-light scattering diagram of figure 7e was also evaluated [63], under the assumption that it can be approximated by u, d, s, and c quark loops with $m_u = m_d = 0.3$ GeV, $m_s = 0.5$ GeV, and $m_c = 1.5$ GeV. In view of the large error in the reported result $[-26(10) \times 10^{-10}]$, we have reevaluated it in two ways. One is under the same assumption as in [63] (except that expansion in m_μ/m_q is not made), and the other is by approximating the diagram of figure 7e by a charged pion loop and various low energy hadron resonances (of which π^0 pole is the most important). The results are [49]

$$a_\mu(\text{had.e}) = \begin{cases} 60(4) \times 10^{-11} & \text{(quark loop)} \\ 49(5) \times 10^{-11} & \text{(pion loop and resonances)}, \end{cases}$$

(34)

(35)

which are consistent with each other. The value of $a_\mu(\text{hadron})$ given in (28) is obtained from (32), (33), and (35).

As was mentioned repeatedly, the primary motivation for improving theoretical prediction and measurement of a_μ is that it provides an opportunity to test the gauge theories of electroweak interaction at the one-loop level. There are several other processes that require inclusion of the one-loop weak interaction effect for a good fit, such as the muon decay, Cabibbo universality, the $|\Delta S| = 1$ semileptonic decays of neutral particles, and $K_L - K_S$ mass difference [66]. Also, the mass shift of weak bosons calculated at the one-loop level is likely to be confirmed shortly by experiment [67]. In comparison, the electroweak effect on a_μ suffers from the fact that it does not become observable until the measurement of a_μ is improved by at least an order of magnitude. According to Hughes [68], however, it may be possible to improve the accuracy of measurement by a factor of 30 using high intensity proton beams that will be available at the LAMPF II or the upgraded AGS and a superconducting muon storage ring. Other approaches are being examined, too [69]. Thus, we may be able to test the electroweak effect on a_μ in the not too distant future. This will test the renormalizability of the gauge theory of electroweak interaction independently of other tests and help us reduce the number of alternative theories. It may even lead to the discovery of a totally unexpected phenomenon.

Acknowledgment

This work was supported in part by the National Science Foundation.

References

[1] The initial news of this conference was brought to Japan, where I was a beginning graduate student, by a science column in the 29 September issue of *Newsweek* magazine.

[2] J. Schwinger, *Phys. Rev.* 73, 416L (1948).

[3] P. Kusch and H. M. Foley, *Phys. Rev.* 72, 1256 (1947).

[4] T. Kinoshita, in Proceedings of the International Symposium of Foundations of Quantum Mechanics, Tokyo, 1983.

[5] See, for instance, a review by L. Lyons, in *Progress in Particle and Nuclear Physics* 10, 227 (1983).

[6] J. Ellis, J. Haselin, and D. V. Nanopoulos, *Phys. Lett.* B113, 283 (1982); R. Barbieri and L. Maiani, *Phys. Lett.* B117, 203 (1982); J. A. Grifols and A. Mendez, *Phys. Rev.* D26, 1809 (1982).

[7] A. Rich and J. C. Wesley, *Rev. Mod. Phys.* 44, 250 (1972).

[8] R. Karplus and N. Kroll, *Phys. Rev.* 77, 536 (1950).

[9] P. A. Franken and S. Liebes, Jr., *Phys. Rev.* 104, 1197 (1956).

[10] C. M. Sommerfield, *Phys. Rev.* 107, 328 (1957), *Ann. Phys.* 5, 26 (1957).

[11] A. Petermann, *Helv. Phys. Acta* 30, 407 (1957).

[12] B. E. Lautrup and E. de Rafael, *Phys. Rev.* 174, 1835 (1968).

[13] A. Petermann, *Phys. Rev.* 105, 1931 (1957).

[14] H. Suura and E. H. Wichmann, *Phys. Rev.* 105, 1930 (1957).

[15] H. H. Elend, *Phys. Lett.* 20, 682 (1966), 21, 720 (1966); G. W. Erickson and H. H. T. Liu, Report UCD-CNL-81 (1968).

[16] C. Bouchiat and L. Michel, *J. Phys. Radium* 22, 121 (1961).

[17] D. T. Wilkinson and H. R. Crane, *Phys. Rev.* 130, 852 (1963).

[18] B. N. Taylor, W. H. Parker, and D. N. Langenberg, *Rev. Mod. Phys.* 41, 375 (1969).

[19] A. Rich, *Phys. Rev. Lett.* 20, 967 (1968); G. R. Henry and J. E. Silver, *Phys. Rev.* 180, 1262 (1969).

[20] S. Granger and G. W. Ford, *Phys. Rev. Lett.* 28, 1479 (1972).

[21] J. C. Wesley and A. Rich, *Phys. Rev.* A4, 1341 (1971).

[22] S. D. Drell and H. R. Pagels, *Phys. Rev.* 140, B397 (1965); R. G. Parsons, *Phys. Rev.* 168, 1562 (1968).

[23] T. Coffin et al., *Phys. Rev.* 109, 973 (1958).

[24] G. Charpak et al., *Phys. Rev. Lett.* 6, 128 (1961).

[25] F. J. M. Farley et al., *Nuovo Cimento* 44, 281 (1966).

[26] J. Bailey et al., *Phys. Lett.* 28B, 287 (1968).

[27] J. Bailey et al., *Phys. Lett.* 68B, 191 (1977); F. J. M. Farley and E. Picasso, *Ann. Rev. Nucl. Part. Sci.* 29, 243 (1979).

[28] This is seen by applying the mass singularity theorem (see [32]) to the muon vertex. As a matter of fact this interesting property was first recognized while I was trying to understand the reason why the coefficient of $\ln(m_\mu/m_e)$ in (7) is equal to $1/3$.

[29] T. Kinoshita, *Nuovo Cimento* B51, 140 (1967).

[30] B. E. Lautrup and E. de Rafael, *Phys. Rev.* 174, 1835 (1968); B. E. Lautrup and E. de Rafael, *Nuovo Cimento* A64, 322 (1969); S. Brodsky and T. Kinoshita, *Phys. Rev.* D3, 356 (1971); R. Barbieri and E. Remiddi, *Phys. Lett.* B49, 468 (1974).

[31] J. Aldins et al., *Phys. Rev. Lett.* 23, 441 (1969); *Phys. Rev.* D1, 2378 (1970).

[32] T. Kinoshita, *J. Math. Phys.* 3, 650 (1962).

[33] B. E. Lautrup and M. A. Samuel, *Phys. Lett.* B72, 114 (1977).

[34] J. Calmet et al., *Rev. Mod. Phys.* 49, 21 (1977); C. Chlouber and M. A. Samuel, *Phys. Rev.* D16, 3596 (1977).

[35] J. A. Mignaco and E. Remiddi, *Nuovo Cimento* A60, 519 (1969); R. Barbieri, M. Caffo, and E. Remiddi, *Nuovo Cimento Lett.* 5, 769 (1972); D. Billi, M. Caffo, and E. Remiddi, *Nuovo Cimento Lett.* 4, 657 (1972); R. Barbieri, M. Caffo, and E. Remiddi, *Nuovo Cimento Lett.* 9, 690 (1974); R. Barbieri and E. Remiddi, *Phys. Lett.* B49, 468 (1974); L. L. DeRaad, Jr., K. A. Milton, and W.-Y. Tsai, *Phys. Rev.* D9, 1814 (1974).

[36] P. Cvitanovic and T. Kinoshita, *Phys. Rev.* D10, 4007 (1974).

[37] M. J. Levine and J. Wright, *Phys. Rev.* D8, 3171 (1973); R. Carroll and Y.-P. Yao, *Phys. Lett.* B48, 125 (1974).

[38] M. J. Levine, H. Y. Park, and R. Z. Roskies, *Phys. Rev.* D25, 2205 (1982).

[39] M. Veltman, CERN Preprint (1967); H. Strubbe, *Comput. Phys. Commun.* 8, 1 (1974), 18, 1 (1979).

[40] A. C. Hearn, "REDUCE 2 User's Manual" Stanford Artificial Intelligence Project Memo AIM-133.

[41] RIWIAD is a version of SHEPPY [See W. Czyz, G. C. Sheppy, and J. D. Walecka, *Nuovo Cimento* 34, 404 (1964)] modified by Lautrup, Sheppy, and Dufner [see B. E. Lautrup, in Proceedings 2nd Colloq. Computer Methods in Theoretical Physics, Marseille, 1971].

[42] G. P. Lepage, *J. Comput. Phys.* 27, 192 (1978).

[43] B. D. Josephson, *Phys. Lett.* 1, 251 (1962).

[44] K. v. Klitzing, G. Dorda, and M. Pepper, *Phys. Rev. Lett.* 45, 494 (1980).

[45] R. Jackiw and S. Weinberg, *Phys. Rev.* D5, 2473 (1972); I. Bars and M. Yoshimura, *Phys. Rev.* D6, 374 (1972); W. A. Bardeen, R. Gastmans, and B. E. Lautrup, *Nucl. Phys.* B46, 319 (1972); K. Fujikawa, B. W. Lee, and A. I. Sanda, *Phys. Rev.* D6, 2923 (1972).

[46] R. S. Van Dyck, Jr. P. B. Schwinberg, and H. G. Dehmelt, *Phys. Rev. Lett.* 38, 310 (1977).

[47] T. Kinoshita and W. B. Lindquist, Cornell University Preprint CLNS-374 (1977).

[48] T. Kinoshita and W. B. Lindquist, *Phys. Rev. Lett.* 47, 1573 (1981).

[49] T. Kinoshita, B. Nizic, and Y. Okamoto (to be published).

[50] C. Itzykson, G. Parisi, and J.-B. Zuber, *Phys. Rev.* D16, 996 (1977).

[51] G. Martinelli et al., *Phys Lett.* B116, 434 (1982); C. Bernard et al., *Phys. Rev. Lett.* 49, 1076 (1982).

[52] T. A. Welton, *Phys. Rev.* 74, 1157 (1948).

[53] Z. Koba, *Prog. Theoret. Phys.* 4, 319 (1949).

[54] Z. Koba, private communication.

[55] P. B. Schwinberg, R. S. Van Dyck, Jr., and H. G. Dehmelt, *Phys. Rev. Lett.* 47, 1679 (1981).

[56]   T. Kinoshita and W. B. Lindquist, Cornell University Report CLNS-510 (1981), and its update (to be published).

[57]   See, for instance, T. Kinoshita, in *New Frontiers in High Energy Physics* (Plenum, New York, 1978).

[58]   E. R. Williams and P. T. Olsen, *Phys. Rev. Lett.* 42, 1575 (1979).

[59]   F. G. Mariam et al., *Phys. Rev. Lett.* 50, 993 (1983); J. R. Sapirstein, E. A. Terray, and D. R. Yennie, *Phys. Rev. Lett.* 51, 982 (1983); J. R. Sapirstein, *Phys. Rev. Lett.* 51, 985 (1983).

[60]   D. C. Tsui et al., *Phys. Rev. Lett.* 48, 3 (1982).

[61]   R. S. Van Dyck, Jr., private communication.

[62]   H. G. Dehmelt, in *Atomic Physics*, Vol. 7 (Plenum, New York, 1981).

[63]   J. Calmet et al., *Phys. Lett.* B61, 283 (1976).

[64]   V. Barger, W. F. Long, and M. G. Olsson, *Phys. Lett.* B60, 89 (1975).

[65]   S. R. Amendolia et al., *Phys. Lett.* B138, 454 (1984).

[66]   See, for instance, M. A. B. Bég and A. Sirlin, *Phys. Rep.* 88, 1 (1982).

[67]   G. Arnison et al., *Phys. Lett.* B122, 103 (1983); M. Banner et al., *Phys. Lett.* B122, 476 (1983); G. Arnison et al., *Phys. Lett.* B126, 398 (1983); P. Bagnaia et al., *Phys. Lett.* B129, 130 (1983).

[68]   V. W. Hughes, talk presented at the Third LAMPF II Workshop, Los Alamos National Laboratory, 1983.

[69]   J. Bailey and F. J. M. Farley, private communications.

[70]   J. E. Nafe, E. B. Nelson, and I. I. Rabi, *Phys. Rev.* 71, 914 (1947).

[71]   D. E. Nagle, R. S. Julian, and J. R. Zacharias, *Phys. Rev.* 72, 971 (1947).

[72]   G. Breit, *Phys. Rev.* 72, 984 (1947).

[73]   J. H. Gardner and E. M. Purcell, *Phys. Rev.* 76, 1262 (1949).

[74]   W. H. Louisell, R. W. Pidd, and H. R. Crane, *Phys. Rev.* 91, 475 (1953).

[75]   H. G. Dehmelt, *Phys. Rev.* 109, 381 (1958).

[76]   A. A. Schupp, R. W. Pidd, and H. R. Crane, *Phys. Rev.* 121, 1 (1961).

[77]   A. Rich and H. R. Crane, *Phys. Rev. Lett.* 17, 271 (1966).

[78]   W. H. Parker, B. N. Taylor, and D. N. Langenberg, *Phys. Rev. Lett.* 18, 287 (1967).

[79]   G. Gräff, F. G. Major, R. W. H. Roeder, and G. Werth, *Phys. Rev. Lett.* 21, 340 (1968).

[80]   J. Calmet and M. Perrottet, *Phys. Rev.* D3, 3101 (1971).

II HISTORICAL PERSPECTIVES

A Short History of Shelter Island I

Silvan S. Schweber

In January of 1948, six months after the 1947 Shelter Island Conference, K. K. Darrow wrote Duncan MacInnes a brief postcard: "... I must quote [you] ... the words of warm commendation used yesterday by I. I. Rabi anent your Shelter Island meeting—he said that it has proved much more important than it seemed even at the time, and would be remembered as the 1911 Solvay Congress is remembered, for having been the starting-point of remarkable new developments...."[1] Similarly, R. P. Feynman many years later recalled, "There have been many conferences in the world since, but I've never felt any to be as important as this."[2]

Shelter Island I was indeed one of the most fruitful of the conferences held right after the end of World War II. Just as the 1911 Solvay Congress set the stage for subsequent developments in quantum theory (see, for example, de Broglie, 1951), Shelter Island provided the initial stimulus for the postwar developments in quantum field theory: effective, relativistically invariant computational methods, Feynman diagrams, and renormalization theory.

The origins of the conference present an interesting case study in the interplay and accommodation of institutional and professional demands, involving the National Academy of Sciences and its leadership, the theoretical physics community, and a particular group of individuals that included MacInnes, Darrow, and Oppenheimer. That story is told in section 1. Section 2 presents the scientific content of the meeting, focusing particularly on the field-theoretical developments, and section 3 describes the impact of this conference and the later ones at Pocono and Oldstone.

1 The Genesis of the Conference

In the fall of 1945, soon after the close of World War II, Duncan MacInnes, a member of the Rockefeller Institute and a past president of the New York Academy of Sciences, went to see Frank Jewett, the head of the National Academy of Sciences (NAS), about an idea he had for specialized science conferences. He felt that scientific and technical societies had grown so large, or had become so narrowly specialized, that no critical discussions were taking place at their meetings.

---

1. Rockefeller University (RU) Archives, box 9-450 M189; Darrow to MacInnes, 16 January 1948. These comments should be contrasted with the remarks that Rabi had made shortly before the conference at a luncheon with Darrow and Willis Lamb: "The last 18 years has been the most sterile of the century in theo. [i.e., theoretical physics]" (K. K. Darrow diaries, Monday, 14 April 1947; the diaries are in the American Institute of Physics—AIP—Archives).
2. Oral interview with C. Weiner, 1966, AIP, Niels Bohr Library.

For a number of years before he became president of the New York
Academy of Sciences, MacInnes had helped arrange a series of small
conferences sponsored by that organization, designed to promote active
discussion of different scientific topics. For the conferences to achieve their
objectives, MacInnes had considered it essential that the topics be in areas
in which actual work was in progress and that participation be restricted to
currently active investigators in the designated fields. Although the early
conference sponsored by the New York Academy had achieved this objec-
tive, MacInnes believed that the effectiveness of the later ones had been
impaired by the fact that they attracted too large a crowd (Shedlovsky,
1970). He felt very strongly about limiting attendence at these conferences.
In October 1944, he had declined to run again for president of the academy
when a decision by the council of the academy on this matter went against
him. In fact, in January 1945 McInnes resigned from the academy over this
issue.

MacInnes suggested to Jewett that much of value would be had from a
two- or three-day conference of 20 or 25 (not more than 30) selected people
brought together to discuss a number of papers prepared and circulated in
advance, covering specific facets of a field in which major problems existed.
He inquired from Jewett what he thought of the idea and, in particular,
whether the NAS would be willing and be in a position to sponsor a series of
such conferences.

Jewett told him that the idea appealed to him provided that the problems
the conferences were to address were real, that their solutions "would be
aided by the concentrated consideration of a small highly qualified group,
meeting in intimate association for discussion after study of the prepared
papers," and that it seemed to him that "the idea of a two or three day
meeting at some quiet place where the men [sic] could live together
intimately seemed ... the best way to obtain the desired results".[3] Jewett felt
confident that the NAS would be willing to sponsor the undertaking
provided that it met these criteria. He suggested that MacInnes pick out one
or two problems that seemed promising and use these as "pilot plants,"
rather than address the problem of a broad program *ab initio*.

Given this encouragement, MacInnes, after discussion with colleagues,
suggested two conferences, "The Nature of Biopotentials" and "The Postu-
lates of Quantum Mechanics." Biopotentials were the current focus of

3. Archives, NAS; F. Jewett to W. Weaver, 23 April 1946.

research of MacInnes's colleague and friend W. J. V. Osterhout. Electrochemistry was the field that MacInnes worked in, and biopotentials presented an interesting area for involving outstanding physicists and physical chemists looking for challenging new problems after the war. The topic of the second conference was an area that intrigued MacInnes, who at the time was studying wave mechanics (see footnote 8). K. K. Darrow—the urbane, by then perennial, secretary of the American Physical Society (APS), a theoretical physicist who had been at the Bell Laboratories in Murray Hill since 1925 and who was a popularizer of science of some stature — offered MacInnes his help in organizing the conference on quantum mechanics. In late December 1945, MacInnes and Darrow consulted Darrow's friend León Brillouin because he had a great deal of experience with Solvay congresses. Reporting on this meeting to Jewett, MacInnes indicated that Brillouin had made a number of worthwhile suggestions, one of which was that they consult with W. Pauli, who had just obtained his Nobel Prize, and was still at the Institute for Advanced Studies in Princeton, where he had been since the outbreak of World War II. Darrow wrote Pauli on 4 January 1946 indicating that MacInnes had thought of the idea "of convoking a congress similar to the excellent Solvay Congresses of old" to discuss problems connected with the foundations and philosophy of quantum mechanics and "subsequently to publish [their] discussions." He also informed Pauli that the National Academy of Sciences was favorably disposed to sponsor the conference, "so that it seems possible to solve the problem of financing." Jewett had indeed explored further the matter of funding and had obtained indications from Fosdick and Weaver of the Rockefeller Foundation, and from Josephs of the Carnegie Foundation that they would be interested in supporting the enterprise.

A meeting in New York of Pauli with Brillouin, Darrow, and MacInnes was arranged and held on 18 January 1946. Pauli, who had just been invited to an international conference that was to convene at the Cavendish Laboratories in the summer of 1946 and was convinced of the importance of such conferences in reestablishing the continuity of the scientific enterprise and reasserting communal values after the terrible ordeal of World War II, suggested a large conference rather than one patterned after the small and elitist Solvay congresses. Moreover, "Pauli was a bit doubtful whether there would be anything very new to consider [in wave mechanics], but got more enthusiastic when the topic was changed to 'Fundamental Problems of Quantum Theory.' He also suggested quite a number of names from abroad for those who might attend.... He ... promised to make up a list of those who would present papers...."[4]

MacInnes did not like Pauli's suggestions and became disturbed "at the way the plans for the quantum theory conference seemed to be going. Pauli was planning for too many of the older men, and [MacInnes thought] that the best results would be to get out the coming generation." On 21 January, MacInnes conferred with Jewett, "who heartily agreed" with him on the matter, and "also suggested having the conference at an inn somewhere."[5]

After his consultation with Jewett, MacInnes wrote Pauli that "Dr. Jewett ... is very much in favor of smaller conferences of the type that we outlined to you." Also, Dr. Fosdick of the Rockefeller Foundation, who "has been thinking along that line for some time," would give financial support only to a small gathering. Additionally, "Dr. Jewett thinks, as I do, that if at all possible we should try to discover and make prominent in such a conference the younger men who are coming along and give them, if possible, the job of preparing and presenting the papers."[6] MacInnes indicated that he had in mind "such men as J. A. Wheeler" and that there had to be others. It was these men "who are most likely to do a good job and to bring in fresh ideas." Thanking Pauli for his help, MacInnes requested of him suggestions as to the names of "such coming men." Soon thereafter, on 15 February, Darrow also wrote Pauli that MacInnes and he were anxious to get his list of talented young physicists "worthy of invitation" to the proposed conference.[7]

On 20 February 1946, MacInnes got a letter, signed jointly by Pauli and Wheeler, in which they indicated "that a proposal for a conference on Wave Mechanics had been made by both Bohr and [by him], the former suggesting that one be held in Denmark early in 1947."[8] Pauli and Wheeler both remembered "the stimulating discussions which have taken place at Bohr's institute and the important consequences for science which have come out of them under his leadership." Since plans were then "underway for a conference in New York in September on 'Physics of the Elementary Particles,'" they suggested that the "second" conference, which had been

4. Duncan MacInnes diaries, 2 January 1946–30 December 1947, page 6 = folder title and page number.
5. MacInnes diaries, entry for 24 January 1946, page 6.
6. MacInnes to Pauli, 25 January 1946. RU Archives.
7. AIP Archives. K. K. Darrow letters. A copy of the letter is also to be found in the MacInnes RU Archives.
8. MacInnes diaries, entry for 20 February 1946.

proposed by both MacInnes and Bohr, be held in Copehagen. Further-more, if MacInnes approved this suggestion, Pauli and Wheeler recom-mended that the funds "which would in any case have been set aside for tra-vel should be used to send scientists from other countries to the conference in Copenhagen. This arrangement might be especially appropriate through the fact that Bohr's Institute has always enjoyed extremely close relations with the Rockefeller Foundation." [9]

Pauli and Wheeler preferred Bohr's conference in Copenhagen to one held in the United States because of their respect and affection for Bohr and because this would be a "means to promote international collaboration in science." Although MacInnes was disposed to go along with them, Darrow convinced him otherwise and suggested that they "get Pauli and Wheeler to NYC." Darrow called Pauli, who, because he was leaving for Europe, delegated Wheeler to act for him in his absence. Darrow then wrote Wheeler to state his disagreement with Pauli that "Bohr's proposed confer-ence should supersede the one which MacInnes is planning" because "(1) Few Americans are likely to be asked to Bohr's conference.... (2) Even among the Americans whom Bohr will invite ... [few will come because of] the difficulty and the cost of the journey.... (3) Bohr did not publish the proceeding of his conferences, whereas MacInnes is in a position to do so." [10] Darrow therefore suggested to Wheeler that he take over "the fulfillment of the request which MacInnes and I made of Pauli, viz. that you select say fifteen of the young and promising Americans in the field of quantum mechanics and send your list to MacInness and me. We will then combine it with the list of established theoretical physicists which MacInnes, Pauli and I drew up some weeks ago, and we will seek an occasion to meet with you and discuss our plans."

Darrow in his letter to Wheeler had made explicit what had been implicit with MacInnes and Jewett: The conference was to be an American one, designed to "bring out" the young American theoretical physicists who had played such a large role in the successful war effort. For Darrow, MacInnes, and Jewett, the conference was to demonstrate that theoretical physics in the United States had come into its own with the "younger men" who had been born and trained in the United States. For Jewett and Darrow, the

9. Pauli and Wheeler to MacInnes, 16 February 1946, MacInnes correspondence. RU Archives.
10. Darrow to Wheeler, 26 February 1946 [copy], Wheeler letters in MacInnes corre-spondence. RU Archives.

conference was to prove American superiority not only in wartime but also in "pure" physics.

MacInnes also wrote Wheeler. He indicated that it seemed to him "that there is room for both the National Academy conference and Dr. Bohr's conference ... since somewhat different groups might be involved. In conversation with Dr. Jewett, he agreed with me that it would be well to include in the group, of say thirty, as many of the younger active producing physicists in the field as possible and have a relatively smaller number of the older men. It is the latter group, of course, that would be most likely to be invited to Dr. Bohr's conference." Moreover, "if the National Academy conference could be held during the coming fall it might prove to be an excellent preliminary to the European meeting...."[11]

Darrow and MacInnes met with Wheeler over lunch on Monday, 11 March 1946, in New York. At that meeting Wheeler suggested to them that their conference deal with "more advanced topics," namely electron-positron theory. He also identified for them some of the "younger men." On 14 March MacInnes wrote Wheeler a letter containing the names he had "obtained from various sources, for the conference to be tentatively called 'Fundamental Problems of Quantum Mechanics,'" as well as those Wheeler had given him. The names on MacInnes's list were

Epstein	Schroedinger
Houston	Van Vleck
Darrow	Oppemheimer [sic]
Kemble	Breit
Rojanski	Debye
Rabi	Tolman
Wigner	Kirkwood
Von Neumann	Einstein
Bethe	Pauli
Eyring	Kimball
Pauling	Bohr
Slater	Born
London	Des Touches
Teller	De Broglie
Dirac	Landau
Frenkel	Onsager

After noting that he had misplaced the memorandum containing the names of "Europeans" that Pauli had given him, MacInnes recorded Wheeler's

11. MacInnes to Wheeler, 27 February 1946. RU Archives.

suggestions:

E. B. Wilson	Robt. Christy
Richard Feyerman [sic]	Leonard Schiff
Chas. Critchfield	E. H. (?) Hill
Wendell Furry	Julian Schwinger
Victor Weisskopf	

He concluded his letter with the statement that his "present idea would be to get together a group consisting mostly of the younger men, who would understand each other's jargon."[12]

After his meeting with Wheeler, MacInnes also wrote to Jewett, "Dr. Darrow and I have come more or less to the conclusion that it would be the most effective thing to get together a group of the younger men who are actively at work or else slowly getting back to their normal work after war research. There should also be some of the older men present as well. Dr. Wheeler has been of great assistance in pointing which of the younger men should be invited and he has outlined the liv[eli]er portions of the topics to be considered. The idea of a long weekend at some country inn seems most desirable for this group."[13]

Having been requested by MacInnes for a program for the contemplated theoretical physics conference, in a letter dated 19 March Wheeler outlined possible topics dealing with "Problems on the Quantum Mechanics of the Electron," and drew up a list of suggested participants. On 28 March MacInnes wrote Wheeler to thank him for his outline and indicated to him that he was particularly pleased that he considered "a group as small as fifteen." In addition he informed Wheeler that Jewett had raised the question whether the contemplated conference "could utilize some of the men who are to be brought over from Europe during the coming October" to attend the Autumn meeting of the National Academy of Science and the American Philosophical Society in Washington and Philadelphia. He also asked Wheeler to "sound out various of the workers as to their willingness to prepare papers as you describe," which were to be published under the auspices of the NAS "after revision in the light of the discussions at the conference."[14]

12. MacInnes to Wheeler, 14 March 1946. RU Archives MacInnes correspondence. The Wheeler letters.
13. Archives, NAS; MacInnes to Jewett: 18 March 1946.
14. D. A. MacInnes to J. A. Wheeler, 28 March 1946. K. K. Darrow paper, AIP Library. A copy of this letter is also to be found in the MacInnes correspondence at RU.

In the middle of April 1946, Wheeler answered him and in his letter detailed both possible problems to be addressed by a three-day conference and likely persons to present papers on these topics. In addition, Wheeler estimated the probable cost of such an undertaking. MacInnes promptly thanked Wheeler and on the same day wrote to Jewett—enclosing Wheeler's letter—to give him his estimate of the cost of the two conferences. The estimate for the biopotential conference amounted to $1,500; that for the quantum mechanics conference was as follows:[15]

Two days at inn (30 people)	420.00
Train Fare for those coming from a distance	750.00
Monograph	1,500.00
Stenographic work (if used)	300.00
Mimeographing, postage, etc.	150.00
	$3,120.00

Jewett in his reply to MacInnes indicated that there ought to be no difficulty in getting the $5,000 or $6,000 to cover the expenses of the two conferences. If necessary, Jewett as president would stretch his authority over the National Academy's Emergency Fund and cover the expenses of the conferences from it. "Further, if these first two prove really successful I think we will have no trouble financing others."[16]

With funding of the conference assured, Darrow and McInnes instructed Wheeler on 7 May 1946 to select "20 or so participants for the conference" to be held in late January or early February 1947 at a country inn and, in consultation with Darrow, choose the authors of the papers to be presented. They also recommended that N. Bohr ought to be got to attend "if possible." Wheeler was also informed that a decision had been made not to have a stenographic recording of the discussion "but to have careful notes made of such discussion by selected members of the conference"[17] and that the conference was to take place at a country inn to be selected by MacInnes.

In his reply to MacInnes, Wheeler informed him that he had talked about the conference to Critchfield of George Washington University and to his

15. AIP; Darrow, MacInnes to Jewett, 16 April 1946.
16. Archives, NAS; Jewett to MacInnes.
17. AIP; MacInnes to Wheeler, 7 May 1946, Darrow papers. The letter formalized the results of a conference between Wheeler, MacInnes, and Darrow held on 5 May. MacInnes diaries, entry for 6 May.

colleague Wigner at Princeton—both suggested participants. Both were very enthusiastic about the choice of subject. But both felt that since George Washington University and the Carnegie Institute of Terrestrial Magnetism were considering sponsoring their first postwar theoretical conference that October "to take advantage of the presence in this country of Bohr and Dirac" (who were to attend the joint NAS and APS meetings), and, since Wheeler's proposed topics "would be suitable for consideration as a program for such a Fall Conference," that "it would be appropriate to unite the [MacInnes] conference with the Washington conference."[18] Wheeler stressed that the presence of Bohr at such a conference was essential. Upon receiving Wheeler's recommendation of a joint sponsorship, MacInnes forwarded the letter to Darrow and requested his reaction to Wheeler's suggestion. His own views were that "the conferences has been developing into something quite different from what I originally intended and now shows signs of being captured by another group." Although he personally would be willing to compromise on a combined conference if it were held, at least partly, under NAS auspices, he confessed that he did not like it. Moreover, the early date of October 1946 did not leave enough time for proper preparation. Finally, "Though the presence of Bohr and Dirac would help, it seems to me that it would be much more profitable to have the younger men in the field do the job and not rely too much on authority of big names."[19] Darrow's reaction was swift and to the point. On 21 June he sent a telegram to MacInnes stating "Deprecate merger of your conference with other."[20]

MacInnes, with Jewett's backing, carried the day and, with Wheeler as the principal consultant, the planning for the conference proceeded on the assumption that it would be held in the spring of 1947 to allow proper preparations. By November 1946, Darrow had prepared a first draft of an invitation to the conference. The announcement stressed that there would be no formal papers—except for three or four brief ones "to invite and excite discussion"—no agenda, and that it was not contemplated to make any record of the proceedings, a practice shared by the Copenhagen and Washington conferences.

The conference on biopotentials was held at the Rockefeller Institute

18. AIP; Wheeler to MacInnes—Darrow papers, 7 June 1946.
19. RU Archives; MacInnes to Darrow, 18 June 1946.
20. RU Archives; Darrow to MacInnes, 21 June 1946.

under the auspices of the NAS on 13 and 14 December 1946. It was in honor
of W. J. V. Osterhout and turned out to be a great success. In contrast with
the biopotential conference, no plans had been made to produce either a
monograph or to make available to the scientific public a record of the
proceedings of the conference on quantum mechanics. This in Jewett's
opinion detracted considerably from the value of the contemplated conference. To meet this objection MacInnes and Darrow decided to postone the
conference on "The Quantum Mechanics of the Electron" to late May of
1947.

Darrow had recommended that Oppenheimer write a paper. When
asked, Oppenheimer, who was coming from California, indicated that he
would only be able to attend if the conference were held "during the two
days just before or the two days just after the Memorial Day weekend."
Darrow recommended to MacInnes that they accept the later date
"because Oppenheimer is so great a man" and also "because it may facili-
tate greatly our task of getting accomodations"[21] (since many inns and
hotels only opened their premises after Memorial Day). They set for the
date of the conference Monday, Tuesday, and Wednesday, 2–4 June, and
decided that K. K. Darrow would be its convener and chairman. The other
persons asked to prepare papers for the conference were Edwin Kemble,
who declined, Victor Weisskopf, and Hendrik Kramers.

On 28 February 1947 Jewett gave MacInnes the green light on the
coming conference. The matter of getting a place for the conference was
attended to by MacInnes, and he turned to the Commerce Commission of
the State of New York for assistance. John Deming, of the commission,
recommended to him "an inn on Ram's Island at the end of Long Island
quite highly ... [that sounded] very good but possibly too expensive".[22] On
10 March 1947 MacInnes visited the inn to judge its fitness and though it
was closed deemed it excellent for the conference. On 26 March he informed
Peter Katavalos, the manager of the inn, that "we would use his Rams Head
Lodge for our conference."[23]

Although Oppenheimer and Weisskopf had initially agreed to prepare
manuscripts of their papers in advance of the conference—which were then
to be copied and circulated and were "to form the backbone of the pub-

21. RU Archives; Darrow to MacInnes, 4 February 1947.
22. MacInnes diaries, entry for 19 February 1947, p. 90.
23. MacInnes diaries, entry for 26 March 1947, p. 108.

Participants at the first Shelter Island conference, June 1947 (left to right): I. Rabi, L. Pauling, J. Van Vleck, W. Lamb, Jr., G. Breit, D. MacInnes, K. Darrow, G. Uhlenbeck, J. Schwinger, E. Teller, B. Rossi, A. Nordsieck, J. von Neumann, J. Wheeler, H. Bethe, R. Serber, R. Marshak, A. Pais, J. Oppenheimer, D. Bohm, R. Feynman, V. Weisskopf, H. Feshbach. (Missing from the picture is H. Kramers.)

lished monograph which [had been] required to [be brought] out in due time"—by late March they proposed that the monograph be written wholly after the conference. Darrow and MacInnes were willing to accept the proposal, but were unwilling to convene the conference without a written agenda. They therefore asked each of them to prepare "an outline of 500 words or there-abouts, of the topics, problems and questions which you propose for consideration" to be delivered by 30 April to Duncan MacInnes.[24] They also asked each of them to draw up a list of people "whom they propose for taking part in the conference." On 7 April Darrow sent MacInnes a list of 25 proposed participants that Rabi and he had drawn up "having before us Wheeler's list from which we made very few deletions."[25] An invitation was sent out by Duncan MacInnes to the 25 proposed participants on 11 April 1947 detailing the plans of the conference and indicating that a grant from the NAS had made it possible to defray the expenses of the participants. "The accomodations will be two in a room but with separate beds",[26] and the participants were later asked for "at least four choices as to roommates." From the original list only Bloch, Einstein, Weyl, and Wigner refused, and of those who accepted, only Fermi did not attend because he developed eye trouble on the way to the conference. The final announcement to the members of the conference detailing travel arrangements was sent out on 24 May. It also included a list of the participants:

Chairman
K. K. Darrow, Bell Telephone Laboratories, New York 14, New York

Discussion Leaders
H. A. Kramers, Institute for Advanced Study, Princeton, N.J.
J. R. Oppenheimer, University of California, Berkeley, Calif.
V. F. Weisskopf,* M.I.T., Cambridge 39, Massachusetts

H. A. Bethe,* Cornell University, Ithaca, N.Y.
David Bohm, Princeton University, Princeton, N.J.
Gregory Breit, Yale University, New Haven, Conn.
Enrico Fermi, University of Chicago, Chicago 37, Ill.

24. Archives, NAS: Darrow to Kramers; RU Archives; Darrow to Kramers, Oppenheimer, and Weisskopf, 1 April 1947 [copy].
25. RU Archives; Darrow to MacInnes, 7 April 1947.
26. RU Archives; Duncan MacInnes, 11 April 1947.
* Editors' note: Participants marked by an asterisk also attended Shelter Island II.

Herman Feshbach,* M.I.T., Cambridge 39, Mass.
R. P. Feynman,* Cornell University, Ithaca, N.Y.
W. E. Lamb, Jr.,* Columbia University, New York 27, N.Y.
D. A. MacInnes, Rockefeller Institute, New York 21, N.Y.
R. E. Marshak,* University of Rochester, Rochester, N.Y.
John von Neumann, Institute for Advanced Study, Princeton, N.J.
Arnold Nordsieck, Bell Telephone Laboratories, Murray Hill, N.J.
A. Pais, Institute for Advanced Study, Princeton, N.J.
L. Pauling,* California Institute of Technology, Pasadena, Calif.
I. I. Rabi,* Columbia University, New York 27, N.Y.
Julian Schwinger, Harvard University, Cambridge 38, Mass.
R. Serber,* University of California, Berkeley 4, Calif.
Edward Teller, University of Chicago, Chicago 37, Ill.
G. E. Uhlenbeck, University of Michigan, Ann Arbor, Mich.
J. A. Wheeler,* Princeton University, Princeton, N.J.
Bruno Rossi, M.I.T., Cambridge 39, Mass.
J. H. Van Vleck, Harvard University, Cambridge 38, Mass.

The conference arrangements proceeded without further incident. The conferees gathered in New York at the American Institute of Physics, 55 East 55 Street, on the afternoon of Sunday, 1 June 1947. From there they were taken "on an old and shaky" bus to Greenport at the tip of Long Island. On the final phase of the trip they were accompanied by a police motorcycle escort and their bus did not stop at any red lights. As they passed each county line a new police escort would meet them. The police escort had been secured by John F. Deming, the regional manager of the Department of Commerce of the State of New York, in charge of the Nassau-Suffolk office, and the prime agent in arranging the "happy sojourn at Ram's Head Inn."

In Greenport the conferees were wined and dined at Mitchell's restaurant —a big oyster and steak dinner!—as guests of the Chamber of Commerce. The dinner, although officially given by the Chamber of Commerce, was actually paid for by John C. White, its president. He had been deeply impressed by the physicists' war efforts and wanted to express his appreciation and indebtedness. In an after-dinner speech he told the audience that he had been a marine in the Pacific during the war and that he—and many like him—would not be alive were it not for the atomic bomb. Deeply touched, Darrow and Oppenheimer graciously replied on behalf of the conferees to express their thanks for the hospitality that Greenport was extending the conference, and for its generous gesture in tendering the dinner. After the dinner they were taken by ferry to Shelter Island, where "they were received by a siren-shrieking police escort to their Inn." [27]

In the late '40s, Shelter Island was sparsely populated. A good deal of its land was still being farmed, and its coastline facing the Atlantic was dotted with great nests of ospreys. During the summer the island's population swelled to about 2,000 as a fair number of summer homes and several summer resorts were located on it. Ram's Head Inn—an unpretentious, but elegant and comfortable, two-storied clapboard structure—had just "been painted and furnished with excellent furniture, and looked most inviting." [28] The conferees were its first guests for that season. In fact, it had opened ahead of schedule to host the conference. The inn can accommodate about 30 people and was just large enough to hold all the participants. Located on a western cove of Shelter Island, it is perched on a promontory, surrounded by lush, green lawns that abut the dunes; and beyond the dunes a beach that can be reached in a leisurely few minutes' walk from the inn. It was an ideal place to come together to address once again the problems of pure science.

Stephen White covered the conference for the *New York Herald Tribune*. His article, dated 2 June, reported on the conference's first day:

Twenty three of the country's best known theoretical physicists—the men who made the atomic bomb—gathered today in a rural inn to begin three days of discussion and study, during which they hope to straighten out a few of the difficulties that beset modern physics.

It is doubtful if there has ever been a conference quite like this one. The physicists, backed by the National Academy of Science, have taken over Ram's Head Inn.... They roam through the corridors mumbling mathematical equations, eat their meals amid the fury of technical discussions and gather regularly in the lobby for blackboard discussion....

The meeting is officially entitled "A conference on the Foundations of Quantum Physics." What it amounts to, is an attempt to clarify, now that war-time work can be set aside, the problems that lie before physicists in the realm of the most advanced portion of the science....

... although quantum physics has yielded ... practical results—the atomic bomb among them—the physicists have always been conscious of the fact that the theories rest on insecure grounds. The science is far too approxi-

27. *New York Herald Tribune*; Stephen White, 3 June 1947.
28. MacInnes Diary entry for 6 June, p. 129.

mate for their taste. In these three days they hope to find, if nothing else, what directions their work should take....

... The conference is taking place with almost complete informality, aided by the fact that the scientists have the inn all to themselves and feel that there is no one to mind if they take off their coats and get to work....

The article gave a list of the participants. Singled out from among them were Dr. J. Robert Oppenheimer, "who directed the construction of the first atomic bomb; Dr. I. I. Rabi, Nobel prize winner; Dr. Hans Bethe, Dr. Julian Schwanger [sic; also identified as Schwainger], who was recently made a full professor at the age of twenty nine" at Harvard University, and Dr. John von Neumann, "man of all science."

The reporting was accurate: The conference was informal, the pace somewhat frenetic, the atmosphere intense. All the participants were aware that the experimental results being presented to them by Lamb and Rabi on the spectrum of hydrogen and by Rossi and others on recent cosmic-ray findings were of exceptional significance.

Darrow chaired the conference, but Oppenheimer was the dominant personality and in absolute charge. MacInnes recorded in his diary that "it was immediately evident that Oppenheimer was the moving spirit of the affair." When, at Darrow's insistence, the evening sessions were terminated at 9:00 P.M., the appointed hour, Oppenheimer adjourned the formal sessions to informal meetings that lasted late into the night. Darrow himself, in his diary, gave a revealing account of Oppenheimer:

As the conference went on the ascendency of Oppenheimer became more evident—the analysis (often caustic) of nearly every argument, that magnificent English never marred by hesitation or groping for words (I never heard "catharsis" used in a discourse on [physics], or the clever work "mesoniferous" which is probably O's invention), the dry humor, the perpetually-recurring comment that one idea or another (incl. some of his own) was certainly wrong, and the respect with which he was heard. Next most impressive was Bethe, who on two or three occasions bore out his reputation for hard & thorough work, as in analysing data on cosmic rays variously obtained (An amusing interchange in which x—I've forgotten who, Teller I think—had put a math'l argument on the board; y said "is there not a logarithm?"; x replied "when I do it decently the logarithm will be there"; Bethe said "When it is done *really* decently, there is no logarithm" (laughter!).

Let me give one further indication of Oppenheimer's stature at the time of the meeting. The conference lasted through the morning of Wednesday, 4 June. After the conference Oppenheimer had to go to Harvard, where he was to receive an honorary degree. Arrangements had been made to fly him, Rossi, Schwinger, and Weisskopf by seaplane from Shelter Island to Boston. However, bad weather forced the plane to come down at the New London Coast Guard Station, which is not open to civilian aircraft. The pilot was very worried since they were not supposed to land there. They were met by a naval officer who was clearly furious and ready to read them the riot act. As they opened up and jumped out, Oppenheimer told his very nervous pilot, "Don't worry." Hand outstretched, he introduced himself to the ranting and raging officer with the statement, "My name is Oppenheimer." The bewildered officer queried, "The Oppenheimer?" To which came the reply, "An Oppenheimer!" After an "official" welcome in the officers' club, they were taken to the New London railway station by car with a military escort, where they boarded a train for Boston.

From Cambridge Oppenheimer wrote Jewett[29]

We have just come away from the Ram Island Conference on the Foundation of Quantum Mechanics, that the Academy has sponsored and aided. On behalf of all participants, and for myself, I should like to thank you, as President of the Academy. The three days were a joy to us, and perhaps rather unexpectedly fruitful; we had a long needed chance for clarification and exchange of views, and came away a good deal more certain of the directions in which progress may lie. We recognized this so generally that we agreed that in our work of the next year, we would wish in publication to acknowledge our indebtedness to the conference and to the Academy explicitly, wherever it touched upon matters there discussed. Since these were indeed many and deep, I expect that the Physical Review will be full of praises of Ram Island. And that is only proper.

We worked rather hard, and could well have spent another day or two. It is my hope to ask the conference to reconvene next winter at the Institute in Princeton.

Very Sincerely,
J. R. Oppenheimer

Similar sentiments were expressed by the other conferees.

Jewett was elated by the results of the conference. Its cost had been minimal: $872 had been expended, even though $1,500 had been appropriated. Both of the conferences that MacInnes had initiated and overseen

---

29. Archives, NAS; Oppenheimer to Jewett, 4 June 1947.

"had paid dividends," and Jewett was gratified by the fact that they had been held under NAS auspices during his presidency. Although his term as president expired on 30 June 1947, he recommended to A. N. Richards, the incoming president, that steps be taken to guarantee the continuance of such conferences under NAS sponsorship. Richards was favorably disposed and he requested MacInnes to prepare a memorandum on his assessment of the two "experimental" conferences, so that a recommendation could be submitted to the Council of the NAS.

MacInnes did so in November 1947. In his report he indicated that a monograph on the bioelectric potentials conference would appear soon and that it would emphasize the direction that future research in that field should take. He also reported that although it was decided not to publish a monograph on the Shelter Island Conference, the conferees agreed to give credit to the conference in papers published as a result of its occurrence. He noted that several papers making such aknowledgement had already appeared: Lamb and Retherford's paper on the fine structure of the hydrogen atom [*Phys. Rev.* 72, 241–243 (1947)], Bethe's nonrelativistic Lamb shift calculation [*Phys. Rev.* 72, 339–341 (1947)], Marshak and Bethe's paper on the two-meson hypothesis [*Phys. Rev.* 72, 506–509 (1947)], and Breit's paper on the hyperfine structure of the hydrogen atom, which was in press at the time of the writing of MacInnes's report. MacInnes's conclusion was that "it would appear that these trial conferences were sufficiently successful to justify the continuation."

Richards presented MacInnes's memorandum to a business meeting of the NAS on 17 November. Those attending endorsed the suggestion that the academy encourage such conferences in the future. In December 1947 Richards wrote Oppenheimer, the now acknowledged leader of the previous conference, that the NAS would be prepared to underwrite a meeting of theoretical physicists similar to that held at Shelter Island—up to a sum of $3,000. Oppenheimer, a few days earlier, had informed him of his plans to convene a second such conference. MacInness also wrote to Oppenheimer to express his satisfaction that his idea of the conferences was proceeding so well, and to pledge his support for continued NAS sponsorship. But he insisted "that the conference be not allowed to grow in size." [30] Although this might cause some feelings to be hurt, he felt very strongly that this was an essential factor for the conferences to be "really successful."

30. National Archives; MacInnes to Oppenheimer, 6 December 1947.

On 10 December Oppenheimer circulated a memorandum containing his plans for a second meeting tentatively scheduled to be held from 30 March to 2 April 1948—a date chosen so as to be able to invite Dirac and Bohr, who would be visiting the United States at that time. Because the Ram's Head Inn was not available for that period, he solicited suggestions for another meeting place. Pocono Manor was the site eventually agreed upon for his second conference. It is located approximately midway between Scranton and the Delaware Water Gap, and it afforded the same kind of setting as had Ram's Head Inn—a place where, as Oppenheimer put it, the conferees could "be together and undisturbed."

The conference was held from 30 March to 2 April 1948. Its size was increased to 28 because of the presence of foreign guests. Of those who had attended Shelter Island, not present at this second conference were H. A. Kramers, who had suffered a heart attack, and MacIness, Pauling, and Van Vleck, whose research interests did not overlap with the topics discussed at the conference. The new participants included Dirac, who was on a visit to the United States, Aage Bohr, Neils Bohr, and Walter Heitler, all of whom were at Columbia University for the academic year 1947/48, Gregor Wentzel, who had just come to the University of Chicago, and Eugene Wigner, who had been unable to attend the Shelter Island Conference.

At the conclusion of the conference Oppenheimer wrote Richards, "We have just come home from the second Academy conference on theoretical physics.... All of those invited attended and, as at the first conference, there was a unanimous and cordial expression of satisfaction and gratitude. We worked quite hard, meeeting all day, and often in the evenings, and turned our attention primarily to the new evidences on the nature of the meson, on the interaction of nucleons with electromagnetic fields, and the magnificent new developments in modern electrodynamics which have so deeply increased our insight.[31]"

Oppenheimer also informed Richards that it had been agreed to attempt another meeting "about a year from now," since both Shelter Island and Pocono had proved so useful, important, and inexpensive—the total cost of the Pocono conference amounting to only $1,550. He added in closing, "I believe that these conferences, quite without publicity, with a minimum of organization, and undertaken only for an exchange of views and for further-

31. Archives, NAS; Oppenheimer to Richards, 2 April 1948.

ing our understanding of the foundations of physical theory, are singularly appropriate for Academy support." [32]

The financing of the conferences that MacInnes had conceived was discussed at a NAS Council meeting in Berkeley in November 1948. It was decided there to follow Jewett's original suggestion to approach private foundations for the funding of future conferences, and Richards was instructed by the Council to do so. When Richards discovered that there still existed a balance of nearly $44,000 in the Academy Emergency Grant— monies originally given by the Carnegie Corporation "to help the Academy realize its usefulness during the early years of the war emergency"—he asked the council to be relieved of the duty of soliciting outside funds, and to be allowed to expend the balance on future conferences.

All in all, 11 such small conferences were held under the auspices of the NAS and financed at a total cost of approximately $48,000. The first 4 had a total budget of less than $4,700; none of the physics conferences on theoretical physics held between 1947 and 1949 expended more than $1,600. All 11 conferences proved to be important, seminal, and successful. They are a fitting tribute to MacInnes's vision and integrity. Incidentally, many of these NAS-sponsored conferences were held at Ram's Head Inn on Shelter Island.

In January 1949, Oppenheimer formally wrote Richards asking support from the NAS for a third meeting. In this third conference, to be held 11–14 April, he indicated that subjects "in so far as they can now be anticipated, will probably again be mesons, field theory, the relations of the elementary particles to nuclear forces, and cosmic rays." He added that "much of the progress of the last eventful years in this field was germinated at Shelter Island and at the Poconos." Moreover, he noted that a sign of the seriousness and nature of these meeting was the mandate he had received to "arrange for four full days instead of the two and a half with which we originally started." At Shelter Island and Pocono the participants had put in "ten and eleven hour days," which may be "some indication of how much we cherish the privilege of these meetings." [33] The site for this third conference, which was held for four days 11–14 April 1949 under the chairmanship of Robert Oppenheimer, was Oldstone-on-the-Hudson in Peekskill,

32. Archives, NAS; A. N. Richards to Council of NAS, 7 January 1949.
33. Archives, NAS; Oppenheimer to Richards, 4 January 1949.

New York. Its cost was $1,150. After the conclusion of the conference Oppenheimer wrote Richards the following letter:[34]

The two years since the first conference have marked some changes in the state of fundamental physics, in large part a consequence of our meetings. The problems of electrodynamics which appeared so insoluble at our first meeting, and which began to yield during the following year, have now reached a certain solution; and it is possible, though in these matters prediction is hazardous, that the subject will remain closed for some time, pending the accumulation of new physical experience from domains at present only barely accessible. The study of mesons and of nuclear structures has also made great strides; but in this domain we have learned more and more convincingly that we are still far from a description which is either logical, consistent or in accord with experience.

The members of the conference this year did not determine positively that they desired to reconvene next Spring, since determination on the fruitfulness of this will in part depend on the developments in physics in the year to come; but from the many expressions of gratitude and appreciation, it was clear that this conference needed to be held.

In fact, no further NAS-sponsored conference on particle physics was held.

The Shelter Island, Pocono, and Oldstone conferences were the precursors of the Rochester conferences. But they differed in important ways from the later Rochester conferences: they were small, closed, and elitist in spirit, in contrast with the more professional and democratic outlook of the Rochester conferences. In a sense they mark the postponed end of an era, that of the '30s, with its characteristic style of doing physics: small groups and small budgets. The Rochester conferences reflected the characteristics of the new era: the large group efforts and the large budgets involved in machine physics. Also, whereas Shelter Island and Pocono looked upon QED (quantum electrodynamics) as a self-contained discipline, the Rochester conferences have witnessed "particle physics" coming into its own with QED as one of its subfields—albeit one in a privileged, paradigmatic position.

The function of the Shelter Island, Pocono, and Oldstone conferences in reasserting the values of pure research should be noted. Conferences—and small ones in particular—have ritual functions. Coming after World War II, the conferences helped to purify and revitalize the american theoretical Physics community. They also asserted the new social reality implied by the

34. Archives, NAS; Oppenheimer to Richards, 15 April 1949.

newly acquired power of theoreticians and helped integrate the most out-standing of the younger theoreticians into the elite—Schwinger, Feynman, Marshak, and Pais at Shelter Island, and Dyson at Oldstone.

The theoretical successes reported at these conferences, as impressive as they were, were undoubtedly further magnified by Oppenheimer, the char-ismatic leader of Los Alamos. Oppenheimer was the dominant figure at these conferences, and it was he who made the assignment of the relative merit of the various contributions reported there. These conferences gave further proof of the intellectual powers of the leading theoreticians. In many ways, they continued the heroic efforts that had been demanded by the war. In fact, most of the theoretical advances were carried out by the same heroic figures who had solved the wartime problems.

2 The Scientific Content of the Conference

By May 1947 Darrow had obtained the "abstracts" that Kramers, Oppen-heimer, and Weisskopf, the discussion leaders, had been asked to write; these 'abstracts' are to be found in an appendix. As these were to form the basis of the discussions at Ram's Head, they were sent to the other partici-pants. Weisskopf's paper outlined the problems in "elementary particle" physics in broad and general terms.

Weisskopf's outline was a succinct statement of what he considered the outstanding problems of physics and an assessment of the methods avail-able for tackling them. It was the work of one of the leading theoreticians who, "returning from the war," had devoted all his energies to solving fundamental problems in pure physics. The frontiers of physics were located in QED, nuclear and meson phenomena, and "high energy particle physics." The theoretical foundations were considered shaky and the mood conveyed was one of frustration: once again, Weisskopf had "knocked his sore head" against the same old wall that he had confronted during the '30s. The proposals that Weisskopf made for the topics of discussion centered on experiments that either had just been performed—the Piccioni et al. and the Bridge-Rossi cosmic ray experiments—or ones that could be performed in the near future, when the accelerators then being built at Berkeley, G. E., Cornell, and Rochester would be running. His emphasis on experiments and what could be learned from them and his advocacy of a much closer cooperation between experiment and theory were undoubtedly a legacy of the Los Alamos experience.

Oppenheimer's outline was more narrowly focused and concentrated on cosmic-ray phenomena. Oppenheimer had returned from the war with his interests changed from physics to statesmanship. But his abilities were such that he could be both an outstanding spokesman for the scientific community and a most impressive theoretical physicist. His presentation reflected his then current interests in physics. His students, S. Wouthuysen and H. Lewis, had just completed some calculations on multiple production of mesons, and with Bethe he had written a paper on infrared divergences to prove that the prescriptions that Heitler had recently given for rendering field theories finite were inadmissible as they disallowed certain necessary processes. All these calculations were based on the Bloch-Nordsieck and Pauli-Fierz methods for handling multiple-particle processes.

Both Weisskopf's and Oppenheimer's outlines reflected their interest— and that of the other theoreticians—in cosmic rays. Since the early '30s experiments exploiting the interaction of cosmic-ray particles with the atoms of the air in their passage through the atmosphere had been an important source of information on "elementary particles." In 1937 cosmic-ray shower experiments and their theoretical interpretation had established the existence of a new (unstable) charged particle, the mesotron, which existed in both positive and negative varieties. Its lifetime was about 10^{-6} seconds. This meson was assumed to be the particle that Yukawa had postulated to account for the nuclear forces and that should therefore interact strongly with nuclei. It was this identification that led to the difficulty that Nordheim had pointed out and that Oppenheimer emphasized in his outline: If—as should be the case—these mesons are copiously produced by cosmic rays in nuclear interactions in the upper atmosphere, why were not more of them absorbed in their traversal through the atmosphere?

These difficulties had been dramatically exhibited in an important and striking experiment carried out by Conversi, Panvini, and Piccioni, the results of which were published early in 1947 in the *Physical Review* (Conversi, 1947, 1983). Their experiment gave the first demonstration that a substantial fraction of the slow negatively charged mesons found at sea level *decayed* in a carbon plate—rather than being absorbed by the carbon nucleus $(Z = 6)$. However, when these mesons were stopped in an iron plate, they were absorbed by the iron nuclei $(Z = 26)$. Conversi el al. also found that all positively charged mesons decayed in both the carbon and the iron plates. Now according to theory (Tomonaga, 1940) positively charged nuclear force mesons should always decay (in agreement with experiment), but negatively charged nuclear-force mesons should never decay (in contradiction with the experiment of Panvini et al.).

Since the mesons are unstable, one must inquire into the time taken by the negatively charged mesons to slow down and be captured into Bohr orbits in an air nucleus (from which nuclear absorption can readily take place) and compare this slowing-down time with the lifetime for decay. If the lifetime for decay is much shorter than the slowing-down time, the mesons will rarely be absorbed by the nucleus and will most often decay. Fermi, Teller, and Weisskopf (Fermi, 1947a, b) in an analysis of the Conversi et al. experiment carried out early in 1947 came to the startling conclusion that "the time of capture from the lowest orbit of carbon is not less than the time of natural decay, that is, about 10^{-6} second." This was in disagreement—by a factor of about 10^{12}—with the estimate of Tomonaga and Araki, which assumed the meson to be responsible for nuclear forces. It was this striking discrepancy that was behind Oppenheimer's and Weisskopf's call for a discussion of the Conversi et al. experiment.

Kramers, for his part, chose to review the difficulties encountered in QED since its inception in 1927 and to indicate one way out of these problems. His outline, which included his own work and that of his students, Serpe and Opechowski, which had been carried out in 1940, presented a theory in which all structure effects had been eliminated. His paper described "how an electron with *experimental mass* behaves in its interaction with the electromagnetic field."

Kramers's essential point was that the infinite shift of spectral lines that resulted from the Hamiltonian describing the interaction of a charged particle with the radiation field that Dirac had written down in 1927 could be "immediately connected with the divergence of the electromagnetic mass for a point electron." In Kramers's treatment the interaction energy, which in Dirac theory was $-(e/c)(\mathbf{v} \cdot \mathbf{A})$, took the form $-(e/c)(\mathbf{v} \cdot \mathbf{Z})$, where $\mathbf{Z}$ is the Hertz vector: $\mathbf{A} = (1/c)(\partial \mathbf{Z}/\partial t)$. And with "the latter interaction term the ultraviolet catastrophy in 2d order perturbation calculation disappears." Kramers's remarks proved to be very important at the conference. Rabi, of course, had known of the progress of Lamb and Retherford's experiment on the fine structure of the 2s–2p levels of hydrogen being carried out at Columbia and had Lamb invited to the conference. That experiment, which first succeeded on 26 April 1947, indicated "that, contrary to [Dirac] theory but in essential agreement with Pasternack's hypothesis, the $2^2S_{1/2}$ state is higher than the $2^2P_{1/2}$ by about 1000 mc/sec" (Lamb, 1947).

By early May rumors of the experimental findings in Rabi's laboratory were spreading. Weisskopf in a mid-May letter to Bethe informed him that he had as yet no news about Rabi's hyperfine structure experiments and inquired from Bethe "whether he had heard rumors that he [Rabi] had measured the spread of the excited level $(2S^{1/2}-2P^{3/3})$ and got a 10% deviation from theory?"[35]

In that letter he also indicated to Bethe that he had written his outline "only because I had no specific idea ready to discuss," adding that "since Kramers, and to some extent Oppenheimer, have something to say, let us give them all the time they need to do so." A few days earlier he had written Bethe that "he would like to hear from Kramers in great detail about his new theory."

By the end of May the preliminary results of Lamb and Retherford were confirmed. Schwinger and Weisskopf discussed the theoretical implication of the experiment on their train ride from Boston to New York to attend the Shelter Island Conference and agreed that the effect was very likely quantum electrodynamic in origin. Schwinger recalls discussing with Weisskopf on the train that since the electron self-energy was logarithmically divergent in hole theory, the $2S_{1/2}-2P_{1/2}$ energy difference would be finite when calculated with that theory. In fact, the matter may have been discussed even earlier over the lunches Schwinger and Weisskopf took together periodically, since this hydrogenic level shift was then one of Weisskopf's research interests. *In the fall of 1946,* Weisskopf had given the problem of a hole-theoretic computation of the 2S–2P level shift in hydrogen to Bruce French. Since that time, French had been working on the problem. It is likely that Weisskopf would have told Schwinger about it, would have elicited his reaction to it, and would have sought his advice on effective computational approaches.

This is how matters stood before the opening of the conference. It was clear that Lamb and Retherford's experiment on the fine structure of hydrogen and that of Conversi et al. on the absorption of sea-level mesons would have an important place in the presentations at the conference. In fact, it had been arranged that Rossi would give a general report on recent cosmic-ray experiments and also that Rabi would report on the other

35. Cornell University Library, Bethe Archives 14/22/76; Weisskopf to Bethe, 19 May 1947.

atomic beam experiments being carried out at Columbia investigating the hyperfine structure of hydrogen, deuterium, and more complex atoms.

The conference opened on Monday morning, 2 June. Unfortunately there exists no record of the various sessions, and the recollections of the participants are vague and hazy. What is known is that on the first day Kramers gave his report and Lamb and Rabi gave detailed presentations of the experiments that were being performed at Columbia.[36] Lamb reported the results of his experiments carried out with Retherford on the fine structure of hydrogen; Rabi told of Nafe and Nelson's and Kush and Foley's experiments on the hyperfine structure of hydrogen, deuterium and more complex atoms, whose findings suggested that a discrepancy might exist between theory and experiment concerning the electron's magnetic moment. According to the article that appeared in the *New York Herald Tribune*, "the subsequent discussion was led by Oppenheimer and Weisskopf. The meeting discussed recent work on the simplest of all atoms, the hydrogen atom...."

The result of the Lamb-Retherford experiment became one of the central and dominant concerns of the conference. Suggestions were made by Weisskopf, Schwinger, and Oppenheimer on how a calculation of this level shift might be attempted in hole theory and why a finite answer might result from the application of Kramers's ideas.

Rossi reported on the results of recent cosmic-ray experiments. The discussion of the Conversi et al. experiment was "animated" (Marshak, 1983, p. 381).[37] Oppenheimer suggested that one ought to consider giving up microscopic reversibility. Weisskopf's suggestion for overcoming the

---

36. Darrow has the following entry in his diary for Monday, 2 June 1947. "Lamb and Rabi gave long papers on the terms of H and D, each embodying experiments of their own: these held to the end of the conf'ce their rank as outstanding contrib.ⁿˢ" (AIP Archives; K. K. Darrow papers). On 17 June 1947 MacInnes recorded in his diary, "After a night's sleep and a good breakfast the conference got down to work, and went off without any delay or explanation after a halting speech from me. It was immediately evident that Oppenheimer was the moving spirit of the affair and the preliminary agenda were hardly followed at all. The subjects seemed to be all of the difficulties of modern physics. The main discussions were by Kremers, Fineman, Lamb, and by Oppenheimer. One evening they continued until eleven thirty. It was all pretty much above my head by I got a bit here and there that was valuable, at least enough to justify my doubts about the quantum theory in its present form" (MacInnes diary, 17 June 1947, p. 129).

37. K. K. Darrow records that on Wednesday, 4 June 1947, "at the morning session, Oppenheimer spoke with his usual quiet grace and Feynman with a clear voice, great rush of words and illustrative gestures sometimes ebullient" (AIP Archives, K. K. Darrow papers).

apparent lack of reversibility between production and absorption was a mechanism whereby a primary cosmic-ray particle when interacting with a nucleus converted one of its nucleons into an "excited nucleon," capable subsequently of emitting mesons. The lifetime of the "pregnant" nucleon could be chosen sufficiently long to account for the subsequent weak interaction between mesons and nucleons. Marshak proposed his famous two-meson hypothesis: There exist in nature two kinds of mesons, possessing different masses. The heavier Yukawa meson (subsequently known as the π meson) is responsible for the nuclear forces and is the one produced with a large cross section in the upper atmosphere. The lighter meson (the μ meson) is a decay product of the heavier one and is the meson normally observed at sea level to interact weakly with matter.

The history of the two-meson hypotheses, including an earlier Japanese version, has been told by Marshak (1952, 1983), and I shall not repeat it here.* I shall only point out that Marshak's hypothesis was put forth before the announcement by Blackett and his collaborators of the discovery of the π meson. It was the discussion of the long lifetime, 2×10^{-6} second, for meson absorption from the carbon K shell, which had been estimated by Fermi, Teller, and Weisskopf in their calculation based on the results of the Conversi experiment, that spawned the two-meson hypothesis at Shelter Island.

As is well known, on the train ride from New York to Schenectady after the Shelter Island Conference, Bethe made a nonrelativistic calculation of the 2S–2P level shift using the suggestions of Kramers, Weisskopf, Schwinger, and Oppenheimer and discovered that with suitable cutoffs such a calculation could account for a major part (1,040 megacycles) of the observed level shift. This had not at all been obvious at Shelter Island. *Bethe's calculation was crucial and confirmed the feeling expressed at Shelter Island that the effect was a quantum electrodynamical one.* Incidentally, until he received Bethe's paper in mid-June, Lamb believed that the effect was of nonelectromagnetic origin.

Bethe's paper (1947) was completed by 9 June, and he circulated it to the participants of Shelter Island. His accompanying letter to Oppenheimer was brief and to the point:[38]

---

* Editors' note: See also the paper in this volume by Marshak.
38. Library of Congress, National Archives, Oppenheimer Papers; Bethe to Oppenheimer.

June 9, 1947

Dr. J. Robert Oppenheimer
Physics Department
University of California
Berkeley, California

Dear Robert:

Enclosed I am sending you a preliminary draft of a paper on the line shift. You see it does work out. Also, the second term already gives a finite result and is not zero as we thought during the conference. In fact, its logarithmic divergence makes the order of magnitude correct. It also seems that Vicky and Schwinger are correct that the hole theory is probably [handwritten insertion] important in order to obtain convergence. Finally, I think it shows that Kramers cannot get the right result by his method.

It was good seeing so much of you during the conference. This should be repeated.

Sincerely yours,
Hans [handwritten]
H. A. Bethe

Bethe's remarks concerning Kramers are worth noting. Although Kramers's work was clearly essential—it indicated the importance of expressing observables in terms of the *experimental* mass of the electron— Bethe's letter suggests that Kramers's presentation at the conference probably stressed the classical features of his approach and that the extension of his approach to the quantum electrodynamic situation was not fully appreciated. Bethe's paper (1947) indicates that the comments by Weisskopf and Schwinger and by Oppenheimer—that a hole-theoretic calculation of energy level *differences* would be finite[39]—had a great and immediate impact on him. Bethe was interested in calculating the level shift *quantum electrodynamically*, and the insight of Oppenheimer, Schwinger, and Weisskopf gave him the justification for cutting off the divergent integrals he encountered in his nonrelativistic approach.

39. It should be recalled that Oppenheimer in 1930, using a *pre* hole theoretic formulation of quantum electrodynamics, had investigated whether energy level *difference* might be finite. The result was negative because of the linear divergence of the electron self-energy in that theory. Lamb and Kroll's calculation of the Lamb shift in 1947/48 is in fact an updated hole-theoretic version of Oppenheimer's paper.

To appreciate the importance of Bethe's calculation, let me indicate Weisskopf's reaction to it. Weisskopf received Bethe's manuscript on 11 June and after studying it wrote Bethe that he was "quite enthusiastic about the result. It is a very nice way to estimate the effect and it is most encouraging that it comes out just right. I am very pleased to see that Schwinger's and my approach seems to be the right one after all. Your way of calculating is just an unrelativistic estimate of our effect, as far as I can see. *I am all the more pleased about the result since I tried myself unsuccessfully to estimate the order of magnitude of our expression. I was unable to do this* [italics mine] but I got more and more convinced that the method was sound. . . . I would like to talk it over with you especially the 'korrespondenz Deutung' of the effect."[40] He added, "I do not quite agree with your treatment of the history of the problem in your note. That the $2S_{1/2}-2P_{1/2}$ split has something to do with radiation theory and hole theory was proposed by Schwinger and myself for quite some time. We did not do too much about it until shortly before the conference. We then proposed to split an infinite mass term from other terms and get a finite term shift, just as I demonstrated it at the conference. Isn't that exactly what you are doing? Your great and everlasting deed is your bright idea to treat this at first unrelativistically. 'Es möchte doch schön sein'[41] if this were indicated in some footnote or otherwise."

A few days later, Bethe answered him in his quiet and thoughtful way, and Weisskopf thereafter responded in a letter acknowledging that Bethe's "abstract" was "harmlöser" (much more harmless) than he initially thought and agreed with Bethe to "Let's forget about patent claims."

Bethe's work made a relativistic calculation of the Lamb shift the next desideratum. French and Weisskopf at MIT continued their calculation, but now made use of Kramers's ideas and thus greatly simplified the subtraction of infinites. Lamb at Columbia started on a hole-theoretic level shift calculation during the summer of 1947. He was soon joined by Kroll, then a graduate student at Columbia. Schwinger, who had married right after Shelter Island and who for nearly two months traveled throughout the United States on his honeymoon, started on such a calculation in late July. All of them used the noncovariant calculational methods developed in the

Discussing physics informally (left to right): standing—W. Lamb, Jr., J. Wheeler; seated—A. Pais, R. Feynman, H. Feshbach, J. Schwinger.

'30s that were to be found in Heitler's *Quantum Theory of Radiation*, whose second edition had appeared in 1944. Bethe, on the other hand, felt that what was needed was a relativistically invariant way of computing quantum electrodynamic effects. On his return to Cornell after Shelter Island he got Feynman interested in this problem. Feynman, making use of his previous work with Wheeler on electrodynamics as an action-at-a-distance theory and of his Lagrangian formulation of quantum mechanics, worked on a formulation of quantum electrodynamics with a relativistic cutoff (Feynman 1948a, b).

3 Later Developments

Lamb and Retherford's experiment on the fine structure of hydrogen performed at Columbia during the fall of 1946 and the spring of 1947 was undoubtedly the major stimulus for the renewed interest by theoretical physicists in quantum electrodynamical problems. As Oppenheimer (1948) indicated in his report to the Solvay Conference of 1948, in their application to level shifts, "these developments [mass and charge renormalization] which could have been carried out at any time during the last fifteen years, required the impetus of experiments to stimulate and verify."

It was the wartime developments of microwave technology that made possible the high precision experiments of Lamb, Foley, Kellogg, Kusch, and others in Rabi's laboratory at Columbia and those of Dicke at Princeton on the spectrum of hydrogen and the g factor of the electron, as well as the search for positronium.

It required the impetus of experiments yielding accurate values for the electromagnetic properties of free and bound electrons to place quantum electrodynamics once again at the center of interest of the theoretical community. Since the mid-'30s, nuclear and high energy phenomena were considered the frontier and exciting fields. The Columbia experiments indicated that much could yet be learned from quantum electrodynamics.

Incidentally, the publication during the war of Wentzel's book on quantum field theory (Wentzel, 1943) and its availability in the United States after the war (it was reprinted by Edwards Bros. in Ann Arbor in 1946) also proved helpful. This book detailed a coherent and unified presentation of relativistic quantum field theories incorporating most of the advances in the field up to 1942. Thus the Shelter Island Conference brought together members of the experimental and theoretical communities after each had either assimilated important new advances or was ready to do so. How rapidly these advances were amalgamated can be gauged from the fact that already in his first paper—the short note on the calculation of the $\alpha/(2\pi)$ quantum electrodynamic correction to the electron's magnetic moment submitted to the *Physical Review* in December 1947—Schwinger (1948) was quite explicit as to the meaning of the renormalization procedure:

Electrodynamics unquestionably requires revision at ultra-relativistic energies, but it is presumably accurate at moderate relativistic energies. It would be desirable, therefore to isolate those aspects of the current theory that involve high energies, and are subject to modification by a more satisfactory theory, from aspects that involve only moderate energies and are thus relatively trustworthy.

This goal has been achieved by transforming the Hamiltonian of current hole theory electrodynamics to exhibit explicitly the logarithmically divergent term proportional to the interaction energy of the electron in an external field [polarization of the vacuum]....

The interaction between matter and radiation produces a renormalization of the electron charge and mass, all divergences being contained in the renormalization factors.

This is essentially also the view propounded by Feynman in the spring of 1948:[42]

The philosophy behind these ideas [mass and charge renormalization] might be something like this: A future electrodynamics may show that at very high energy our theory is wrong. In fact we might expect it to be wrong because undoubtedly high energy gamma rays may be able to produce mesons in pairs, etc., phenomena with which we do not deal in the present formulation of the electron-positron electrodynamics. If the electrodynamics is altered at very short distances then the problem is how accurately can we compute things at relatively long distances. The result would seem to be this; the only thing which might depend sensitively on the modification at short distances is the mass and the charge. But that all observable processes will be relatively insensitive and we are now in a position to be able to compute these real processes fairly accurately without worrying about the modifications at high frequencies. Of course it is an experimental problem yet to determine to what extent the calculations we are now able to make are in agreement with experience.

In other words, it seems as though with these methods of mass and charge renormalization we have a consistent and definite electrodynamics for the calculation of all possible processes involving photons, electrons, and positrons.

By January 1948, when the American Physical Society met in New York, it was apparent to Weisskopf, Lamb, and Schwinger that it was imperative to use relativistically invariant and gauge invariant calculational techniques. To obtain finite answers, all of them had made use of the ideas of Kramers on mass renormalization and the earlier ones of Heisenberg and Weisskopf on charge renormalization. But these finite answers were plagued with inconsistencies. The ambiguities that Schwinger encountered in calculating the anomalous magnetic moment of a bound electron and the Lamb shift convinced him that a new approach was needed.

By April 1948, when the Pocono Conference was held, both Schwinger and Feynman had calculational schemes that were relativistically invariant. Schwinger, going back to the early work of Dirac, Fock, and Podolski (Dirac, 1932) on the interaction picture for QED (which incidentally was

42. CIT Library, Feynman Archives.

outlined in Wentzel's book), proceeded to work out a systematic approach based on a series of canonical transformations designed to make descriptions of the vacuum and of the one-particle states simple. The vacuum was to be the relativistically invariant state of zero charge, energy, and momentum; the one-electron states had to describe correctly a particle of experimental charge e_{exp} and experimental mass m_{exp} whose energy and momentum were related by $E_p = c\sqrt{\mathbf{p}^2 + m^2_{exp}c^2}$. Feynman's computational scheme, on the other hand, was much more intuitive. It was based on his integral-over-path quantization procedure, applied to a *cutoff* version of Wheeler-Feynman theory, which had been altered, "mapped," and cataloged to give results in a perturbative expansion identical to those found in the perturbative expansion of hole theory.

One root of Feynman's diagrammatic approach is to be found in this cataloging procedure. Feynman's hurried presentation at Pocono, coming on the heels of Schwinger's day-long presentation, was met with skepticism and even some hostility. Bohr, in particular, objected to his viewing positrons as electrons going backward in time and to his use of diagrams in general. He insisted that the uncertainty principle precluded the assignment of trajectories to particles. He also objected to Feynman's belief "that it is possible to get a consistent theory without using loops," i.e., without vacuum polarization terms. Bohr's remarks on Schwinger and Feynman were abstracted as follows in the unpublished notes of the Pocono Conference taken by Wheeler and Wigner:

It was a mistake in the older days to be discontented with field and charge fluctuations. They are necessary for the physical interpretation. Schwinger's treatment is an advance in bringing the theory into a regular order where effects can be interpreted. As regards the infinities he had other views than those just discussed. He objected to Feynman's view of the electron going backward in time. It is also unreasonable to cut things out.

No sound argument yet exists as to where to cut off electron theory, if at all. It may however be an extravagant wish to hope to have electron theory stand alone.

Despondent, though not despairing, Feynman realized that the only way that he would be able to convince his audience was to publish a detailed presentation of his theory. After Pocono he embarked on writing his classic papers, "The Theory of Positrons" and "Space-Time Approach to Quantum Electrodynamics (Feynman, 1949a,b).

Right after the Pocono Conference, Oppenheimer received a letter from

Sinitiro Tomonaga informing him of the parallel work on quantum electrodynamics that was being carried out in Japan. Tomonaga's influential article on a relativistically invariant formulation of quantum field theory—an article he had written in 1943 and published in Japanese during the war—had been translated into English and had appeared in the first issue of the *Progress of Theoretical Physics* in the winter of 1946. In fact, Oppenheimer called it to Schwinger's attention after the latter's talk at the Physical Society meeting in New York in January 1948. But no one had been aware of the extensive research program in QED that was being carried out in Japan since the results of the Lamb experiment had been learned from a *Newsweek* article in the summer of 1947.

The history of the Shelter Island, Pocono, and Oldstone conferences chronicles the development of QED from 1947 to 1950 and thus also tells the story of how Schwinger, Feynman, and Dyson worked out their respective formulations.

Shelter Island indicated the problems and possible solutions, primarily by virtue of the focus that Kramers's insight had given. At Pocono, Schwinger and his methods were at the center of the stage. Oldstone was Feynman's show—and also Dyson's. By then the great power of Feynman's calculational schemes had become patently clear. By April 1949 Karplus and Kroll were well on their way to completing their calculation of the fourth-order α^2 corrections to the magnetic moment of the electron using Feynman-Dyson methods. The deeper differences between Schwinger and Feynman were also apparent at Oldstone. As the unpublished notes from the conference indicate, Dyson, in his report stressed that " ... Feynman [has] more interest in solutions than in equations. Most though not all problems calculable as scattering problem.... Largely from Feynman one has systematic technique for writing down solutions of scatter problem." By the time of the Oldstone Conference it had also become clear that QED in its renormalized version was probably able to account for all the then available data on the properties of an electron—its anomalous moment and its energy level sprectrum in H and He$^+$. In fact, by Oldstone the interest of the theoretical community had already shifted to "π meson physics."

The initial accomplishment of Schwinger in the 1947–1949 period was primarily the formulation of a somewhat unwieldy but coherent and systematic apparatus for doing relativistic field theoretical calculations and the proof that these methods could be successfully applied to interesting problems (e.g., the computation of the electron g factor and the radiative corrections to Coulomb scattering). Feynman provided deep new insights by *visualizing* the processes in a manner that translated these intuitive representations into simple, *extremely* efficient and effective calculational methods for computing observable quantities. For the first time one could conceive of doing higher order calculation routinely. But it could be argued that after Kramers's suggestion for mass renormalization and the earlier one of Dirac, Heisenberg, and Weisskopf for charge renormalization had been adopted, both Feynman's and Schwinger's formulations, although technically much more manageable and much superior, were not major advances in terms of *physical content* over the quantum field theory of the '30s. In fact the first *correct* calculation of the Lamb shift by French and Weisskopf and by Kroll and Lamb still used the computational apparatus of the '30s. The question to be answered, therefore, is why the post-World War II generation, unlike the generation of the '30s, was successful in bypassing the obstacles posed by quantum field theory. For an answer we must look to the broader context in which these developments took place. The workers of the '30s—Heisenberg, Bohr, Rosenfeld, Pauli, Dirac—all sought solutions of the problems in terms of *revolutionary* departures (e.g., nonpositive metrics in Hilbert space and new fundamental lengths). Special relativity and quantum mechanics had been advanced by such revolutionary steps, and in the '30s Bohr constantly advocated and encouraged such attitudes for the solution of the field-theoretic problems. The solution advanced by Feynman, Schwinger, and Dyson was in essence conservative; it asked to take seriously the received formulation of quantum mechanics and special relativity and to explore the content of this synthesis. A generational conflict manifested itself in this contrast between the revolutionary and conservative stances of the pre- and post-World War II generations (Dyson, 1965).

It was Dyson who provided a major contribution to the understanding of the *structure* of field theories by his formulation of the concept of renormalizability and his analysis of the higher order contributions and of the structure of quantum field theories to *all orders* of perturbation theory. In this connection it should be remembered that by mid-1948 neither Feynman nor Schwinger had considered the problems connected with either the sufficiency or the consistency of the renormalization procedure in higher orders of perturbation theory. Their primary interest had been in obtaining predictions from QED that could be compared with experiments (e.g., the calculation of the magnetic moment of an electron, the radiative corrections to scattering processes, and the level shifts in H and D). It was Dyson's contribution to indicate how Feynman's visual insights could be translated into answering the question whether charge and mass renormalization were sufficient to remove all the divergences in QED to all order of perturbation theory and what renormalizability implied for other field theories. Feynman diagrams (and their analogs in many-body theory), in addition to their intuitive appeal, can also be viewed as a representation of the logical content of field theories (as stated in their perturbative expansions). It is this aspect of diagrammatics that enabled Dyson to gain deep insights into the *structure* of quantum field theories and thus to make his important contributions to the concepts of renormalization. Moreover, it is precisely the recovery of the possibility of "visualizing [atomic] physical phenomena in terms of the concepts of daily life"[43] that made Feynman's contribution so telling. Starting with his doctoral dissertation in 1940, but particularly after 1946, when he started teaching, Feynman's revealing reinterpretation of the formal and abstract rules of quantum mechanics—which allowed transition amplitudes to be interpreted as being made up of contributions from alternative paths that a particle could take in space-time—and his insights into the nature of the probability calculus relevant to quantum mechanics resulted in a totally novel visual way of describing atomic phenomena. Since the ability to visualize is for most people the most important way of understanding the world, Feynman made it possible to conceptualize the unseen.

Advances in quantum field theory (and in QED in particular) have always been marked by an amalgamation of perturbation-theoretic with nonperturbation-theoretic insights. The ideas of mass renormalization in quantum field theory can be traced back to Pauli and Fierz's seminal 1938 paper in which solutions of a Hamiltonian describing a nonrelativistic extended charged particle interacting with the quantized electromagnetic field were obtained. Their work in turn was based on the nonperturbative

---

43. Heisenberg (1938, p. 99): "The passage from the prerelativistic to the relativistic theory or from the prequantum to the quantum theory was not so much a correction of the older theories as the recognition that upon ascribing finite values to c and h, the possibility of visualizing physical phenomena in terms of the concepts of daily life must in part be abandoned. Thus in the relativity theory the finite velocity of light precludes the introduction of an absolute time independent of the state of motion of the observer, while in the quantum theory the finite reaction of the observer upon the objects observed makes illusory every attempt to measure the position and velocity of a particle simultaneously with any desired accuracy."

insights of Bloch and Nordsieck's analysis of the infrared divergences. The successes in the period from 1945 to about 1949—the papers of Dyson (1949a,b), Salam (1950, 1951), and Ward (1950a,b) mark the end of the period—were based on the perturbative expansion of the solutions of quantum field theories. Renormalizability was one such inference. By 1949, the emphasis had turned to nonperturbative covariant formulations—the work of Salpeter and Bethe (1951), Nambu (1950), Källén (1952), Gell-Mann (1951), and Schwinger (1951) mark the beginning of this period— partly because bound state problems were attracting attention (e.g., the relativistic calculations of the Lamb shift and hyperfine structure of hydrogen and the structure of positronium) but also because by then π meson physics was becoming the center of interest and it was clear that meson-nucleon interactions were much "stronger" than electrodynamic ones.

One consequence of the emphasis on relativistic covariance is worthy of comment. One of the legacies of the 1947–1952 period emphasizing covariant methods was to impose too rigid an interpretation of relativistic invariance. The stress on relativistically invariant computational methods led to a tacit assumption that the Heisenberg state vector describing the vacuum state of a relativistic quantum field theory was always a *nondegenerate* vector and that this vector, which corresponded to the ground state of the system, always posessed the full invariance properties of the Lagrangian. Because of this, the advances brought about by the insights regarding symmetry breaking represented a shift in outlook that can only be compared to the "breaking of the circle" by Kepler's ellipses.

Since the '50s many of the important advances in quantum field theory have been the results of nonperturbative methods. Oftentimes, these were first developed in many-body theory (e.g., superconductivity and ferromagnetism).

In retrospect, perhaps the most important lesson that could have been gleaned from the period 1945–1952 was to appreciate how rich and malleable a structure relativistic quantum field theory was and how easily it could accommodate the dramatic advances of that period (and subsequent ones!).

Acknowledgments

I have received valuable source materials, recollections, criticisms and suggestions from H. A. Bethe, R. H. Dicke, F. Dyson, R. P. Feynman, R.

Finkelstein, M. Fierz, R. Jost, P. Galison, M. Goldberger, W. E. Lamb, F. Manuel, C. Morette-de Witt, A. Pais, H. Pendleton, J. Schwinger, R. Serber, E. P. Wigner, A. S. Wightman, and J. A. Wheeler. I thank them.

I would also like to acknowledge the help received from numerous librarians: in particular, Carolyn Kopp, the archivist at Rockefeller University; Judith Goodstein, the archivist at the Caltech Millikan Library; Joan Warnow at the AIP; as well as those at Brandeis, the National Academy of Sciences, the Neils Bohr Library at the AIP, Churchill College, Cambridge, and Cornell University. I appreciate obtaining their permission to quote from materials in their archives.

This research was supported in part by a grant from the National Science Foundation.

Appendix

Foundations of Quantum Mechanics
Outline of Topics for Discussion
Victor F. Weisskopf

The theory of elementary particles has reached an impasse. Certain well known attempts have been made in the last fifteen years to overcome a series of fundamental problems. All these attempts seem to have failed at an early stage. An agenda for a conference on these matters contains, necessarily, a list of these attempts. After returning from war work, most of us went through just these attempts and tried to analyze the reason of failure. Therefore, the list which follows will be well known to everyone and will probably invoke a feeling of knocking a sore head against the same old wall. The success of the conference can be measured by the extent it deviates from this agenda.

It may perhaps be useful to divide the discussion into three parts:

A. The difficulties of quantum electrodynamics.
B. The difficulties of nuclear and meson phenomena.
C. The planning of experiments with high energy particles.

These topics can be subdivided:

A. Quantum electrodynamics
    1) Self Energies. Attempts to remove infinite self energies.
        a) Classical attempts: theories using advanced minus retarded potentials. Non linear theories.
        b) Attempts which change formalism after quantization: λ-process,

Dirac negative light quanta, Heitler-Peng, Riess-method. Why do logarithmic divergencies defy most of these methods?

2) How reliable are the "finite" results of quantum electrodynamics derived by means of a subtraction formalism? Polarization of the vacuum and related effects. Is there a high energy limit to quantum electrodynamics?

3) Infra-red catastrophe and related topics.

B. Nuclear forces and mesons.
1) Present knowledge of nuclear forces, range, type, form, singularities, saturation.
2) Present knowledge of meson properties.
3) Resume of meson theories and their representation of nuclear forces.
4) Discrepancies in the interaction of mesons with matter (Piccioni experiment). Shower formation by primary cosmic radiation (Bridge-Rossi experiment).
5) Discussion of new cosmic ray experiments.
6) Problems connected with the theory of β-decay. Inconsistencies. Connection with meson decay.

C. Proposed experiments
Discussion of fundamental experiments to be done with the high energy machines.
1) Electron accelerators.
2) Proton accelerators.
3) The relative importance of different energy regions and the planning of new machines.

In view of the failure of the present theories to represent the facts, and the small probability that this conference may produce a new theoretical idea, Part C of this agenda could become the most useful part of the conference. A number of very powerful accelerators in the energy region of 200–300 MeV are near completion, and it is time to inaugurate a symmetric program of research, with experiments which can be reliably interpreted. All too often much more thought is given to the production of the beam and too little to a reasonable instrumentation.

The Foundations of Quantum Mechanics
Outline of Topics for Discussion
J. R. Oppenheimer

It was long ago pointed out by Nordheim that there is an apparent difficulty in reconciling on the basis of usual quantum mechanical formalism the high rate of production of mesons in the upper atmosphere with the small interactions which these mesons subsequently manifest in traversing matter. To date no completely satisfactory understanding of this discrepancy exists, nor is it clear to what extent it indicates a breakdown in the customary formalism of quantum mechanics. It would appear profitable to discuss this and related questions in some detail.

We might start this discussion by an outline of the current status of theories of multiple production. Some illuminating suggestions about these phenomena can be worked out in a semi-quantitative way, for instance on the basis of the neutral pseudo-scalar theory of meson couplings. The suggested results appear to agree reasonably well as to energy dependence of multiplication, energy and angle distribution with the experimental evidence, which is admittedly sketchy. However, no reasonable formulation of theories along this line will satisfactorily account for the smallness of the subsequent interaction of mesons with nuclear matter. Similar difficulties appear when one attempts to make a theory involving couplings of meson pairs to nuclear matter. There are two reasons for these apparent difficulties. One is that in all current theory there is a formal correspondence between the creation of a particle and the absorption of an anti-particle. The other is that multiple processes are in these theories attributable to the higher order effects of coupling terms which are of quite low order, first or second, in the meson wave fields. The question that we should attempt to answer is whether, perhaps along the lines of an S matrix formulation, both these conditions must be abandoned to accord with the experimental facts.

It would be desirable to review the experimental situation with an eye to seeing how unambiguous current interpretations are.

The calculation of the multiple production of mesons is in some ways an extension of the treatment given by Block and Nordsieck of the radiation of electrons during scattering. The difficulties of a complete description of these phenomena appear in exaggerated form in the problem of meson production. It would therefore be profitable to review the present status of the theory of radiation reaction and of certain recent suggestions for improving the theory.

The Foundations of Quantum Mechanics
Outline of Topics for Discussion
H. A. Kramers

At almost every important stage of the development of quantum mechanics, not only were new positive results added to what had already been achieved, but also certain "defects" revealed themselves. In some cases such defects were removed by the next step; in some cases they just stayed or even were emphasized more strongly.

When in 1926 the field of free Maxwell radiation was quantized, the infinite zero-point energy revealed itself. We may consider this as a defect which even now has by no means been removed in a satisfactory way.

When in 1927 Dirac gave his non-relativistic quantum theory of the interaction between radiation field and charged particles, the emission, absorption and scattering of photons were, on the whole, described in a satisfactory way. Several defects could be noticed, however, pertaining mainly to the divergencies involved in many perturbation-calculations of the second order. We mention the infinite shift of spectral lines, the impossibility of describing a steady state of an atomic system in the radiation field, and—connected therewith—the impossibility of arriving at an exact dispersion formula in which the phase of the scattered wave was duly accounted for, and finally the difficulty, if not impossibility, of describing the reaction of the radiation on the atomic particles.

The year 1928 brought the Dirac theory of the spin electron and its incorporation in a tentative relativistic quantum mechanics of electrons in the radiation field. Second order divergencies stayed, the negative energy-states gave new troubles, but the relativistic energy levels of the H-atom and the Klein-Nishina formula were significant achievements. They were both obtained, it should be remembered, without basing oneself explicitly on a quantization of the radiation field.

The hole theory of 1930, 1931 brought new significant results. Those which could immediately be connected up with experiments: existence of the positron, materialization and annihilation of electrons, did not—as far as I can see—require explicitly the introduction of the concept of the infinite sea of negative-energy states. Calculations which needed this concept led on the one hand to the not yet observed interaction of light with light; on the other hand they led, as regards the self energy, to results which threw a new light on the divergencies resulting from second and higher order perturbations (logarithmnic divergency discussed by Weisskopf and by Pais).

The meson theory of nuclear forces showed encouraging features, but also brought new divergence sorrows.

With regard to this general situation two points seem to me worth closer attention. The first refers to the fact that the non-relativistic Dirac theory of 1927 showed, in its fundaments, by no means the necessary correspondence with classical electron theory. If we ask for a system of formulae which, by means of a variational principle, describes how an electron with experimental mass behaves in its interaction with the electromagnetic field, we get a Lagrangian different from that on which Dirac's 1927 theory is based. The latter gives, even in purely classical problems, often divergent results where the former automatically leads to convergent results. As an example, it can be shown that the infinite shift of spectral lines, with the Dirac Lagrangian, is immediately connected with the divergence of the electromagnetic mass for a point-electron (Serpe 1941). As a second example we mention that the main interaction term in the Hamiltonian expressing the coupling of field and electron, which is Dirac's theory, took the form $-(e/c)(\bar{V}A)$, in the correspondence-theory referred to above takes the form $-(e/c)(\bar{V}\bar{Z})$, where Z is the Hertz potential of the electromagnetic radiation field

$(\bar{A} = (1/c)\ (\partial\bar{Z}/\partial t)$. With the latter interaction term the ultraviolet catastrophy in 2nd order perturbation calculations disappears.

The second point refers to an—in my opinion—unsatisfactory feature which is already shown by the relativity-treatment of *free* particles. It consists in this, that with an arbitrary initial choice of the symbolic wave functions, which are supposed to describe the presence of such particles, the knowledge of those initial wave functions in a finite domain of space does not allow precise prediction of the probable existence of those particles within that finite domain, as was the case in non-relativistic quantum theory. This situation appears unsatisfactory with respect to the exigencies of the general idea of "Nahewirkung" in physics.

It is proposed, if there is time, to go into more specific details as regards the first of the points mentioned.

Bibliography

Bethe, H. A.

1947 "The Electromagnetic Shift of Energy Levels," *Phys. Rev.* 72, 339–341.

Broglie, Maurice de

1912 with P. Langevin. *La Théorie du Rayonnement et les Quanta*. Rapports et Discussions de la Reunion venues a Bruxelles du 30 Octobre au 3 Novembre 1911. Sous les auspices de M. E. Solvay. Paris: Gauthier-Villars.

1951 *Les Premiers Congrès de Physique Solvay et l'Orientation de la Physique depuis 1911.* Paris: Editions Albin Michel.

Cochrane, R. C.

1978 *The National Academy of Sciences. The First Hundred Years: 1863–1963.* Washington DC: National Academy of Sciences.

Conversi, M.

1947 with E. Pancini and O. Piccioni. "On the Disintegration of Negative Mesons," *Phys. Rev.* 71, 209–210.

1983 "The Period That Led to the 1946 Discovery of the Leptonic Nature of the Mesotron," in *The Birth of Particle Physics*, L. N. Brown and L. Hoddeson, editors. Cambridge: Cambridge University Press.

Dirac, P. A. M.

1932 with V. A. Fock and B. Podolsky. "On Quantum Electrodynamics," *Phys. Zeits. der Sowjet Union* 3, 64–72.

Dyson, F. J.

1949a "The Radiation Theories of Tomonaga, Schwinger and Feynman," *Phys. Rev.* 75, 486–502.

1949b "The S-Matrix in Quantum Electrodynamics," *Phys. Rev.* 75, 1736–1755.

1965 "Tomonaga, Schwinger, and Feynman Awarded Nobel Prize for Physics," *Science* 150, 588–589.

Fermi, E.

1947a with E. Teller and V. Weisskopf. "The Decay of Negative Mesotrons in Matter," *Phys. Rev.* 71, 314–315.

1947b with E. Teller. "The Capture of Negative Mesotrons in Matter," *Phys. Rev.* 72, 399–408.

Feynman, R. P.

1948a "A Relativistic Cut-Off for Classical Electrodynamics," *Phys. Rev.* 74, 939–946.

1948b "Relativistic Cut-Off for Quantum Electrodynamics," *Phys. Rev.* 74, 1430–1438.

1949a "The Theory of Positrons," *Phys. Rev.* 76, 749–759.

1949b "Space-Time Approach to Quantum Electrodynamics," *Phys. Rev.* 76, 769–789.

French, B.

1949 with V. Weisskopf. "The Electromagnetic Shift of Energy Levels," *Phys. Rev.* 75, 1240–1248; reprinted in Schwinger (1958).

Gell-Mann, M.

1951 with F. Low. "Bound States in Quantum Field Theory," *Phys. Rev.* 84, 350–354.

Heisenberg, W.

1938 Report to the Congress of the Institute of Intellectual Cooperation, Warsaw, 1938.

Heitler, W.

1944 *The Quantum Theory of Radiation*, 2nd edition. Oxford: The Clarendon Press.

Källén, G.

1952 "On the Definition of the Renormalization Constants in Quantum Electrodynamics," *Helv. Phys. Acta* 25, 417–434.

Karplus, R.

1949 with N. Kroll. "Fourth-Order Corrections in Quantum Electrodynamics and the Magnetic Moment of the Electron," *Phys. Rev.* 76, 846(L), *Phys. Rev.* 77, 536–549.

Kroll, N.

1948 with W. E. Lamb, Jr. "On the Self-Energy of a Bound Electron," *Phys. Rev.* 75, 388–398.

Lamb, W. E.

1947 with R. C. Retherford. "Fine Structure of the Hydrogen Atom by Microwave Method," *Phys. Rev.* 72, 241–243; reprinted in Schwinger (1958).

Marshak, R.

1947 with H. A. Bethe. "On the Two-Meson Hypothesis," *Phys. Rev.* 72, 506–509.

1952 *Meson Physics.* New York: McGraw-Hill.

1970 "The Rochester Conferences," *Bulletin of Atomic Scientists* 26, 92.

1983 "Particle Physics in Rapid Transition: 1947–1952," in *The Birth of Particle Physics*, L. M. Brown and L. Hoddeson, editors. Cambridge: Cambridge University Press.

Nambu, Y.

1950 "Force Potential in Quantum Field Theory," *Prog. Theo. Phys.* 5, 614–633.

Oppenheimer, J. R.

1930 "Note on the Theory of the Interaction of Field and Matter," *Phys. Rev.* 35, 461–477.

1948 "Electron Theory," in *Rapport du 8ᵉ Conseil de Physique Solvay, Sept. 27–Oct. 2, 1948.* Bruxelles: R. Stoop, 1950; reprinted in Schwinger (1958).

Pauli, W.

1937 with M. Fierz. "Zur Theorie der Emission langwelliger Licht Quanten," *Nuovo Cimento* 15, 167–188.

Penick, J. L.

1972 with C. W. Pursell, M. B. Sherwood, and D. C. Swain, editors. *The Politics of American Science: 1939 to the Present.* Cambridge, MA: MIT Press.

Rojanski, V. B.

1938 *Introductory Quantum Mechanics.* New York: Prentice-Hall.

Salam, A.

1950 "Differential Identities in Three-Field Renormalization Problem," *Phys. Rev.* 79, 910–911.

1951 "Overlapping Divergence and the S-Matrix," *Phys. Rev.* 82, 217–227.

Salpeter, E.

1951 with H. Bethe. "A Relativistic Equation for Bound-State Problems," *Phys. Rev.* 84, 1232–1242.

Schwinger, J.

1948a "On Quantum Electrodynamics and the Magnetic Moment of the Electron," *Phys. Rev.* 73, 416–417.

1948b with V. Weisskopf. "On the Electromagnetic Shift of Energy Levels," *Phys. Rev.* 73, 1272.

1951 "On the Green's Function of Quantized Fields," *Proc. Nat. Acad. Sci. USA* 37, 452–459.

1958 editor. *Selected Papers on Quantum Electrodynamics.* New York: Dover Press.

Shedlovsky, T.

1970 *Duncan Arthur MacInnes: Biographical Memoirs.* Washington, DC: National Academy of Sciences.

Tomonaga, S.

1940 with G. Araki. "Effect of the Nuclear Coulomb Field on the Capture of Slow Mesons," *Phys. Rev.* 58, 90–91.

Ward, J.

1950a "The Scattering of Light by Light," *Phys. Rev.* 77, 293.

1950b "An Identity in Quantum Electrodynamics," *Phys. Rev.* 78, 182.

Wentzel, G.

1943 *Einfuhrung in der Quantentheorie der Wellenfelder.* Wein: Franz Deutiche; reprinted by Edwards Bros., Ann Arbor, 1946.

Panel on Shelter Island I

H. A. Bethe, V. F. Weisskopf, and W. E. Lamb, Jr.

BETHE

I was asked to arrange for a discussion of Shelter Island I. It was announced as a panel, but it is not; it is a sequence of five speakers, who will speak for different lengths of time about some of the parts of Shelter Island I that had influence on the later development of physics.* One of the things that stands out, in my opinion, is how simple physics was, and how beautiful it was that we could at that time compare theoretical results with experiments. The first of the speakers will be Viki Weisskopf....

WEISSKOPF

I'm supposed to speak for 10 minutes about Kramers's influence on renormalization, which I think was quite great. What I'm telling is, however, my own personal experience and is almost that story only. Up to Shelter Island, the existence of the Lamb shift was doubtful. There was some indication, called the Pasternack effect, and that excited some theorists like myself, and Kramers and Euler. The decisive moment of Kramers's influence, as far as I am concerned, was the summer school of 1940 in Ann Arbor (I think it was the last Ann Arbor summer school before the war). There it happened that Kramers and myself were lecturers, and Kramers said, "You younger people should calculate the difference between the energy of the hydrogen 2s state and the free electron. Both are infinite, and you should take the difference, which is hopefully finite, and that may be the Pasternack effect."

That was a very nice suggestion. Only there are several reasons as far as I was concerned why it was not done right away. First of all, there was a lot of very exciting nuclear physics going on at that time, and there was also a war going on. But I and many others were in Los Alamos and had other things to do. Soon afterward some of us took it up again. I had a very gifted graduate student, Bruce French, and we tried it but with not too much enthusiasm; after all it was not clear whether the Pasternack effect was real or not.

Then came Shelter Island.... There are many important aspects of that conference (you'll hear about the meson problem from Bob Marshak), but as far as this problem is concerned, it was the first time that I heard of

---

* Editors' note: Of the five panelists—Bethe, Weisskopf, Lamb, Kinoshita, and Marshak— only the first three appear here; however, papers read to this conference by the last two appear in this volume.

Retherford and Lamb's experiment. It may be that other people heard of it before; for a lot of us, it was here that we were told by Lamb that the 2s-state of hydrogen differs from the Sommerfeld value by about 1,000 megahertz.

This gave us, of course, a tremendous push, and right away when I came home we went on with more energy to calculate this difference, and so did Kroll and Lamb. It was not so simple as Kramers had said. You cannot just subtract the infinite mass of the electron from the infinite result you get from a bound state, because the bound state is a distribution of momenta, whereas a free electron has a definite momentum. You have to be careful, but I don't want to go into this; it could be done (Kroll and Lamb did it, and French and I did it).

Shortly after Shelter Island, before we finished our calculation, Hans Bethe made an estimate that right away put the whole thing on the map in an almost quantitative way. He saw right away that the subtractions with which we had so many difficulties would probably make everything above mc^2 (m is the electron mass) unimportant, and indeed it did: it enters only as the upper limit of a logarithm, which is not very much. The main point was to calculate the lower limit of a logarithmic integral, and that lower limit depended actually only on nonrelativistic properties of the hydrogen atom.

Here, of course, Hans Bethe was in a much better position than anybody else, because he knows all the matrix elements of the hydrogen atom by heart. Or at least he did at that time. As Murph Goldberger reminded me a minute ago, the second reason why Hans could do it was that he took a train from here to Schenectady. He had time enough, knowing all the matrix elements by heart, to do the calculation, and he then published his famous paper that showed for the first time that the Lamb shift indeed is a consequence of quantum electrodynamics.

All we had to do afterward was to calculate this not very important, less than 10% addition to Hans's expression that comes from cutting off at mc^2. That was done, of course, by several people, most elegantly by Feynman and Schwinger. Probably everybody knows this story. French and I had the result all right, and I called up Schwinger and Feynman, but both of them had the same result, which differed from ours. In my judgment, the probability that Feynman and Schwinger had made exactly the same mistake was zero. Unfortunately, it wasn't. They did make exactly the same mistake, and that's why our publication was a little retarded. But compared with 36 years, it's a negligible time difference.

I would like to make one general remark, which perhaps weakens a little

Hans's statement that at that time we could really compare things with experiment.... It speaks not very favorably of the theorists that they began to calculate the effect seriously only after Lamb and Retherford had measured it. We should have and could have calculated this already in 1935 or '36; indeed, Heitler and Euler and even myself speculated about these things, but there was no experimental result and therefore no calculation. Nowadays the situation is somewhat different; the fewer experimental results, the more calculations. Both procedures are wrong, and maybe we'll find some middle ground.

Perhaps you would be interested in the proposals brought forward when the meeting was planned. It shows what kinds of silly theories were discussed at that time and reminds me a little of the other sessions that we heard today.... I'll only read the things connected with QED. The following items were supposed to be discussed: Self-energies: attempts to remove infinities—that's OK. Then: Classical attempts, theories using advanced minus retarded potentials (of course that's Feynman); nonlinear theories; attempts to change the formalism of quantization, such as the lambda process (you may vaguely remember what that was); Dirac negative light quanta (A. Pais's method of introducing an additional massive scalar field); the Heitler-Penz method; why do logarithmic divergences defy most of those methods? I mention these just to show you how many crazy theories there were. Then: "How reliable are the finite results of QED derived by means of subtraction formalism?... polarization of the vacuum and related effects... is there a high-energy limit of QED?" You see, there was a mixture of not very fruitful attempts and attempts that turned out later to be correct. It is hoped that this is also true today. Thank you.

BETHE

... I also heard of Kramers's renormalization procedure for the first time at that time, namely, the idea that the self-energy of a free electron is simply part of its mass, and you have to subtract that self-energy from the self-energy that you get for a bound electron. So after Shelter Island I took that famous train ride to Schenectady and tried to write down what this difference of self-energies might be, and it turned out that you could fairly easily subtract one from the other. I'll write at least a couple of equations. The self-energy according to the most elementary theory of radiation interaction, which you find, e.g., in Heitler's book, is

$$W_{bound} = -C \int k \, dk \sum_n \frac{\langle n|v|m \rangle^2}{E_n - E_m + k}, \tag{1}$$

where C is some constant, k is the energy of the light quantum, and $\langle n|v|m \rangle$ the matrix element of the velocity of the electron between states m and n of the atom.

Now, if you have a free electron, then it can go only to one momentum state, and if I consider the light quantum's momentum small, then for the free electron I get simply that same constant and the same integral $k \, dk$, but now simply the matrix element of v^2 for the state itself, the state of the free electron, divided by k, the energy of the light quantum, this being now the only energy difference; thus

$$W_{free} = -C \int k \, dk \frac{\langle m|v^2|m \rangle}{k}. \tag{2}$$

The analog for the atom is the expectation value of v^2 for the ground state. Now using the completeness relation, (2) becomes

$$W_{free} = -C \int k \, dk \sum_n \frac{\langle m|v|n \rangle^2}{k}. \tag{3}$$

By subtracting (3) from (1), the difference between the self-energy of a bound electron and a free electron is then

$$W_{bound} - W_{free} = C \int dk \sum_n (E_n - E_m) \frac{\langle m|v|n \rangle^2}{E_n - E_m + k}. \tag{4}$$

Thus taking the difference has reduced the divergence by one power of k. I knew from Weisskopf's previous theories that once you have electron-positron hole theory, then relativistically the self-energy diverges only logarithmically with k, so if I get one additional k in the denominator, then it won't diverge at all. That's the whole story.

My equation (4) still diverges logarithmically with k, but then borrowing from the previous theory, I said the integral goes up to a maximum momentum K, which is about the mass of the electron. I calculated this and found the result to be 1,040 megahertz, which happened to agree well with the observed quantity.

Then, of course, the real work began, namely, to do the relativistic theory and find the actual number, and that's what made for the progress in physics. My calculation could only show that the subtraction method could give results in agreement with experiment, and so it may have encouraged people to go on and do the actual calculations. As you know, and as Viki already mentioned, the main progress was made by Schwinger and Feynman. Neither of them will talk about it, because it is assumed that everybody knows that....*

LAMB

I shall first make some extraneous remarks bearing on previous discussions. (1) I was at Ann Arbor in the summer of 1940, and do not believe that Kramers was there that year.** (2) When I heard Kramers talk at Shelter Island I it seemed to me that he mainly said that we should be applying the methods used by Lorentz for the classical electron. I could not see that he indicated how such a program could be carried out, and hence did not derive great inspiration from his talk....*** Unlike Weisskopf, who was working on these problems actively before this wonderful time, Kroll and I were only inspired when we found out how Bethe had subtracted the self-energy of a free electron.

The earliest idea that microwaves might be used for measurements of hydrogen fine structure came to me in the summer of 1945. Construction of the first apparatus was begun by Retherford in the last quarter of 1946. The first resonance signals were seen on a Saturday (in those days people sometimes worked in the laboratory on Saturday) in April 1947. Rabi was out of town that day and only heard about the results on the following Monday. I don't know whom Rabi told about it; obviously he told Oppenheimer, or I probably wouldn't have been invited to either of the Shelter Island conferences.

The first slide shows three old illustrations that deal with the fine structure of atomic hydrogen. The first [figure 1] shows the levels of the $n = 2$

---

*Editors' note: The following interchange precedes Lamb's discussion. Seitz: "Incidentally, Kramers' visit must have been before 1940, because Holland was invaded that spring." Weisskopf: "We'll have to check that; it may have been '39. By the way, perhaps it's interesting to add that the presence of Kramers in June of 1947 was due to the fact that he was chairman of the United Nations international commission of the internationalization of nuclear weapons and nuclear power, an attempt that as you all know unfortunately did not succeed." Frederick Seitz, a solid-state physicist, is a former president of Rockefeller University and a former president of the National Academy of Science.
**Editors' note: Weisskopf makes the following interjection: "It must have been '39."
***Editors' note: Bethe makes the following interjection: "He did inspire me."

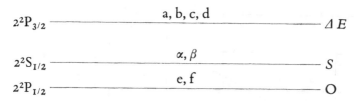

Figure 1
Fine structure of $n = 2$ levels of hydrogen. According to the Dirac theory the $2^2S_{1/2}$ and $2^2P_{1/2}$ levels coincide exactly. The letters a, b, c, d, e, f, α, and β denote the magnetic sublevels that are split in a magnetic field.

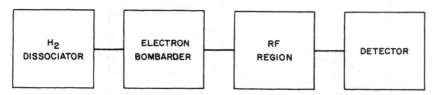

Figure 2
Modified schematic block diagram of apparatus.

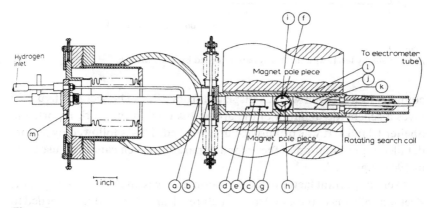

Figure 3
Cross section of second apparatus: (a) tungsten oven of hydrogen dissociator; (b) movable slits; (c) electron bombarder cathode; (d) grid; (e) anode; (f) transmission line; (g) slots for passage of metastable atoms through interaction space; (h) plate attached to center conductor of r-f transmission line; (i) d.c. quenching electrode; (j) target for metastable atoms; (k) collector for electrons ejected from target; (l) pole face of magnet; (m) window for observation of tungsten oven temperature.

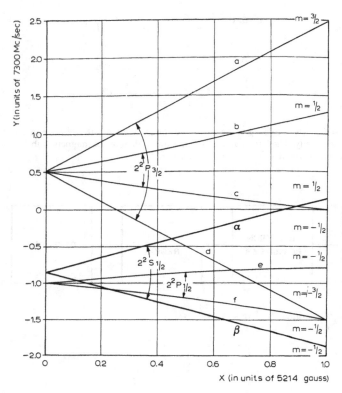

Figure 4
Zeeman energy levels with the $2^2S_{1/2}$ pattern raised by 1,000 Mc/sec.

complex in hydrogen with 2s displaced from $2P_{1/2}$. The second [figure 2] gives a schematic diagram of the apparatus used in the work reported at Shelter Island I. It was primitive, but it worked. This is followed by a more detailed representation of the second apparatus, which came into use early in 1948 [figure 3].

A very important factor in these experiments was an understanding of the Zeeman splitting of the $n = 2$ fine structure. I learned everything I needed to know about that from the 1933 edition of the Geiger-Scheel *Handbuch der Physik* article by Bethe. The second slide [figure 4] shows the energies of the $n = 2$ levels as functions of magnetic field. Bethe's article also contained a discussion of perturbing electric fields on a two-level system of decaying atomic states.

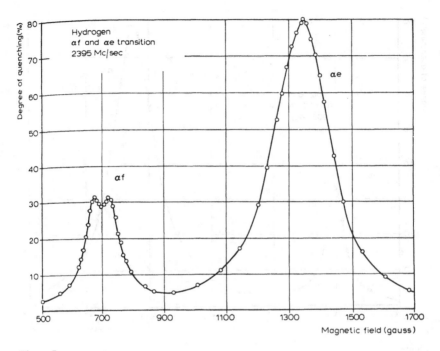

Figure 5
Observed resonance curves for hydrogen.

The third slide shows two of the resonance curves that we obtained [figures 5 and 6]. The 2s metastable atoms were detected by their ejection of electrons from a metal plate. We induced transitions from the 2s to the 2p states by microwave radiation obtained from a klystron oscillaor. The 2p atoms promptly underwent Lyman-alpha radiative decay to the 1s ground state in which the atoms were no longer able to give a detector signal.

I know you are all very eager to hear about the comparison between theory and experiment, but I'm going to delay that for a bit while I tell what happened to me after 1947.... During the time I was at Columbia, my colleagues Retherford, Dayhoff, and Triebwasser had done the $n = 2$ states of hydrogen and deuterium, while Skinner, Novick, Yergin, and Lipworth observed fine structure transitions in $n = 2$ of $Z = 2$ singly ionized helium. As has been mentioned before, Kroll and I made an excursion into QED field theory to bring Bethe's 1947 calculation into relativistic form.

In '51 I moved to Stanford (it's too long a story to explain this move right

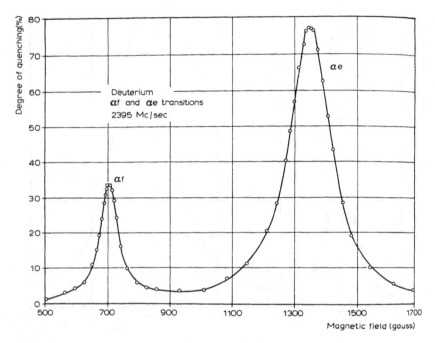

Figure 6
Observed resonance curves for deuterium.

now), and there my colleagues Sanders and Wilcox managed to make measurements on the $n = 3$ and $n = 4$ states of hydrogen. Maiman and Wieder studied the $n = 2$ and $n = 3$ triplet fine structure of the helium atom, also a subject discussed very clearly in the Bethe *Handbuch* article of 1933. I also began to work on maser theory in 1955. In '56 I moved to Oxford, where I was supposed to be a professor of theoretical physics. Having no experimental facilities, I began work on laser theory.

In 1962, I went to Yale, and continued to work on laser theory. I again had the opportunity to be involved in experimental work. Our previous work had been plagued by small signal-to-noise ratios. I thought I could find a way to make a relatively enormous yield of metastable hydrogen atoms. This succeeded in the work with Lea, Leventhal, and Kaufman. Unfortunately, however, the experimental conditions became so complicated that we could not analyze them well enough to learn anything. The work was essentially a failure. Mader and Jacobs made useful determinations of the fine structure of the $n = 3$ and 4 states of singly ionized helium. This work did lead, as a kind of lucky accident, to the discovery of a way to do microwave spectroscopy on very high Rydberg states of helium. I also began work on the quantum theory of measurement.

In Arizona, where I am now, I have been involved in high resolution laser spectroscopy on simple molecules like HD^+, and more complicated ones like D_3^+. Also I continued to work on the quantum theory of measurement, and did computational physics of one sort or another, such as integrating time-dependent Schrödinger equations and dealing with many-body problems in certain ways.

Except for the measurements on hydrogenic fine structure, our experiments have not had much impact on elementary phenomena. The accuracy was not great enough. The helium atom is a very good thing to study, but you have to make rather accurate calculations before you can begin to look for the possible consequences for elementary particle physics. The HD^+ molecule has radiative corrections to the Born-Oppenheimer vibrational potential, which are just barely visible. It also offers a rather unlikely possibility that two nucleons have additional forces betweem them of a long range nature. Future measurements by my colleague Wing will have much higher precision.

I might point out that before the fine structure experiments were done I knew about the Uehling effect of vacuum polarization (that's about 27 MHz); I thought there might be some additional interactions between the electron and proton spins, so I was expecting the experimental results to bear more on hyperfine structure than on the fine structure. However, as soon as the first results were obtained, we could tell immediately that we had to do with the Pasternack explanation of Houston's spectrocopic data.

The fourth slide shows a table of experimental results for the level shift of $n = 2$ in hydrogen [table 1]. (There are already results for many other values of Z. Probably they don't provide a severe test for QED now, but will do so in the future.) Each of the values consists of the value of 1,057 MHz— "1,057 and all that"—plus an indicated decimal fraction, with error estimate shown in parentheses. TDL stands for Triebwasser, Dayhoff and Lamb; RS for Robiscoe and Shyn; LP for Lundeen and Pipkin; and NAU for Newton, Andrews, and Unsworth. The theoretical results of Erickson and Mohr have both been modified by Lundeen and Pipkin to allow for a new and larger rms proton radius, $r_p = .862(12)$ fm.

Experimental techniques have advanced quite a bit since the days I was

Table 1[a]

Experiment	Calculation
.77(6) (TDL, 1953)	.930(10) (Erickson, 1971)[b]
.90(6) (RS, 1970)	.884(13) (Mohr, 1975)[b]
.892(20) (LP, 1975)	
.862(20) (NAU, 1979)	
.845(9) (LP, 1981)	

a. $\mathscr{S}(n = 2)$; H, 1,057.+ MHz.
b. $r_p = .862(12)$.

involved in this work. For one thing, instead of using electron bombard-
ment of an atomic beam to excite the atom, one can use ion beam foil
techniques. One can work in zero magnetic field and tune the oscillator. We
had to vary the magnetic field to Zeeman tune through the resonance. The
corrections for the effects of motional electric fields were an important
source of possible error. The Ramsey method of two separated rf transition
regions, applied to decaying states as first suggested by Hughes, offers gret
benefits. Finally, computers are enormously helpful in the analysis of the
data and the application of the many corrections required to obtain the
final results.*

* Editors' note: The following discussion takes place. Kinoshita: "The Erickson and Mohr
difference has been resolved in favour of Mohr by the work of Fegelstein, so the agreement
between theory and experiment is much closer. That was published a year or so ago." Lamb: "I
missed it." Bethe: "The radius of the proton is mean-squared...?" Lamb: "Yes, rms. I would
worry a little bit myself, being very naive, that the effective size of the proton for the 2s atomic
electron might not be exacly the same as that indicated by high energy scattering experiments.
One curious thing: Pipkin was once involved in some electron scattering work in which he
thought he was getting discrepancies from QED. At that time, the agreement between
calculation and experiment was pretty good. Now the latest Lundeen and Pipkin result is
higher than Mohr's value by 0.039(16) MHz, which is rather good, if not perfect, agreement."

Origin of the Two-Meson Theory

R. E. Marshak

When Shelter Island I convened in early June 1947, the Yukawa meson was in serious trouble despite a brilliant start only a a decade earlier. Yukawa's meson hypothesis of nuclear forces was put forward in 1935 and it took only two years before it seemed to be spectacularly confirmed by the discovery of a particle of intermediate mass in the cosmic radiation. However, the Yukawa theory soon ran into quantitative difficulties when it was confronted with the known properties of the two-nucleon interaction and the small observed scattering cross section of sea-level cosmic-ray mesons by nucleons [1]. It was found that any postulated meson-nucleon interaction leading to the required tensor force between two nucleons gave an unacceptable $1/r^3$ behavior in the weak coupling approximation; this could be patched up by accepting a mixture, say, of pseudoscalar and vector mesons, either with equal masses [2] or unequal masses [3]. But this patched-up weak coupling theory could not explain the observed reduction by at least a factor of 10^2 below the predicted meson-nucleon scattering cross section. Some progress was made [4] in resolving the dilemma of the small scattering cross section by inventing a strong coupling version of meson theory in which the charge and spin inertias of the nucleon in the presence of the meson field led to excited states of nucleons whose interference with the ground state could suppress the matrix element for scattering. But unfortunately, the strong coupling theory predicted incorrect properties for the nucleon (e.g., equal and opposite magnetic moments for the neutron and proton). The situation became increasingly confused as World War II put a halt to serious work on testing Yukawa's theory.

As the war began to wind down and physicists went back to work, the mystery of the Yukawa meson reached crisis proportions. In a paper entitled "On the Disintegration of Negative Mesons" published in the 1 February 1947 issue of *Physical Review*, Conversi, Pancini, and Piccioni [5] reported that a substantial fraction of negative (sea-level) mesons decayed in a carbon plate but were absorbed in an iron plate. All positively charged mesons decayed in both carbon and iron plates. According to the theory of Tomonaga and Araki [6], Yukawa mesons carrying positive charge should always decay (in agreement with experiment) and those carrying negative charge should never decay (in disagreement with experiment). Analysis of this experiment by Fermi, Teller and Weisskopf [7] led to the conclusion that "the time of capture from the lowest orbit of carbon is not less than the time of natural decay, that is, about 10^{-8} second. This is in disagreement with the previous estimate by a factor of about 10^{12}. Changes in the spin of the mesotron or the interaction form may reduce this disagreement to 10^{10}...." This tremendous discrepancy of a factor of 10^{10-12} between theory and experiment deduced from meson absorption now replaced the disturbing discrepancy of a factor of 10^2 found in meson scattering. It was no longer a question of calculating the meson-nucleon interaction with a weak coupling, strong coupling, or some other sort of coupling approximation, but rather something more fundamental was at issue.

As this dilemma was being discussed during the spring of 1947, Oppenheimer was busy organizing Shelter Island I, a small informal conference to be held 2–4 June. Fully cognizant of the Italian experiment and its analysis, he sent a memorandum to each participant in the forthcoming conference entitled "The Foundations of Quantum Mechanics: Outline of Topics for Discussion." I quote from this memorandum [8]:

It was long ago pointed out by Nordheim that there is an apparent difficulty in reconciling on the basis of usual quantum mechanical formalism the high rate of production of mesons in the upper atmosphere with the small interactions which these mesons subsequently manifest in traversing matter. To date no completely satisfactory understanding of this discrepancy exists, nor is it clear to what extent it indicates a breakdown in the customary formalism of quantum mechanics. It would appear profitable to discuss this and related questions in some detail....

We might start this discussion by an outline of the current status of theories of multiple production.... However, no reasonable formulation of theories along this line will satisfactorily account for the smallness of the subsequent interaction of mesons with nuclear matter.... There are two reasons for these apparent difficulties. One is that in all current theory there is a formal correspondence between the creation of a particle and the absorption of an anti-particle. The other is that multiple processes are in these theories attributable to the higher order effects of coupling terms which are of quite low order, first or second, in the meson wave fields. The question that we should attempt to answer is whether, perhaps along the lines of an S matrix formulation, both these conditions must be abandoned to accord with the experimental facts.

While Oppenheimer took the trouble to alert us to the importance of the Italian experiment, Shelter Island I naturally started with a thorough discussion of the significance of the newly discovered Lamb shift. After an exhaustive and historic discussion of this subject, the attention of the conferees focused on the Italian meson experiment. There was little inclination during the spirited discussion that ensued to support Oppenheimer's suggestion that one should consider surrendering microscopic reversibility.

At one point, Weisskopf [9] proposed to overcome the apparent lack of reversibility between "creation" and "absorption" of mesons by having the primary cosmic-ray proton convert a normal nucleon in the "air" nucleus into an "excited" nucleon, capable of emitting mesons; the lifetime of the "meson-pregnant" state could be chosen sufficiently long, Weisskopf argued, to account for the subsequent weak interaction between mesons and nucleons. This hypothesis seemed inelegant and I proposed an alternative solution of the difficulty, to wit, that two kinds of mesons exist in nature, possessing different masses: the heavy (Yukawa) meson is produced with large cross section in the upper atmosphere and is responsible for nuclear forces, whereas the light meson is a decay product of the heavy meson and is the normal particle observed at sea level to interact weakly with matter. It is to be noted that this two-meson hypothesis was not to be confused with the Moller-Rosenfeld or Schwinger mixture of two strongly coupled mesons. In my two-meson theory the heavy meson was *strongly* coupled to the nucleon and the light meson *weakly* coupled to the nucleon and, a fortiori, the two mesons were weakly coupled to each other.

When I left the Shelter Island Conference, I decided to publish a brief note on the two-meson hypothesis and to use the observed decay rate from the K mesic orbit in carbon to estimate the lifetime for the decay of the heavy (Yukawa) meson into the light (sea-level) meson. Before I could do so, the 24 May 1947 issue of *Nature* arrived (it did not reach the United States until mid-June—after the Shelter Island Conference—because in 1947 journals were not sent airmail) carrying an article by Lattes et al. [10] showing two pictures (in Ilford plates) of a meson coming to rest in the emulsion and decaying into a lighter meson of fixed range. I was immediately convinced that the two beautiful $\pi \to \mu$ decays (as they came to be known) were small in their statistics but clear-cut in their support of the two-meson hypothesis. I decided to enlist Bethe's help in writing the paper because of his extensive knowledge of the cosmic-ray data. Our joint paper was sent to *Physical Review* during July and published in September [11]. [There was no reference in our paper to Pontecorvo's important observation (see following) concerning the Italian experiment [12] because his paper was only published in August—preprints were uncommon in those days.]

There is no point going into the details of the paper by Bethe and myself except to note that the most important result was an estimate of the lifetime τ_H for the decay of the heavy to the light meson. From the deep penetration of cosmic-ray mesons underground (to the equivalent of 1,000 meters of water), we estimated an upper limit on τ_H of 2×10^{-8} sec. If we then related τ_H to the strength of the heavy meson-nucleon interaction and the observed decay of the light meson in carbon in the Italian experiment, we deduced $\tau_H \sim 10^{-8}$ sec, very close to the upper limit derived from the cosmic-ray data. In making this calculation, it was necessary to assign spins to the heavy and light mesons. We acknowledged in our paper that "the best qualitative theory of nuclear forces so far worked out is based on a pseudoscalar meson field with pseudo-vector coupling," and realized that the spin of the light meson could not exceed $\frac{1}{2}$ from the well-known analysis of the frequency of burst production associated with cosmic ray mesons [13]. "For the sake of a model," as we put it, we computed τ_H with spin $\frac{1}{2}$ for the heavy meson and spin 0 for the light meson, although it was clear that the converse choice of spins would have little effect on τ_H [14]. In any case, the large value of τ_H (on the nuclear time scale) indicated that the light meson—despite its intermediate mass between that of the electron and the nucleon—was very different from the Yukawa meson.

In our paper, Bethe and I did not directly pin down the light meson as a heavy (second-generation) lepton. This was done by Pontecorvo within the framework of a *one-meson theory*. Pontecorvo's paper [12] was published in the 1 August 1947 issue of *Physical Review* and was sent 21 June 1947. He was reacting to the Italian experiment and to the analysis of Fermi et al. [7] and made this brilliant observation: "We notice that the probability ($\sim 10^6$ sec^{-1}) of capture of a bound negative meson is of the order of the probability of ordinary K capture processes, when allowance is made for the difference in the disintegration energy and the difference in the volumes of the K shell and of the meson orbit. We assume that this is significant and wish to discuss the possibility of a fundamental analogy between β processes and processes of emission or absorption of charged mesons...." He obviously had not seen the *Nature* article by the Bristol group and understandably was not aware of our two-meson theory. He realized the need to understand the large cross section and multiple production of mesons and appealed, of all things, to my early meson pair (sometimes called heavy electron) theory of nuclear forces [15]. Specifically, he said in his paper, "The hypothesis that the meson decay is not a β process, while the meson absorption is a β process, does not require that hypothetical particles such as neutral mesons are invoked to account for nuclear forces. In fact, a heavy electron pair theory of nuclear forces was successfully developed by

Marshak. Moreover, a pair theory is capable of accounting, at least in principle, for the existence of processes in which several pairs of mesons are produced in a single act, as suggested by Heisenberg in connection with a different problem...." This was a curious twist but does not diminish the importance of Pontecorvo's statement that the muon is a heavy electron (i.e., the second-generation lepton). Indeed, a reading of Pontecorvo's paper immediately persuaded me that the choice of spin 0 for the heavy (Yukawa) meson and spin $\frac{1}{2}$ for the light (sea-level) meson was the highly likely choice in the two-meson theory.

In November 1947, I visited Powell's cosmic-ray laboratory in Bristol. The nuclear emulsion technique was the center of attraction and many visitors had arrived to join in the quest for new discoveries in meson physics. Armed with the basic ideas of the two-meson theory and the likelihood that the Yukawa meson was the boson and that its decay product (the muon) was a heavy electron, the Powell group and I engaged in a fascinating discussion of the consequences that could be tested with nuclear emulsions: e.g., π^-s should be captured by all nuclei, whereas μ^-s should not; in nuclear capture π^- should convert an appreciable fraction of its rest energy into "star" energy, whereas μ^- should not. I do not recall speaking to a single theorist in Powell's Laboratory, but it did not matter because there were still so many exciting discoveries to make in nuclear emulsions and so many improvements in technique still possible (e.g., development of electron sensitive plates and proper correction for fading). As a theorist, I almost felt as if I was intruding on the private domain of the experimentalist.

After Bristol, during that European trip in November 1947, I visited Blackett's laboratory in Manchester, where I was briefed by his colleagues Rochester and Butler (Blackett by that time had transferred his interests to the new field of paleomagnetism). Rochester and Butler showed me their first two cloud chamber photographs (in contradistinction to Bristol, the cloud chamber technique dominated at Manchester at that time) of V particles [16]—the first serious evidence for the existence of particles heavier than pions (with masses about 1,000 times the electron mass) decaying into charged particles. Here were two excellent examples of what later came to be known as "strange particles," and I should have "spread the good word" when I returned to the United States. But unfortunately, Rochester and Butler also showed me some other cloud chamber photographs in which they said they were observing muons produced in the metal plates inside the cloud chamber. They insisted that they could distinguish between the pion and muon masses and that they were seeing the direct production of muons. I told them that this was in direct contradiction with my two-meson theory (in which I naturally had great confidence) and that I could not accept their interpretation of the photographs. Since I was confident that they were wrong about the direct production of muons, I was hesitant—unfairly it turned out—to accept their early discovery of the V particles. It later turned out that the source of the difficulty was the incorrect calibration of the π-to-μ mass ratio that Rochester and Butler had received from Bristol, which initially claimed a value 1.7 instead of the correct value, 1.3. (The high value was connected with a fading problem that was soon resolved.) It is clear now that with the substantially smaller mass ratio, Rochester and Butler would not have argued for the direct production of muons [17].

Two months after my return from the European trip, at the American Physical Society meeting in New York in January 1948, Oppenheimer shared with me a copy of the very interesting paper by Sakata and Inoue [18] that had just reached him at Princeton. This paper contained an earlier version of the two-meson theory, and I must try to set the record straight, by appealing to a book containing Sakata's collected papers [19], and to my own recollection. The editor of the book, O. Hara, in listing Sakata's major contributions in the preface, states, "The next is the proposition of the two-meson hypothesis with Y. Tanikawa, in which he introduced the μ meson, which is intimately related to but is distinct from the π meson originally predicted by Yukawa. This hypothesis was verified by C. F. Powell after the war, and removed the confusion of the meson theory in its early stage." The published paper, "On the Correlations between Mesons and Yukawa Particles," by Sakata and Inoue [19], contains the footnote, "The contents of this paper was read before the Symposium of the Meson Theory held on September 1943. The printing was however delayed owing to the war circumstance." A second paper, with another version of the two-meson theory, was published by Tanikawa the following year [20]. Apparently, fairly intense discussions of meson theory were held in Japan during World War II (at least until 1943), and it was at such symposia that these two papers were first presented.

The two-meson models of Sakata and Inoue and of Tanikawa were evidently motivatd by the attempt to explain the small observed meson-nucleon scattering cross section in the cosmic radiation. In the Sakata-Inoue paper, the heavier (Yukawa) meson is assigned spin 0 or 1, decaying into the lighter (sea-level) meson with spin $\frac{1}{2}$ and having a weaker coupling with the nucleon by a factor of 10 compared with the heavy meson. In the Tanikawa paper, the spin of the light meson was taken as 0. Sakata and Inoue were thus trying to explain the factor 10^2 discrepancy in the scattering cross section; they predicted $\tau_H \sim 10^{-21}$ sec. This simply means that the light meson of the Japanese authors was not the second-generation lepton (muon) at all but a meson that was less strongly coupled (by a factor of 10) than the Yukawa meson. Until the Italian experiment was done, it would have been difficult to decide between the "strong coupling" explanation of the small scattering cross section and the two-meson approach of the Japanese workers. I have no doubt that had Sakata and Inoue learned about the Italian experiment when it was first published, they would have predicted the correct value of τ_H. (The difference between our estimate of 10^{-8} sec and their 10^{-21} sec is essentially the factor 10^{12} discrepancy deduced by Fermi et al. from the Italian experiment.) It is unfortunate that the same postwar disruption of communications that was responsible for the late arrival of the Sakata-Inoue paper prevented the timely arrival of the Conversi et al. paper in Japan.

Before concluding the two-meson story, I should like to mention one final flap that occurred in connection with the alleged observation of the direct production of muons in nuclei. In early 1948, Lattes came over from Bristol to show the physicists using the Berkeley synchrocyclotron how to detect mesons with Ilford plates. The 340 MeV alpha particles in the Berkeley synchrocyclotron soon became a copious source of pions. But then one of the Berkeley groups began to see muons in the Ilford plates that seemed to come from the target, supposedly providing evidence for the direct production of muons in alpha-nuclei collisions. I fought this "threat" to the two-meson theory with great conviction at several conferences in the spring of 1948. By the summer, the Berkeley effect vanished when it was determined that the apparent direct production of muons was due, in fact, to the direct production of pions, decaying to muons that were returned to the target area under the influence of the internal magnetic field. The Yukawa meson was unequivocally differentiated from the second-generation lepton and a new chapter in meson physics began.

References

[1]   J. G. Wilson, *Proc. Roy. Soc.* 174, 73 (1940).
[2]   C. Möller and L. Rosenfeld, *Proc. Inst. for Theor. Phys., Copenhagen* 17, No. 8 (1940).

[3]   J. Schwinger, *Phys. Rev.* 61, 387 (1942).

[4]   Cf. W. Pauli and S. M. Dancoff, *Phys. Rev.* 62, 85 (1942); W. Heisenberg [*Zeits. f. Physik* 113, 61 (1939)] first worked out the strong coupling theory for vector mesons. Cf. also J. R. Oppenheimer and J. Schwinger, *Phys. Rev.* 60, 150 (1941).

[5]   M. Conversi, E. Pancini, and O. Piccioni, *Phys. Rev.* 71, 209 (1947).

[6]   S. Tomonaga and G. Araki, *Phys. Rev.* 58, 90 (1940).

[7]   E. Fermi, E. Teller, and V. Weisskopf, *Phys. Rev.* 71, 314 (1947).

[8]   Oppenheimer was not alone in searching for a singular explanation of the Italian experiment. Apparently, Niels Bohr ventured the (unpublished) suggestion that the effect could be due "to a quite unusual mechanism of capture of the slow negative meson by atoms of small charge number: there would in fact be a large probability of its being retained on an "orbit" of *large* angular momentum (and consequently with small chance of capture by the nucleus) during a time sufficiently long to permit its decay into leptons". See page 17 of *Nuclear Forces* by L. Rosenfeld (North-Holland, Amsterdam, 1948).

[9]   I urged Weisskopf to publish a note on his hypothesis [V. F. Weisskopf, *Phys. Rev.* 72, 510 (1947)].

[10]   C. M. Lattes, H. Muirhead, G. P. S. Occhialini, and C. F. Powell, *Nature* 159, 694 (1947).

[11]   R. E. Marshak and H. A. Bethe, *Phys. Rev.* 72, 506 (1947).

[12]   B. Pontecorvo, *Phys. Rev.* 72, 246 (1947).

[13]   R. F. Christy and S. Kusaka, *Phys. Rev.* 59, 414 (1941).

[14]   This was soon established by R. Latter (Caltech thesis, 1949, unpublished).

[15]   R. E. Marshak, *Phys. Rev.* 57, 1101 (1940).

[16]   D. Rochester and C. Butler, *Nature* 160, 855 (1947).

[17]   Private communication from D. Rochester (10 December 1980).

[18]   S. Sakata and T. Inoue, *Prog. Theor. Phys.* 1, 143 (1946).

[19]   *Shoichi Sakata Scientific Works*, O. Hara, editor (Horei Printing Co., Tokyo, 1977).

[20]   Y. Tanikawa, *Prog. Theor. Phys.* 2, 220 (1947).

The Activity of the Tomonaga Group up to the Time of the 1947 Shelter Island Conference

Y. Nambu and K. Nishijima

With the timely announcement of the Lamb shift, the Shelter Island Conference of 1947 turned out to be a remarkable event for the development of quantum electrodynamics (QED), especially with respect to renormalization theory, which is today regarded as a standard framework in quantum field theory and statistical mechanics. As far as the research in the United States is concerned, certainly the Shelter Island Conference marked the beginning of QED in the form known today. Around the same time, however, a similar development had been reaching maturity in Japan. S. Tomonaga of the University of Education in Tokyo (predecessor of today's University of Tsukuba) was leading his collaborators there and at the University of Tokyo in an effort to tackle the problem of renormalizability, without the benefit of direct experimental support such as the Lamb shift. The following is a brief account of the motivation and the achievements of Tomonaga and his group up to that time, and of the eventual interaction that ensued between the United States and the Tokyo groups.

Tomonaga came to Tokyo in 1932 as an assistant to Y. Nishina at the Institute for Physical and Chemical Research (RIKEN). He chose QED for his research topic, and got interested in two particular problems. One was to formulate QED in a completely covariant manner by generalizing Dirac's many-time formalism [1]. The other was to find a way to handle the problem of divergent infinities characteristic of field theory.

Regarding the covariant formulation, apparently Tomonaga was motivated by the proposal [2] of Yukawa concerning the general transformation functional (depending on an arbitrary closed hypersurface in spacetime). During World War II, he had already completed a so-called supermany-time theory, and published the results in Japanese in 1943 [3]. Its English translation [4] appeared in *Progress of Theoretical Physics*, which was founded by Yukawa immediately after the war. It seems that Tomonaga was driven to this formulation not only for its esthetic values, but also with an expectation that it would actually be useful someday.

Tomonaga's second problem, i.e., the problem of divergences, has to do with the fact that in field theory the higher order corrections to a physical quantity are always divergent, a fact that was already known at the time of Lorentz's classical theory of the electron. In QED, on the other hand, a detailed analysis by Weisskopf [5] of the electron's self-energy revealed that, in contrast with the classical case, the divergence was only logarithmic when computed according to Dirac's positron theory.

As the next problem of interest, theorists then turned to the higher order

corrections accompanying scattering processes by external fields. It was
Dancoff [6] who performed the first relativistic calculations.

   Various divergences show up in the scattering amplitude of an electron.
Some people held that in quantum field theory, higher order corrections, or
the effects of field reaction, were actually meaningless and to be disregarded.
On the other hand, Tomonaga, during his stay at Leipzig, had been strongly
influenced by W. Heisenberg, who had emphasized the importance of field
reaction. It was thus natural for him, after returning to Tokyo, to devote
himself to the question of whether and how the divergences in scattering
and those in self-energy were related to each other.

   In this study, Tomonaga found it useful to apply the method of contact
(canonical) transformation that had been utilized by Bloch and Nordsieck
[7] and by Pauli and Fierz [8] in connection with the problem of infrared
catastrophe. This method brings out clearly the fact that at least part of the
divergences in scattering has the same origin as the self-energy. Examination of some simple soluble models confirmed in fact that the highest
divergences in scattering processes are of the same type as the mass divergences, and hence the former can be absorbed into the latter. Yet Dancoff
had concluded in his relativistic calculations that the two divergences are
different, contrary to Tomonaga's results. (Besides the divergences of the
self-energy type, there are also divergences of the vacuum polarization type,
which induce a divergent renormalization of the electron's charge. But
Dancoff's could not be attributed to either of them.)

   In 1946, Sakata proposed a mechanism [9] by which the divergent self-
energy of the electron could be eliminated. The idea was to introduce a new
"cohesive" field associated with the electron so that its contribution to self-
energy, with a proper adjustment of the coupling constant, will cancel that
of the electromagnetic field. He showed in fact that a neutral scalar field
(called C meson) could do the job. (This idea is identical with that proposed
independently by Pais [10].) Consequently, Tomonaga decided to investigate how far the method would work. It was already known that the
divergences of the vacuum polarization type could not be disposed of by
this mechanism, but the question was whether or not the mechanism was
effective in the scattering problem. For this purpose one had only to replace
the electromagnetic field in Dancoff's formulas with the C field and add the
results to Dancoff's. The upshot, however, was negative, meaning that one
was still left with uncanceled divergences.

Tomonaga thereupon felt it necessary to redo completely Dancoff's calculations. To this end he made use of his covariant formulation [4] and the method of contact (canonical) transformation [7, 8], thereby separating the mass divergences from the beginning. Being simpler and more transparent than Dancoff's original method, it soon revealed an oversight in the latter's calculations. Inclusion of the overlooked term immediately restored the cancellation between the electromagnetic and the C field effects [11].

Thus the results of Tomonaga's recalculation of the scattering process brought about an important change in his conclusions concerning renormalizability. The divergences could now be attributed only to self-energy and vacuum polarization, with no new divergences appearing in scattering processes. In this way he arrived at the idea of renormalization of mass and charge. It was at the very end of 1947 [12].

The term "renormalization" is already found in Dancoff's work, but the actual validity of this working hypothesis had to be demonstrated by a correct calculation. Such a recalculation was also carried out in the United States by Lewis and by Epstein [13].

By the time of the Shelter Island Conference in mid-1947, Tomonaga's group was equipped with a theoretical framework for calculating higher order effects. The news of the Lamb shift [14] was first brought to Japan through the science columns of American weekly magazines [15]. Only a few copies of *Physical Review* were available in Tokyo at that time, but the ingenious and quick estimates of the Lamb shift by Bethe [16] became known without much delay. So Tomonaga's group launched a concerted effort for systematic calculations [17] of the Lamb shift, the anomalous magnetic moment, and the radiative corrections to electron scattering, making use of a normal product expansion that Tomonaga had developed. The results of their calculations agreed with those done by the physicists in the United States [18]. A brief summary of the activities of the Tomonaga group through the end of 1947 was quickly published as a *Physical Review* letter [19]. (The interested reader is also referred to his 1965 Nobel Lecture [20].)

Acknowledgment

This work was supported in part by the National Science Foundation under grant PHY-83-01221.

References

[1]   P. A. M. Dirac, *Proc. Roy. Soc.* 136, 453 (1932).

[2]   H. Yukawa, *Kagaku* 12, 251, 282, 322 (1943) (in Japanese).

[3]   S. Tomonaga, *Bull. I.P.C.R.* (*Riken-Iho*) 22, 545 (1943) (in Japanese).

[4]   S. Tomonaga, *Prog. Theor. Phys.* 1, 27 (1946).

[5]   V. Weisskopf, *Phys. Rev.* 56, 72 (1939).

[6]   S. M. Dancoff, *Phys. Rev.* 55, 959 (1939).

[7]   F. Bloch and A. Nordsieck, *Phys. Rev.* 52, 54 (1937).

[8]   W. Pauli and M. Fierz, *Nuovo Cimento* 15, 30 (1947).

[9]   S. Sakata and O. Hara, *Prog. Theor. Phys.* 2, 30 (1947).

[10]   A. Pais, *Phys. Rev.* 68, 227 (1945).

[11]   D. Ito, Z. Koba, and S. Tomonaga, *Prog. Theor. Phys.* 2, 216 (1947), and errata to this paper, ibid., 217. For the full paper see *Prog. Theor. Phys.* 3, 290 (1948).

[12]   Z. Koba and S. Tomonaga, *Prog. Theor. Phys.* 21, 218 (1947).

[13]   H. W. Lewis, *Phys. Rev.* 73, 173 (1948). S. T. Epstein, *Phys. Rev.* 73, 179 (1948).

[14]   W. E. Lamb and R. C. Retherford, *Phys. Rev.* 72, 241 (1947).

[15]   The 29 September issue of *Newsweek* magazine.

[16]   H. A. Bethe, *Phys. Rev.* 72, 339 (1947).

[17]   T. Tati and S. Tomonaga, *Prog. Theor. Phys.* 3, 391 (1948); H. Fukuda, Y. Miyamoto, and S. Tomonaga, *Prog. Theor. Phys.* 4, 47, 121 (1949).

[18]   N. M. Kroll and W. E. Lamb, *Phys. Rev.* 75, 388 (1949); J. B. French and V. Weisskopf, *Phys. Rev.* 75, 1240 (1949).

[19]   S. Tomonaga, *Phys. Rev.* 74, 224 (1948).

[20]   S. Tomonaga, *Nobel Lecture* in *Scientific Papers of S. Tomonaga*, Vol. 1, T. Miyazima, editor (Misuzu Shobo Publ. Co., Tokyo, 1971), p. 712.

Contributors

Stephen L. Adler
Institute for Advanced Study
Princeton, NJ 08540

H. A. Bethe
Department of Physics
Cornell University
Ithaca, NY 14850

M. J. Duff
Blackett Laboratory
Imperial College
London SW7 2BZ
England

Murray Gell-Mann
Department of Physics
California Institute of Technology
Pasadena, CA 91125

Alan H. Guth
Center for Theoretical Physics
Massachusetts Institute of Technology
Cambridge, MA 02139

S. W. Hawking
Department of Applied Mathematics and Theoretical Physics
Cambridge CB3 9EW
England

Roman Jackiw
Center for Theoretical Physics
Massachusetts Institute of Technology
Cambridge, MA 02139

Nicola N. Khuri
Department of Physics
Rockefeller University
New York, NY 10021

Toichiro Kinoshita
Newman Laboratory
Cornell University
Ithaca, NY 14853

W. E. Lamb, Jr.
Department of Physics
University of Arizona
Tucson, AZ 85721

T. D. Lee
Department of Physics
Columbia University
New York, NY 10027

A. D. Linde
Lebedew Physical Institute
Moscow 117924
USSR

R. E. Marshak
Department of Physics
Virginia Polytechnic Institute and State University
Blacksburg, VA 24061

Y. Nambu
Enrico Fermi Institute and Department of Physics
University of Chicago
Chicago, IL 60637

K. Nishijima
Department of Physics
University of Tokyo
Tokyo, Japan

John H. Schwarz
Department of Physics
California Institute of Technology
Pasadena, CA 91125

Silvan S. Schweber
Department of Physics
Brandeis University
Waltham, MA 02254

Steven Weinberg
Theory Group, Physics Department
University of Texas
Austin, TX 78712

V. F. Weisskopf
Center for Theoretical Physics
Massachusetts Institute of Technology
Cambridge, MA 02139

P. C. West
Department of Mathematics
King's College
London WC2
England

Edward Witten
John Henry Laboratories
Princeton University
Princeton, NJ 08544

Bruno Zumino
Department of Physics
University of California
Berkeley, CA 94720